Frontiers in Computational Chemistry

(*Volume 5*)

Edited by

Zaheer Ul-Haq

Dr. Panjwani Center for Molecular Medicine & Drug Research,
International Center for Chemical & Biological Sciences,
University of Karachi,
Karachi,
Pakistan

&

Angela K. Wilson

Department of Chemistry,
Michigan State University,
East Lansing, MI,
USA

Frontiers in Computational Chemistry

Volume # 5.

Editors: Zaheer Ul-Haq and Angela K. Wilson

ISSN (Online): 2352-9458

ISSN (Print): 2352-944X

ISBN (Online): 978-981-14-5779-1

ISBN (Print): 978-981-14-5777-7

ISBN (Paperback): 978-981-14-5778-4

©2020, Bentham Books imprint.

Published by Bentham Science Publishers Pte. Ltd. Singapore. All Rights Reserved.

need for a court order if at any point you breach any terms of this License Agreement. In no event will any delay or failure by Bentham Science Publishers in enforcing your compliance with this License Agreement constitute a waiver of any of its rights.

3. You acknowledge that you have read this License Agreement, and agree to be bound by its terms and conditions. To the extent that any other terms and conditions presented on any website of Bentham Science Publishers conflict with, or are inconsistent with, the terms and conditions set out in this License Agreement, you acknowledge that the terms and conditions set out in this License Agreement shall prevail.

Bentham Science Publishers Pte. Ltd.
80 Robinson Road #02-00
Singapore 068898
Singapore
Email: subscriptions@benthamscience.net

BENTHAM SCIENCE

CONTENTS

PREFACE

Computational chemistry is an important partner to experiment in understanding a very broad range of chemical problems, providing insight not possible or not easily possible to obtain via experiment, and enabling a greater understanding of experiment when it is possible. The span of computational chemistry approaches in terms of both method and applicability is significant – with methods including electronic structure calculations (*e.g.*, density functional theory (DFT)) and free energy relationships (*e.g.*, QSAR, QSPR), with applications spanning from in-depth description of the spectroscopic properties of the smallest of atoms and molecules to the design of new molecules and materials in medicine.

The focus of *Frontiers in Computational Chemistry* is on the application of computational chemistry approaches to biological and organic processes.

In this fifth volume, the six chapters address a diversity of topics including:

Chapter 1 **"Recent Advances and Role of Computational Chemistry in Drug Designing and Development on Viral Diseases"** Amit Lochab, Rakhi Thareja, Reena Saxena, Sangeeta D. Gadre.

This chapter outlines a number of approaches that are commonly used in drug design, including structural-based and ligand-based computational strategies, and the role of quantum mechanics and molecular mechanics. A brief overview of how these methods have been utilized in the development of drugs against viral disease, addressing ebola, zika, hepatitis C, and coronavirus is provided.

Chapter 2 **"Molecular Modeling Applied to Design of Cysteine Protease Inhibitors – A Powerful Tool for the Identification of Hit Compounds Against Neglected Tropical Diseases"** Igor José dos Santos Nascimento, Thiago Mendonça de Aquino, Paulo Fernando da Silva Santos-Júnior, João Xavier de Araújo-Júnior, and Edeildo Ferreira da Silva-Júnior.

The impact and importance of computational chemistry in drug development is significant. However, the drug discovery process is truly a fine art, with numerous methods and strategies available. In this chapter, the authors consider a number of different molecular modeling techniques, and demonstrate their use in the development of cysteine protease inhibitors. Cysteine proteases are known to play important roles from growth and development of plants to bone development in humans and animals. In this study, the design of cysteine protease inhibitors against a number of tropical diseases is considered.

Chapter 3 **"Application of Systems Biology Methods in Understanding the Molecular Mechanism of Signalling Pathways in the Eukaryotic System"**, Aditya Rao S.J. and M. Paramesha.

Signalling pathways are critical cascades of reactions that can impact metabolic functions from cell division to cell death. Understanding the underlying mechanisms of signalling pathways is critical, as this can provide insight about how abnormalities can impact the activation or deactivation of signalling events. The authors provide an overview of computational routes that can be used to understand signalling path mechanisms, including systems biology and data mining. Wnt signalling pathways are the focus of this chapter due to their role in growth to cancer.

Chapter 4 "**Implementation of the Molecular Electrostatic Potential over GPUs: Large Systems as Main Target**" J. César Cruz, Ponciano García-Gutierrez, Rafael A. Zubillaga, Rubicelia Vargas and Jorge Garza.

Electrostatic interactions are vital to non-covalent interactions which are prevalent in biological systems. A very useful means to gain insight about these electrostatic interactions is via the molecular electrostatic potential (MEP). MEP is generated using quantum mechanical methods, which represents a significant computational challenge for all but the smallest of molecules. This chapter provides two routes to extend the utility of MEP to larger molecules, utilizing graphical processing units (GPUs) to generate the MEP. The theoretical details are provided, as are a number of useful examples of the application of the methods.

Chapter 5 "**Molecular Electron Density Theory: A New Theoretical Outlook on Organic Chemistry**" Luis R. Domingo, Nivedita Acharjee.

This chapter highlights molecular electron density theory (MEDT), which was introduced by the co-author, Luis Domingo, in 2015. MEDT is based upon the philosophy that changes in electron density rather than molecular orbital interactions drive molecular reactions. The authors discuss a broad range of quantum mechanical principles and approaches that are a part of MEDT and facilitate the understanding of molecular interactions and reactions. The authors provide a long list of reactions, and conclusions that can be drawn from MEDT about these reactions.

Chapter 6 "**Frontier Molecular Orbital Approach to the Cycloaddition Reactions**", Anjandeep Kaur.

Cycloaddition reactions, more specifically, 1,3-dipolar cycloaddition reactions, play a critical role from drug discovery to materials design. Controlling the regioselectivity, enantioselectivity, and diastereoselectivity of these reactions is a major challenge. In this chapter, the reactivity and selectivity of a wide variety of 1,3-dipolar cycloaddition reactions is overviewed. A Frontier Molecular Orbital (FMO) approach is considered.

We hope that the readers will find these reviews to be valuable, and that they may inspire trigger further research in the field. We are grateful for the timely efforts made by the editorial personnel, especially Ms. Mariam Mehdi (Assistant Manager Publications), Mr. Obaid Sadiq (Manager Bentham Books), and Mr. Mahmood Alam (Director Publications) at Bentham Science Publishers.

Zaheer Ul-Haq
Dr. Panjwani Center for Molecular Medicine and Drug Research
International Center for Chemical and Biological Sciences
University of Karachi
Karachi
Pakistan

&

Angela K. Wilson
Department of Chemistry
Michigan State University
East Lansing, MI
USA

List of Contributors

Amit Lochab	Department of Chemistry, Kirori Mal College, University of Delhi, Delhi, India
Aditya Rao S.J.	Department of Plant Cell Biotechnology, CSIR-Central Food Technological Research Institute, Mysuru, India
Anjandeep Kaur	Department of Chemistry, Government Mohindra College Patiala, Patiala, India
Edeildo Ferreira da Silva-Júnior	Chemistry and Biotechnology Institute, Federal University of Alagoas, Maceió, Brazil Laboratory of Medicinal Chemistry, Pharmaceutical Sciences Institute, Federal University of Alagoas, Maceió, Brazil
Igor José dos Santos Nascimento	Chemistry and Biotechnology Institute, Federal University of Alagoas, Maceió, Brazil
João Xavier de Araújo-Júnior	Laboratory of Medicinal Chemistry, Pharmaceutical Sciences Institute, Federal University of Alagoas, Maceió, Brazil
J. César Cruz	Departamento de Química, Universidad Autónoma Metropolitana-Iztapalapa, México
Jorge Garza	Departamento de Química, Universidad Autónoma Metropolitana-Iztapalapa, México
Luis R. Domingo	Department of Organic Chemistry, University of Valencia, Valencia, Spain
M. Paramesha	Department of Plant Cell Biotechnology, CSIR-Central Food Technological Research Institute, Mysuru, India Department of Food Technology, Davangere University, Karnataka, India
Nivedita Acharjee	Department of Chemistry, Durgapur Government College, West Bengal, India
Paulo Fernando da Silva Santos-Júnior	Chemistry and Biotechnology Institute, Federal University of Alagoas, Maceió, Brazil
Ponciano García-Gutierrez	Departamento de Química, Universidad Autónoma Metropolitana-Iztapalapa, México
Rakhi Thareja	Department of Chemistry, St. Stephens College, University of Delhi, Delhi, India
Rafael A. Zubillaga	Departamento de Química, Universidad Autónoma Metropolitana-Iztapalapa, México
Reena Saxena	Department of Chemistry, Kirori Mal College, University of Delhi, Delhi, India
Rubicelia Vargas	Departamento de Química, Universidad Autónoma Metropolitana-Iztapalapa, México
Sangeeta D. Gadre	Department of Physics, Kirori Mal College, University of Delhi, Delhi, India
Thiago Mendonça de Aquino	Chemistry and Biotechnology, Federal University of Alagoas, Maceió, Brazil

Frontiers in Computational Chemistry, 2020, *Vol. 5*, 1-62 **1**

Recent Advances and Role of Computational Chemistry in Drug Designing and Development on Viral Diseases

Amit Lochab[1], Rakhi Thareja[2], Sangeeta D. Gadre[3] and **Reena Saxena[1,*]**

[1] *Department of Chemistry, Kirori Mal College, University of Delhi, Delhi, India*

[2] *Department of Chemistry, St. Stephens College, University of Delhi, Delhi, India*

[3] *Department of Physics, Kirori Mal College, University of Delhi, Delhi, India*

Abstract: The growing number of contagious viral diseases among different geographic regions has become a threat to human health and the economy on a global scale. Various viral epidemics in the past have caused huge casualties due to lack of effective vaccine, the recent outbreak of COVID-19 is a good example of it. Drug designing and development is a lengthy, tedious and expensive process that is always associated with a high level of uncertainty as the success rate of their approval as a drug is very low. Computer-aided drug designing by utilizing *in silico* methods has shown prominent ways to develop novel drugs in a cost-efficient manner and has evolved as a rescue in the past few years. Interestingly, the highest FDA approval reached a maximum (59 drugs) in 2018 for which a lot of credit goes to the successful development of computational chemistry tools for drug designing in the last two decades. These methods provide better chances of getting hit compounds in a far more accurate and faster way. Drug designing is a cyclic optimization process that involves various steps like creating a molecule, selecting the target for this molecule, analysing the binding pattern and estimating the pharmacokinetics of the molecule. The final development of a drug candidate is cumulative of positive results obtained in each aforementioned step. Various computational techniques/approaches such as molecular dynamic studies, homology modelling, ligand docking, pharmacophore modelling and QSAR can be utilized in each phase of the drug discovery cycle. In this chapter, we aim to highlight the recent advances that have taken place in developing tools and methodologies that lead to *in silico* preparation of novel drugs against various viral infections like Ebola, Zika, Hepatitis C and Coronavirus.

Keywords: Computational chemistry, Homology modeling, *In Silico*, Ligand-based drug designing, Ligand docking, Multi target drug designing, Pharmacophore modeling, Protein target, Quantum mechanics, Structure-based drug designing, Viral infection, Virtual screening.

---

* **Corresponding author Saxena Reena:** Department of Chemistry, Kirori Mal College, University of Delhi, Delhi, India; E-mails: rsaxena@kmc.du.ac.in; reenasax@hotmail.com

Zaheer Ul-Haq and Angela K. Wilson (Eds.)

INTRODUCTION

There is a huge global effort indulged in wiping out the infectious viral diseases. Viral infections include various contagious diseases like Ebola, Hepatitis, HIV-AIDS, Rabies, Zika and Corona viruses. These ailments cause a huge impact on both the economy and health. Methods in controlling these diseases like vaccination, public awareness through advertisement and campaigns cause a reduction in the budget which is not that effective also. The current available drugs for the disease have their own limitations of being toxic, less potent and high cost. There are several viral microbes that gain resistance toward these drugs and there is always a continuous need for developing new effective drugs. Studies show that the traditional path for discovering drugs and bringing it to market costs around 2 billion USD. In addition, they require a long time, and the process is highly laborious to establish their safety and effectiveness. At the starting of the 20th century, the drug industry was used to screen out various natural and synthetic compounds experimentally in search of therapeutic characteristics for a particular target. Then the compound was optimized for better pharmacological properties having less toxicity which after clinical trials used to take on an average of ~15 years to come in the market [1]. The concerns over various incurable diseases and an inadequate number of potent drugs have forced us to develop innovative drugs with high specificity and potency for the respective target. The ways in which these microbes are mutating their genes to make a come back in our world have proved to be hazardous in an irreparable fashion as is evident from the outbreak of the pandemic of Covid-19 in December, 2019. This has further added the interests of the researchers in fast and efficient drug discovery tools.

Drug discovery using computational chemistry is established very well from the past few years due to development in combinatorial chemistry with computational screening and optimizing tools, with enhanced, fast and efficacious results. The computational methods help in predicting the conformational interactions of active drugs with the target sites. High throughput screening (HTS) and Computer Aided Drug Discovery (CADD) techniques have helped in suggesting favourable drugs out of huge libraries in a short time by understanding the interaction between the target molecule and the proposed drug. The drug discovery process includes several computational approaches before the clinical trials, right from the beginning in which identification of target and their association with a particular disease is considered for studies. The second step is to investigate the interaction of proposed drug molecules with validated target which is followed by the optimization of lead molecules for the improvement in their potency and biological toxicity [2, 3].

This chapter aims to give an overview of different computational approaches and tools for the development in drug designing based on explanation from quantum mechanics. This covers various optimization procedures for enhancing potency of lead compounds. Finally, recent applications of CADD in designing drugs for viral diseases such as Ebola, Zika, Hepatitis C and Coronavirus are discussed.

The first modelling approach in computational drug development method is to identify the probable target related to particular disease. Generally, these targets can be proteins, enzymes or complex bio molecules having specific bioactivity. CADD can be divided into Structure based drug design (SBDD) and Ligand based drug design (LBDD) based on the availability of the structures of the above bio molecule targets as shown in Fig. (**1**). Both approaches are complimentary to each other as SBDD employs known structure of the target moiety for the screening of active new compounds. The structural information of target is used to find new lead compounds by suggesting a design of potent molecule or through screening from virtual libraries and databases. Whereas LBDD is a suitable approach, when the crystal structure of drug target is not available. However, one must clearly understand that SBDD is based on the drug-target structure where in the binding efficiency by a specific ligand/drug is given major importance which may be studied using several docking tools. On the other hand, LBDD makes use of ligands of the target *i.e.* potential drugs shortlisted to bind to the biological target of the drug [4].

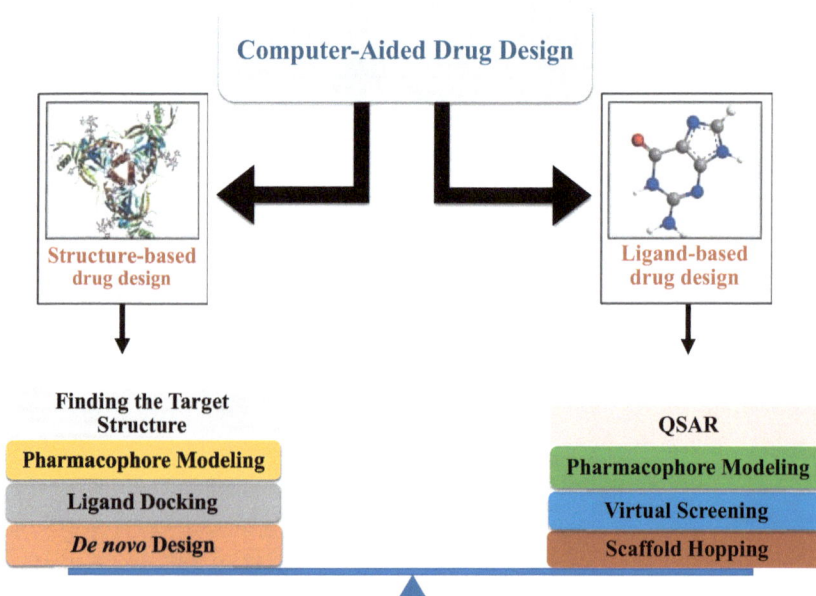

Fig. (**1**). Computer-Aided Drug Design.

STRUCTURE-BASED COMPUTATIONAL METHODS IN DRUG DESIGNING

The process of drug designing is truly very challenging and costly both in terms of currency as well as time. The role of computational tools in structure based drug designing acts as a shorter route in the adventurous journey by potentially decreasing the revenue involved in research and its development further. Today it has become one of the most applicable and requisite tools for the development in this field. This section lays emphasis on the different steps involved in structure based computational methods in drug designing, which shall be discussed in detail step by step.

Finding the Target Structure

Whether the drug-discovery method is based on structure or it is based on the ligand, the first and foremost work of utmost importance is the identification of the target. It is the most important process as it involves classification of the direct molecular target which may be of biological origin *e.g.* a protein or a nucleic acid. Finding the target is primarily done for finding an efficient target of a drug. There are several techniques which are involved in finding the target structure which may be based on principles of several scientific disciplines like biochemistry, genomics, biophysics and chemical biology. Recent research articles lay emphasis on the importance of the target structure for a particular class of chemical compounds. The idea is that once a potential target has been identified, it simultaneously becomes important to work out and capture the entire clinical spectrum of the biological problem and the role of the drug in the treatment of disease. It is very important for the drug to bind to the molecular target productively [5].

Once the target is identified, it can be changed or modified by a chemical molecule in the form of a drug without bothering about its size. There may be different types of targets *e.g.* targets of biological origin like proteins, nucleic acids, *etc*. The latter *i.e.* the target's activity should be susceptible to change when made to interact with a drug of chemical or biological origin. Therefore, just like a wave function, we expect the target to be well behaved. A ***Well-behaved target*** or a 'decent' target must possess some key characteristics:

• The target selected must have some direct or indirect connection to some disease.

• It must play an important role in modification of the disease under consideration.

• The target must have a favourable profile of toxicity which helps in effective prediction of side-effects or harmful effects, with the help of data of phenotype available for the target.

• The target should be easily examinable which would enable it to be screened vigorously through any latest technology available.

• The structure of target should be available to assess its interaction with the drug.

• The market value of the target specified should be high with great relevance to the pharmaceutical industry.

For a good drug designing, the broad-spectrum analysis of the disease with which the target is associated should be carried out and captured. The role of the potential biological target in affecting a particular disease is of great significance. No disease will get cured or prevented unless the drugs of chemical or biological origin get bounded to the target firmly [6].

There are a number of ways, which may be employed for identification of target. One of the most convenient methods, though requires a dedicated study, is to carry out an elaborate literature survey. Another method based on similar lines is to go through the public databases available. These methods are available in an open domain and hence have chances of great competition in the day-to-day research and development. In order to be in the lead to perform research in drug designing, it sometimes becomes mandatory to consider those targets also for which the data available is insufficient; hence, would pose fewer competitions. Other strategies for identification of potential target would include either of the following two: Designing of an effective drug followed by identification of target or *vice-versa* method which implies looking for a new target followed by virtual screening to establish a drug that binds to the target efficaciously. In all ways, the main motto is to establish a drug design which should be impactful [7].

There is a term used for identification of the target specifically which is known as 'target deconvolution' which has an approach based on phenotype. According to this method, models of animals, tissues or cells are exposed to molecules of variable sizes (preferably small) to check the efficacy of the drug's impact onto the former which is primarily checked on the basis of change in its phenotype. Small molecules can be comfortably characterized using several models of animals but still the use of cells especially those of mammals are often favoured. The reason is their compatibility with virtual screening and hence, relevance to their physiological properties. There are a number of methods that fall under this term *e.g.* chromatographic techniques based on affinity, microarray of proteins, biochemical suppression, *etc.* [8]. This approach is anytime suited as it extensively involves study of signalling routes or pathways, which goes beyond the normal research, regarding the basic structure of individual proteins or nucleic acids. The benefit of approaches based on phenotype of the drug is that the drug's effectiveness can be easily established with regard to the environment in which it has to prove itself. The interaction of drug with the target site gives a clearer picture in one go rather than waiting to identify the target in its pure form on biological display. However, there are several challenges associated with it like cost factor, complicated methodology based on assays. Nevertheless, improvements are taking place both in terms of scalability and relevance with respect to physiology and visualization of 3D models of cells. Screening based on phenotype has provided motivation to many advances in technology that include tools for gene-editing, assays for detection and imaging [9].

After having discussed identification of target based on target convolution, now we may throw some light on drug designing based on target specifically which may also be referred to as the 'Discovery of Target'. Targets of biological origin are already known even before the discovery of the lead begins. Therefore, screening done on the basis of target search has 'target discovery' as the main cornerstone. At the very onset of the process of drug designing, it becomes pertinent to know about the target's characteristics in details so as to get a clear picture of its role in a particular disease functioning. The target identified then will be used to develop assays based on justified systems. Virtual screening of large compound libraries is the next step to look out for the best hit which will be considered as the drug with the best candidature. This is the best suited method to understand the mechanism involved through which the drug would interact with the target and hence, is considered better than method discussed earlier for target identification. These methods offer a simple and a cheaper approach to design and develop a drug through a quick approach.

Drug discovery based on target identification first can be done using several methodologies such as biochemical sciences, binding kinetics, binding thermodynamics, molecular modeling and crystallographic studies. All the mentioned approaches will help in understanding the chemical interaction between the target and the drug in a deeper way so that it enables us to develop QSAR (Quantitative Structure Activity Relationship), biomarkers and even drugs of the present day and those that belong to future that act directly at the target specified [10]. The following sections in the present chapter now discuss the structure and ligand based drug designing methodologies that are employed suitably after identification of the target is done.

Pharmacophore Modeling

Pharmacophore modeling plays a very important role in both ligand as well as structure based drug designing. It is a distinct and truly a unique subfield of computer aided drug designing. This concept is not only studied for designing of novel drugs but also as an important tool for computational/molecular modeling in the process of drug discovery. The pharmacophores created are used to display and recognize molecules on a two as well as three-dimensional platform by visualization of important features related to identification of the molecule. The research based on pharmacophore modeling started and developed during late 1800s when it was initially established by Paul Ehrlich [11]. In the earlier days it was believed that the biological effects were a result of the presence of functional groups in the chemical molecule and the ones having groups from homologous series were expected to show similarities in their action. But today, pharmacophores are believed to be molecules related to patterns of abstract features and not concerned with functional groups in particular. The IUPAC defines it in a specific fashion to be an ensemble of steric as well as electronic features pertinent for minimal supramolecular interactions to block the target site and hence, its biological response [12]. This is what the ultimate goal of structure based drug designing is. All types of atoms or particular groups of molecules which are known to display some key feature properties that concern with recognition of molecular target are eligible to be shortlisted for pharmacophore properties. These may be classified in various forms such hydrophilic or hydrophobic, hydrogen acceptor/donors, aromatic, anionic, cationic, or any other permutation or combination of these features and more [13]. To have a quick understanding of structure based pharmacophore modeling, one may now remember that here one starts with the three-dimensional structure of a protein or any other biological macromolecule target or a ligand complex with a complicated molecule. The format of this method deals with first doing a research of the important chemical reactions and bindings possible at the target site both directly and in a complementary fashion, followed by formation of an assembly of pharmacophore model with specifically selected features for the desired application. Pharmacophore models based on structure can be divided into two categories of macromolecules: (i) without ligand (ii) with ligand. Pharmacophore models with ligand are more approachable as it is useful in detection of the active site where the ligand can bind at the macro-biomolecular target structure and is also beneficial in determination of the principal interaction sites between the drug molecules or ligands [14]. When the specified models are available with a decent number of ligands, then several programs can be employed to carry out Pharmacophore structure based drug designing [15]. LigandScout is the best-suited example for this [16]. Other examples for complex of bimolecular–small molecule based pharmacophore modeling programs are GBPM and Pocket v.2

[17]. When many molecules of different origin are compared in pharmacophores modeling, then this practice refers to "Pharmacophore fingerprints". This representation is what reduces the chemical moiety to a combination of all features at two or three-dimensional level. In case, only a few properties/key features of pharmacophores are taken into account during a three-dimensional model study then pharmacophores are often called a 'query' [18].

One of the most commonly used applications of pharmacophores is Virtual Screening. There are many strategies possible for the same. This concept of pharmacophores modeling is not only used for target identification but is also employed for ADME-tox modeling and off-target prediction. When these pharmacophores are clubbed with MD Simulations (Molecular Docking Simulations), virtual screening gets improved. Now pharmacophores modeling is not only used for Structure based drug designing but is also significant in ligand based drug designing that will be discussed later in this chapter, for designing of proteins and also to study protein-protein interactions [19].

The key properties of pharmacophores models describe the standard interfaces between the structure and ligand or protein and ligand. This can definitely be mapped to form a chemically and biologically active conformation of the small molecule which is serving as a ligand. Generally, the model of protein is obtained from XRD studies or NMR Studies, but homology modeling and other similar tools/databases are also being used nowadays. It is quite important to know at least one ligand structure but it is more advantageous to have a three-dimensional picture of several ligands to have a deeper understanding of common interactions. This approach is used by many pharmacophores modeling techniques. The first software package that could generate a query automatically from pdb files based on interactions between proteins and ligands is 'LigandScout'. Queries generated based on structure based drug designing have a wide range of applications. Ligand binding pose prediction, virtual screening, and comparison of target sites of interactions, are examples to name a few [20].

Future perspectives on pharmacophore modeling can be defined infinitely. Ranging from target identification to scaffold hopping, from similarity metrics to virtual screening, from structure modification to ligand optimization, the list of its applications to computer aided drug designing is endless [21]. Scientists world-wide are known to generate models (query) based on pharmacophores that encodes the rightly ordered three-dimensional pattern required for interaction. The information about the structure becomes mandatory and the multiple routes available for construction of a query - all depends upon the information available about the target of a particular protein. Recent developments have shown that queries have been generated from important features known for an active site of protein. The idea initially focused on ligand based drug designing but now, it has also been reversed and pharmacophore models now represent key feature of the structure of proteins/queries as well [22].

Structure based drug designing based on pharmacophores however have limitations too. It needs to have the three-dimensional structure in a mandatory fashion. This implies that in case there are no ligands to target the active site, then one cannot apply the structure based drug design. Then in order to overcome this problem, macromolecule based approach will have to be used. Structure based pharmacophores method can be employed that will convert interaction maps of internal protein binding site to features related to pharmacophore like hydrophobicity, hydrophilicity, H bond acceptor, H bond donor *etc.* Other approaches like GRID can also be used [23, 24]. Tintori *et al.* have also reported another apoprotein-based approach [25 - 27]. The development, formation and utilization of the pharmacophore models is based on an important supposition that all the ligands created must bind in a similar manner to the target of the biomolecule.

Ligand Docking

The basic principle behind protein-ligand docking is the study of association of the ligand with a protein of known structure. There are several applications that help us in doing ligand docking. For example an application known as AutoDock Vina is freely available for everyone to study molecular modeling and drug designing [28]. Through this application we can predict the best fit for ligand and the protein molecule and also come up with a ranking for all the combinations formed. It is important for a researcher to visualize the interactions between protein and ligand before running the simulation on screen. There are several Molecular Modelling tools that help us in visualizing interactions occurring between the ligand and the biomolecular target with high clarity. The software's that support these visualization tools include Argus Free Lab, AutoDock, Schrodinger and Discovery Studio to name a few. While the first two are freeware, the last two are paid software's [29]. A graphical overview of ligand docking with a protein target is shown in Fig. (**2**). This needs to be done in order to make sure that the final outcome of docking *i.e.* the product in the form of a complex that gets formed is stable and strong enough for further applications. This would pave the way for an early screening for an efficient ligand-protein docking.

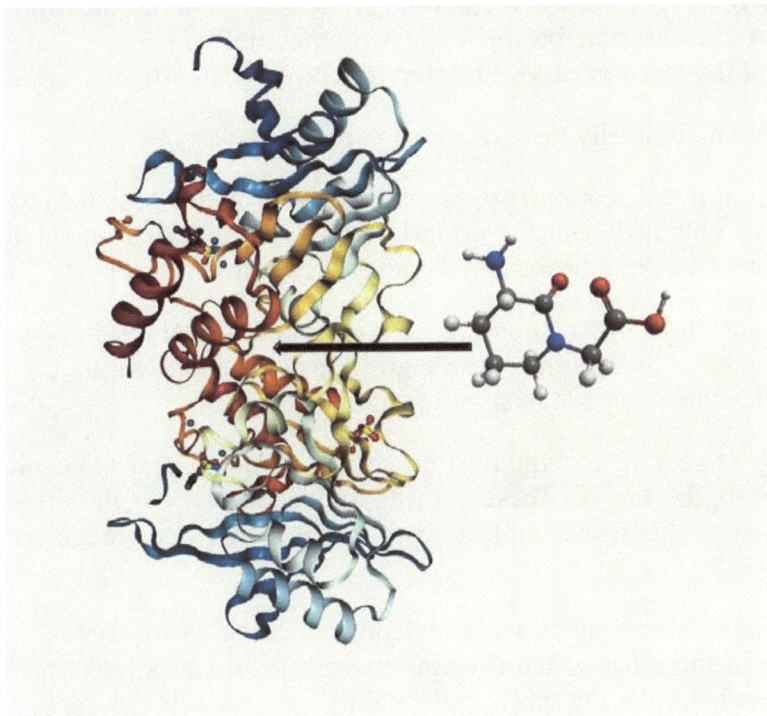

Fig. (2). Protein-Ligand Docking.

In this regard, structure based drug designing has not only helped in framing efficient analogues in the form of best possible macromolecules for the known as well as unknown drugs that have been known but do not get to showcase minimal output for a particular disease. Nowadays there is an increasing demand of data scientists who can collect and combine data in a synchronized fashion from the constantly increasing database that facilitates future study in the area.

For ligand docking, the structure of the protein sometimes poses a major drawback owing to its inflexibility. This poses a hindrance to the aim of achieving maximum efficiency. But steps and measures are being looked for to overcome this problem. One of the structure based virtual screening method is Protein-ligand docking. High throughput screening is employed for protein ligand docking and crystallographic data is exploited by many leading programs. It is not an easy to study the interactions between two or more molecules. The solution gets complicated due to multiple forces which are involved in bonding like Van der Waals forces, stacking forces, hydrophobic and hydrophilic forces. There are still many unanswered queries due to the complexity involved in studying the ligand-protein complex and also inadequate information and knowledge about the impact of the solvent. The entire methodology is based on mimicking the natural pathways of interaction between the two moieties *via* lowest energy profile. Structure of the receptor plays a primary role to have an efficient ligand docking.

Ligand docking basically involves three methodologies:

• First step in this direction is to prepare suitable compounds for analysis of the database so obtained. Here, appropriate tautomeric and protonated forms are prepared to avoid electrostatic and steric clashes between ligand and binding site.

• Second step includes computing of conformational algorithm by accounting various degrees of freedom. These algorithms help in exploring conformational spaces inside the target site in an efficient manner.

• Last step is scoring of binding free energy of the ligand molecule inside the active site of the target. These scoring functions can be classified into three groups such as empirical, molecular force field and knowledge based scoring functions [30].

To summarize, one may talk about ligand docking as a process that involves prediction of the most suited orientation with regard to interaction with ligand when it gets bound to the macromolecule to form a stable complex. This is most importantly studied to develop an understanding of protein-ligand interactions which are important to the development of computer aided drug designing. Binding energies are then calculated for small molecules to predict the best bound conformation. Experiments based on ligand docking are very helpful in exploring the role of target and even the ligand. It is also useful for virtual screening where thousands of compounds/ligands/small molecules are docked and are given priorities. These may then be employed for identification of new inhibitors for designing of modern day drugs [31 - 34].

De Novo Design

For the drug designing, *De novo* design offers a unique approach. The optimization of the system can't just be restricted to the performance of a pre-designed or earlier known system but rather it must be extended to design a minimal ligand that can effectively work as a drug. This implies that our aim is to carry out search for an effective drug rather than an efficient drug. *De novo* design is not based on optimizing an ill-configured structure system which may prove to be wasteful and ineffective but it's based on designing of minimally selected systems that hardly require any further refinements or geometry optimizations [35].

How a particular protein relates to a specifically exclusive structure remains as an untold story to the intrigued chemists which ultimately leads to the designing and studying of those molecules having already well-established structures. Recent studies have shown that certain regularly repeating systems have proved to be effective for certain foldamers that have been designed especially to accept a tertiary structure with helices' bundles. Designing of the single moiety or the monomer geometry is done for creation of non-living or abiotic foldamers. Such kind of new designing approaches definitely will pave way for several more applications.

To summarize the *de novo* design patterns, one may suggest growing a particular template computationally for candidate molecules for highly specific materials. These may be microporous in nature too. Simulations have been able to rationalize templates that are known for chabazite and EU1 and *de novo* design has led to the suggestion for making new template molecules for structure synthesis for the same [36].

Both structures based and ligand based drug designing techniques are used in *de novo* designing techniques. The former is essentially known when a well-established structure of the macromolecule like protein is known beforehand. The latter is described in the sections to follow. When the information is available in abundance, then these methods can be used in combination too to enhance the efficiency of the method and increase the effectiveness of the drug designing method. The improvements and developments of these methods for *de novo* drug design have been discussed elaborately in several reviews too [37].

There are computer programs that are freely available for structure based *de novo* drug design too like LigBuilder that help the researcher develop better insights into understanding the concepts in a better fashion [38]. The *in silico* methods used now in wide range of projects involving drug designing proves its success in latest biological, chemical, medicinal and pharmaceutical research [39 - 41].

LIGAND BASED COMPUTATIONAL METHODS IN DRUG DESIGNING

The discovery of novel drugs using *in silico* approach is being widely used in various preclinical steps. The screening and prioritization of lead compounds through huge databases is done by Ligand and Structure based drug designing methods. Ligand based drug design (LBDD) approach is applicable only when the 3D structure of the target is unknown. Different LBDD methods employ physicochemical characteristics and pharmacophore models in order to define variation in the structure activity relationships (SAR). The SAR thus obtained can be used to predict the novel compounds with improved potency. The general belief behind this approach is that the structure of ligand contains all the required information that explains the action of mechanism, lipophilicity and other pharmacokinetic characteristics. Then by exploiting the properties of known active ligands, similar structured new ligands are developed with improved pharmaceutical properties. The advantage of LBDD over SBDD is that the later one is more costly and suffers from poor prioritization of binding orientation. SBDD is also associated with other limitations like it cannot identify 'activity cliffs', which can be achieved by specific LBDD approach known as Quantitative Structure Activity Relationships (QSAR) through measurement of physicochemical characteristics. This ligand based approach allows quick database filtration and predict novel drugs on a large scale [42]. However, LBDD have certain drawbacks such as the prioritization of model through statistics can itself produce false results. Also, this approach is beneficial only when sufficient amount of information about active conformation of compounds is available [43].

Generally, LBDD method uses two elements mathematical tools to predict models and molecular descriptors [44]. Some of the widely used methods included in LBDD are discussed in the following sub sections to give a broad understanding for the same.

Quantitative Structure Activity Relationships (QSAR)

QSAR methodology is very efficient in drug designing as they are cheaper alternative for predicting physicochemical properties of unknown compounds. The basic concept behind this methodology is that the difference in the structure of molecules can be quantitatively related to the difference in their biological activity. Thus, all the biological activities or other functions of molecules can be related to molecular descriptors which can be used to measure the contribution towards biological effect. The generation of a QSAR model is possible only after designing a suitable experiment to derive a model that utilizes least number of descriptors to explain maximum activities and characteristics as possible. This serves as a tool for reducing the number of steps in synthesis as well. One must have more than five compounds for each descriptor and each descriptor must undergo a verification analysis. The verification can be done on the basis of whether the change in descriptor's value is bringing about any change in activity or not. If no, then there is no point in including that descriptor. The descriptors included must have correlations amongst themselves.

Development of QSAR requires various software such as Concord, CORINA, and Frog *etc.* to draw 3D structures while Dragon, PaDEL and Molconn are used for descriptor calculations in following steps:

• Generation of 3D model for geometry-based descriptor calculations.

• Molecular structure descriptors are generated with help of previous step.

• Feature selection tools are used to select most significant descriptors.

• Formation of QSAR model through the help of prioritized descriptor sets.

• Validation of the model is done by predicting activity of the compound in external prediction set in the last step.

The numerical interpretation of molecular information is done by using molecular descriptors. There are different QSAR model that can be differentiated based upon the dimensions of molecular descriptor:

I. 0D QSAR depends upon the descriptors based on molecular formula and hence, gives information related to the molecular weight, number of atoms and their respective properties.

II. 1D QSAR consist descriptors derived from substructures of the molecule such as rings, functional groups, substituents. It in addition gives the data related to fragment counts and their presence like the number of primary carbon atoms and secondary carbon atoms).

III. 2D QSAR contains graphical representation of the molecule and provides information about how the atoms are connected in the molecule with interatomic interactions.

IV. 3D QSAR uses three-dimensional geometrical representation of the molecule. The descriptors involved contain surficial and geometrical properties of the molecule.

V. 4D QSAR further includes freedom of alignment to prepare the training sets and also includes conformational data by incorporating calculations related to averaging of ensemble which adds the fourth dimension.

VI. 5D QSAR evolves after the simulated evolution is carried out and the fifth dimension is attributed to the probability of introducing an ensemble of different type of induced fit models which are at least six in number.

VII. 6D QSAR include calculations of every solvation model specifically and simultaneously.

The properties or the basic definitions mentioned for the respective QSAR approaches give us the best suited model according to each methodology and also opens up the prospects of understanding the concept of drug designing in a better fashion [45, 46]. Nowadays, Partial Least Squares method and Multiple linear regression are used as the methods for deriving QSAR, QSPR and QSTR models (A=Activity, P=Property, T=Toxicity).

QSAR plays a crucial role in designing chemical structures having good potency, target specificity and low toxicity with reduced computational cost. QSAR models in combination with other methods can be helpful in evaluation of inhibitory activity. These studies are also helpful in predicting antioxidating properties of Coumarin derivatives. Additionally, CoMFA approach of 3D QSAR studies has shown good results in predicting models based on steric and electrostatic field as descriptors. This approach aligns ligand on the basis of 3D structure of a specific grid. Another approach of 3D QSAR, CoMSIA takes into account different terms such as hydrogen bond donor-acceptor, columbic and steric interactions. It also uses special Gaussian function to describe electrostatic and steric components.

However, the accuracy obtained through QSAR studies need to be improved by encoding measured molecular structure descriptors with structures. The reliability of the developed QSAR models can be enhanced by combining it with other techniques such as molecular dynamics simulations along with the improvement in hardware and software of the computers [47]. There are many other methods now in practice employed for generation of models, many of which are suited for attending to certain specific types of issues and problems which will be discussed later.

Pharmacophore Modeling

Pharmacophore modeling utilizes efficient methods in order to filter lead compounds from huge databases by virtual screening. According to IUPAC, pharmacophore is an ensemble of electronic and steric properties which is necessary for molecular interaction with the selected target. As already discussed, a pharmacophore is a portrayal of molecular features that are essential for recognition of molecule by biological molecule. These molecular features include aromatic rings, anions, cations and hydrogen bonds which are also known as pharmacophoric points that can be present on either the ligand or the receptor. A well-established pharmacophore model contains both hydrogen bond vectors and hydrophobic volumes with spatial arrangement of different molecular features that are necessary to interact with the target. This method is highly effective in classifying ligands into inactive and active groups. There are various commercial software available for generation of pharmacophore automatically such as MOE, GLIDE, SLIDE and PHASE. Usually, the pharmacophore generated through training set of ligands include two steps:

i) **Creation of conformational space** of every ligand of the training set in order to represent flexibility of different conformation of the ligands.

ii) **Ligands are aligned** from the training set and essential molecular features are determined to prepare resourceful pharmacophore models [48].

The different processes that are included in pharmacophore modeling are:

• **Dataset preparation** – A set of similar structured ligands with good biological activity are converted into their pIC_{50} (-log IC_{50}) that is used as an input data for the software.

• **Ligand preparation** – ligands having best suited conformations and chirality are developed which somewhat similar to the molecular docking approach.

• **Pharmacophore sites preparation** – different molecular features such as aromatic rings, hydrophobic sites, hydrogen bond donor-acceptor are used to produce pharmacophoric sites for creating pharmacophore model.

• **Arranging and alignment of ligands** – pharmacophores having similar molecular features and spatial arrangements from the dataset are identified separately with the help of software. Finally, a scoring function is employed to give the best suited alignment of the ligands [49].

The use of specific features to create pharmacophore models both according to SBDD or LBDD is nothing but a simple extrapolation of the 'bioisosterism' concept that lays emphasis on identification of common biochemical and biophysical characteristic present in certain functional groups [50]. After implementation of database searching, swiping through theoretical databases, generation of programs for development of structures, search and analysis of conformations, it is that 'Pharmacophore mapping' is done. Some methods employed for the same include Constrained Systematic Search, Clique Detection, Likelihood method, genetic algorithm, *etc.* Each of the method is associated with some limitations too and hence, owing to these limitations, newer approaches are getting developed for obtaining the best pharmacophores [51]. Therefore, what this section defines is basically how ligands can be overlaid based on similar characteristic features when related to their binding abilities at the target site [52].

Virtual Screening

Virtual screening is considered as computational analogue technique of High Throughput Screening. It has been extensively utilized in the various steps for the development of novel drugs. More specifically they help in identification of lead compounds and in their optimization step. Virtual screening basically filters the active compounds by ranking or scoring with the help of various computational tools. It is a fast and cheap alternative for high-throughput screening in order to discover potent drugs. The two main branch of computational drug designing for virtual screening are structure and ligand-based drug designing. The structural based virtual screening can be used only when the 3D structure of the target is known. Whereas, ligand based virtual screening (LBVS) is a complementary technique that works when there is no 3D structure of the target is available. LBVS is based on the fact that molecules having similar structure with an active compound are more likely to have drug like property as compare to random molecules. This method uses the available information of known potent drug for filtering lead compound in place of the 3D structure of the target. This approach is highly effective in case of apo form of protein and G-protein coupled receptor where actual structure is not clear [53].

The process involved in LBVS is the evaluation of new compounds having similar structure with the known active molecule. There are various criteria for scoring and filtering assessment in the virtual screening. The similarity measurements are further carried by different techniques like molecular shapes and pharmacophores. Molecular shape technique attempts to overlap shapes maximally by assigning the similarity value based upon the degree of shape overlap. Pharmacophores technique on the other hand uses distances between different molecular features such as hydrogen bond donor-acceptor, aromatic rings to generate patterns. Then similarity value is calculated by comparing the respective patterns [54].

Even after huge development in various combined computational approaches the improvement in the development of drugs is not appreciable. The issue forced the researcher to introduce more effective concepts such as 'drug likeness' in virtual screening. It attempts to distinguish the particular moiety of the active molecule which is responsible for their biological activity. Also, filters based on ADMET properties, descriptors and Lipinski rule of five are being utilized to screen lead compounds from huge libraries. However, there is still a huge scope in the improvement in the field of scoring and development of reliable models which can be enriched with advanced algorithms and improved hardware [52].

Scaffold Hopping

Scaffold Hopping was first employed by Gisbert Schneider in 1999 to recognize different functionalised core molecules having similar functionality. The aim of scaffold hopping is to find isofunctional characteristics in structurally different molecules. In the case where 3-D structure of the target is absent, pharmacophores play a vital role for drug development. However, pharmacophore doesn't provide sufficient activity towards the target. Additionally, selectivity and ADMET properties require further improvement. So, there is requirement of methods that can keep the beneficial properties and eliminate the bad ones of the screened compounds. Since huge number of compounds in databanks and libraries have poor pharmacokinetic properties that further results in inheriting their unfavourable characteristics. The appropriate solution to this problem can be achieved by replacing the parts of the molecule responsible for the undesired properties. The scaffold hopping is used to provide information about the additional changes required to replace the undesired parts of the core structure that are responsible for poor activity. Sometimes, there is need of replacing the whole core structure or scaffold of parent compound, whereas in some situation small modification in side chains are also sufficient to eliminate the undesirable properties. Various virtual screening methods have been predicted for scaffold hopping and ligand based similarity searching is one of them. Though, virtual screening attempts to figure out novel structures whereas, scaffold hopping focuses only on the particular portion of the molecule that needs to be replaced or in other words we can say that scaffold hopping emphasis on already existing chemical moiety. Later the scaffolding target compound is converted into scaffold query that is appropriate for database searching [55].

The classification of scaffold hopping approach is based on the idea that how much variation is introduced in parent molecule to provide different derivatives. The small modification such as replacing particular atoms in the parent chain or molecule comes in the category of 1° hopping. Generally, MORPH software used for 1° hopping that is useful for creating ring structures systematically without altering the coordinates of the ring. 2° hopping includes ring opening and closure, as most drug compounds have ring moiety, so ring closure and opening are efficient methods to develop new scaffolds. However, there is no particular software to provide ring closing and opening strategies but Cambridge Structural Database (CSD) can be helpful in validating the design in this field. 3° hopping includes mimicking of various features of peptides with the help of templates. Development of peptides is useful in situation where there is imbalance of the peptides in the body that leads to various diseases. ReCore and CAVEAT software are used to design scaffolds in order to substitute peptide parts. 4° hopping which is also known as topology/shape based scaffold hopping, can be created by using virtual screening. Virtual screening mainly focuses on filtering whole new compounds whereas scaffold hopping focuses on discovering new core structures. CSD software can also be used in this case to define various topological needs [56].

Scaffold hopping, though a new technique is gaining attention of the scientific community for discovering novel and effective drugs. It plays a crucial role in optimization of hit compounds and enhances the chances of developing compounds having more activity by replacing undesirable core structures. It can also be utilized to jumpstart project from naturally occurring substrates to effective drug like compounds [57].

MULTI TARGET DRUG DESIGNING

There has been a huge advancement in the field of different screening and optimizing tools for the development of drug. Still, there is stagnant growth in the availability of successful drugs in last few years. It was observed that single target drugs are not always able to inactivate the target successfully.

Specially, in case of complex diseases which can affect the host in compensatory manners cannot be effectively treated using single target drugs. Scientists found that the compensation can be made by introducing the concept of multi target drug design. The best example to simplify the concept can be provided by cancer diseases whose targets are scattered widely and require diverse chemotherapy methods. A Multi target drug plays a vital role in such scenarios where inhibition potency against multiple active sites is required in comparison to single target as depicted in Fig. (**3**). It was observed that the different targets related to a complex disease have networked connections and multi target drugs strategy can be easily applied against them. Additionally, it was suggested that partial inhibition of multiple targets through multi target drugs approach can be more beneficial than total inhibition of single target [58].

This strategy has been extensively strengthened by both branches of computer aided drug designing *i.e.*, Structure and Ligand based drug designing. The various tools involved in structure and ligand based methods as discussed above, have made the multi target drug design approach faster and easier to handle [59].

This approach has been used by various researchers in order to develop efficient drugs against various disease such as Vincetti *et al.* attempted to develop drug against dengue virus by virtual screening. He found that series of purines derivatives showed inhibition properties against virus replication at a very low concentration with no toxicity. The compound 16i among the various identified purines found to have much higher potency than ribavirin. It was able to target both host kinases c-Src/Fyn and NS5-NS3 interactions thus, exhibiting multi target characteristics [60]. Yang *et al.* on the other hand chose ementine (anti-protozoal drug) against Ebola and Zika virus infection. The drug showed inhibition activity against NS5 polymerase of Zika virus and lysosomal functioning at nano-molar concentration. Also, it showed similar efficiency against Ebola virus infections and portrayed itself as a novel multi target drug [61].

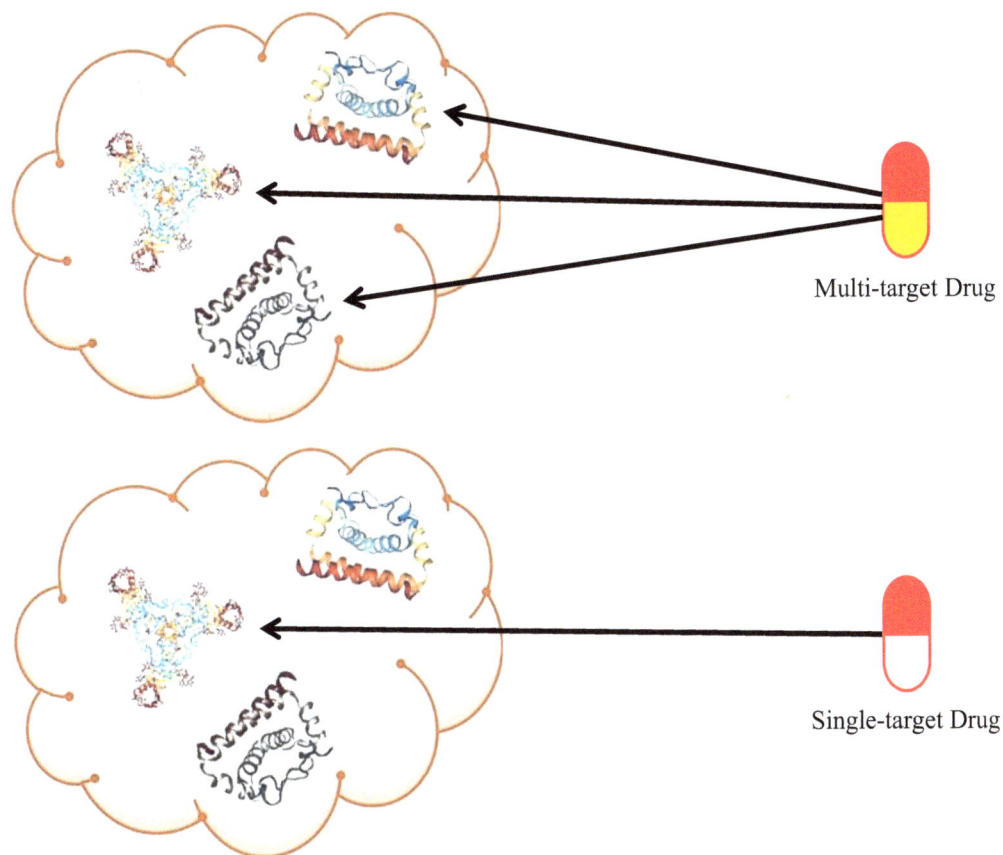

Fig. (3). Comparison between Multi-target and Single-target drugs.

However, this approach is associated with side effects and toxicology but rational drug designing can give better clinical results with safety. Thus, focusing on multiple targets by developing drugs with good safety profile rather than inhibiting single target can give better outputs in the medicinal field.

ROLE OF QUANTUM AND MOLECULAR MECHANICS IN DRUG DESIGNING

Homology Modeling

Earlier, development of novel drugs to target active site of the protein structures were used to undergo screening process *via* experiments called high throughput screening, which was both expensive and tedious. Even after the huge development in carrying automated large number of experiments simultaneously at optimized conditions with the help of laboratory robotics, the method suffers from poor identification of potent drugs. As discussed earlier, SBDD employs the interactional information of target with known structure to the compounds from various databases in order to screen out potential hits. The structure available for different macromolecules and proteins as target is limited which is the biggest limitation of this approach. However, prediction of structure of such targets can be made through computational techniques, which is known as homology modeling. The principle of prediction in homology modeling is based on the fact that similar structured proteins have similar sequence alignment. This means that 3-D models can be predicted for unknown targets that shares structural similarity or have homologous sequence. Homology modeling of proteins basically follows four steps. The first step is the formation of template based on the known 3-D structure of correlated proteins. This is followed by the second step that is to align the sequence of target and in next step a model is prepared based on both alignment and the structure of template. The last step comprises refining, validation and evaluation to build an adequate model as given in Fig. (**4**). Validation of the prepared model generally starts by energy minimization with the help of proper molecular mechanics force fields [62].

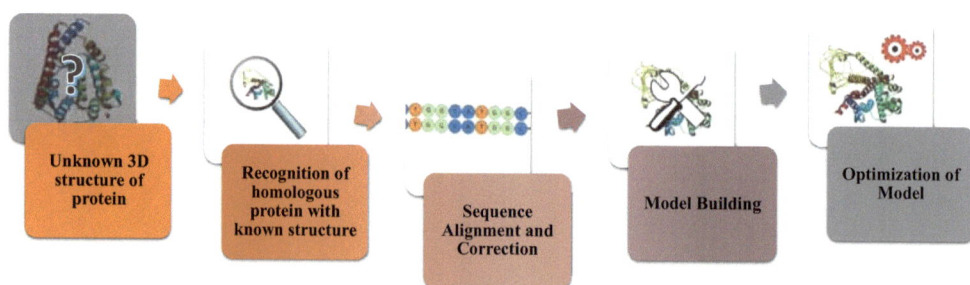

Fig. (4). Homology Modeling.

The accuracy of the predicted model may vary based upon the degree of likeness in sequence alignment and structural similarity. Usually, the models with more than 50% sequence similarity are ideal for the drug development, while the sequence similarities in between 25 to 50% are good enough in order to assess druggability of a target. Multiple template approach is followed in case where sequence alignment is low and single template fails to give whole structural knowledge of the target. The multiple template technique uses dynamic programming in order to align several sequences simultaneously to predict various functional groups of the target structure.

The application of homology modeling is not just limited to predict the 3-D structure of the target but extends to explain the interaction patterns and their functions. The target with known physiological role can perform as ideal target for the discovery of drug. Although this method is highly applicable to screen out hits from libraries in the absence of target structure but development in refinement tools is very essential to have more precise models with active binding sites and conformation [63]. Several studies have been reported based on the application of homology in the treatment of various diseases *via* protein-ligand interactions and molecular docking. For example, Fernando *et al.* and Ekins *et al.* have used the homology structures in order to develop lead compounds against Zika virus that will be discussed in more detail in later section of this chapter [64, 65].

Molecular Dynamic Simulations

The research based on many body problems becomes important and hence, is carried out with computational techniques to reduce the complexity involved in solving simultaneous equations of motion. Molecular dynamics simulations predict the time dependent interactions and movements of each atoms present in a protein or other macromolecule system. It is based on Newton's laws of motion and one can speculate the position of individual atom with respect to time. Once the spatial information of each atom in different environment is available one can measure the different interactions exerted by other atoms on them. Firstly, molecular dynamics simulation was carried out in 1970s for a protein structure and has become very popular in recent years [66]. Generally, the molecular actions of interests are happening at a time scale of macro to nanoseconds and for the same reason simulations require very small-time steps of the order of 10^{-15} seconds which makes the calculations difficult to interpret in the beginning. However, developments in both software and hardware have made such interpretations very easy and economically favourable. The main advantage is that molecular dynamics simulations offer real time environment by subjecting the entire system to temperature and pressure conditions by offering it to evolve in time according to the interactions between the many bodies that take place during the simulations. Also, this method is highly compatible with modern experimental techniques for structure elucidation like Cryo-electron microscopy; X-ray crystallography, Nuclear Magnetic Resonance *etc.* [67]. There are many experiments designed for its testing in Critical Assessment of Protein Structure Prediction (CASP).

This method has gained a lot of popularity over the past few years especially in the field of material sciences, biochemical sciences and biophysics. The simulations can provide insight of various molecular processes like protein folding and conformational variation. The various applications of molecular dynamic simulations are:

• They play an important role in predicting how the macromolecules will behave at atomic level when they are subjected to any perturbation like ligand addition-removal, mutation and protonation.

• To analyse the dynamic structural fluctuations that are happening due to presence of external environment such as salt ions and water molecules that plays a vital role in ligand binding.

• Simulations also helps in revealing several biological processes and transportation steps happening across membranes as shown in Fig. (**5**).

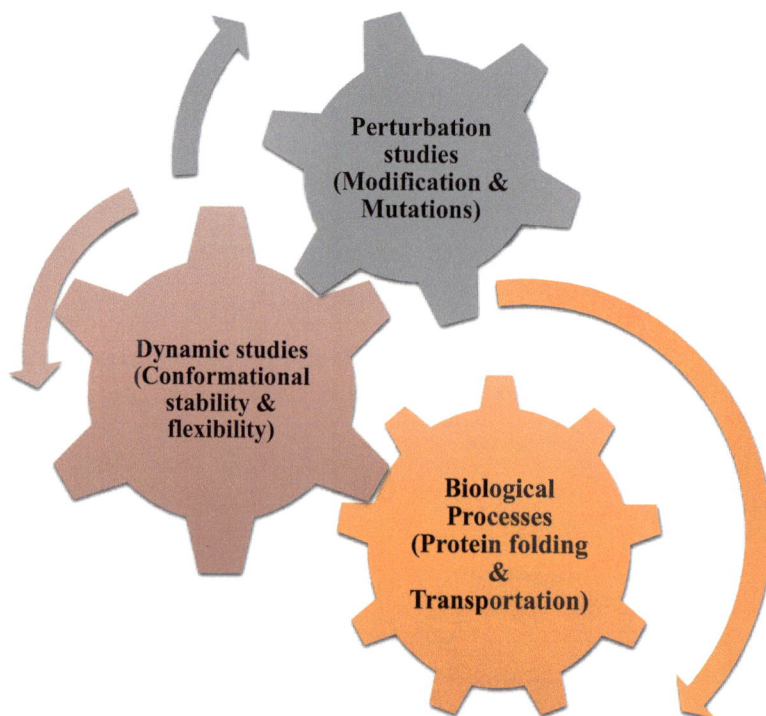

Fig. (5). Molecular Dynamic Simulations.

The binding free energy values provided by simulations are quantitatively more precise in predicting ligand binding interactions than other *in silico* methods like docking.

Carrying molecular dynamics simulations practically require optional hardware systems like Graphics Processing Units (GPUs) and supercomputers to measure forces using molecular mechanics force field. The term 'force field' refers to the potential energy which is calculated on the basis of bonded as well as non-bonded interactions. Hence, it forms the basics of Molecular Mechanics calculations as it needs to be set first before even planning out a run to be given for optimization or simulations. The parameters included are related to bond stretching, angle bending, out of plane bending, improper and proper torsions. Whereas electrostatic and Van der Waals interactions are included for the non-bonded interactions. The concept is vast and has several aspects to it, which is beyond the scope of the present chapter but still, to get a quick gist, a generalized Molecular Mechanics force field may be represented as follows:

$$E = \Sigma_{bonds} K_b (b - b_0)^2 + \Sigma_{angles} K_\Theta (\theta - \theta_0)^2 + \Sigma_{dihedrals} K_\Phi (1 + \cos(n\varphi - \delta) + \Sigma_{improperdihedrals} K_\Phi (\phi - \phi_0)^2 +$$

$$\Sigma_{UreyBradley} K_{UB} (r_{1,3} - r_{1,3;0})^2 + \Sigma_{non\text{-}bonded} \frac{q_i q_j}{4\pi D r_{ij}} + \varepsilon_{ij}[(\frac{Rmin.ij}{r_{ij}})^{12} - (\frac{Rmin.ij}{r_{ij}})^6]$$

In which the first five terms are for bonded interactions or internal interactions and the last two are for non-bonded interactions or external interactions. The main challenge in choosing the force field is to select the right parameters which ensure sufficient coverage of the chemical space occupied by the molecule of interest to generate the best suited model for the same. Depending upon these parameters and the suitability of the calculations for a specific type of a molecule different force field can be picked, a few of them are mentioned here:

Molecular Mechanics Force Fields (MM2, MM3 and MM4)

Initially, MM2 force field was developed to utilize structural data and heat of formation in 1976. Then MM3 was used to cover rotational and conformational energies parameters of hydrocarbons. Later to introduce vibrational energies and to model physical characteristics of compounds more appropriately MM4 force field was introduced.

Merck Molecular Force Field (MMFF)

It was developed to deal with large molecules such as biomolecules and drugs. It was therefore a preferential choice to study drug – target interactions to an extent.

Universal Force Field (UFF)

It was used to estimate the bond angles, connectivity and hybridisation of the molecules. However, it was not able to define strained compounds as it doesn't include dipole interactions.

Chemistry at Harvard Molecular Mechanics (CHARMM)

This force field was designed to treat complex structures *e.g.* lipids of different membranes and other macromolecules. The advanced versions of HAFMM are good enough to do simulation study and modeling of sterols and nucleic acids.

Assisted Model Building with Energy Refinement (AMBER)

It was initially designed for optimization of nucleic acids by including all degrees of freedom. Additionally, it was able to include parameters like partial charges and hydrogen bonding in the overall calculations.

Similarly, other mentionable force fields are Class II Force Field (CFF), SHAPES, GROMOS which have their own strengths and drawbacks. The big challenge in recent research is to introduce polarization effects in the force field calculations to carry out the modelling of molecules especially of organic and biological origin. This is especially needed for computer aided drug designing as the drug-ligand interactions should preferably be studied in water as the solvent medium. The challenge enhances on considering a polar molecule in water. Poisson followed by Onsager developed an equation so as to predict the permittivity's of solvents to develop a better understanding of polarizabilities and dipole moments. So, an extension from the partially charged model that was fixed is required to attain the goal of computational drug design approaches. Thus, CHARMM force field was modified to develop these calculations by addition of polarizability terms. There are different types of models to introduce polarizability into calculations for force fields. Like the point dipole model and fluctuating charge model. Point polarizabilities are also added in AMBER. Owing to inaccuracies in results where polar residues are involved, there was an urgent need to develop a polarizable force field that would treat polarization effects specifically. Many researchers, to develop a polarizable force field in this regard, have already initiated an innovation drive [68]. The limitations of these methods owe to the force field based on molecular mechanics techniques that are used and also to the parameters sets that are shortlisted for calculations. Molecular dynamics calculations lay more emphasis on optimization of potential energy and not on the protein's free energy which implies that the contributions of entropy that result in the polypeptide chain and protein's thermodynamic stability are ignored. Hydrophobic effects and hydrogen bonding between different molecules are also not included as modern force fields don't include them explicitly. Moreover, the Lennard Jone potential is applicable for systems in vacuum to calculate the van der Waals forces interaction which are actually of electrostatic origin and hence, have dependency on dielectric constant. Therefore, in Molecular Dynamics, the dependence on immediate environment of the macromolecule is often neglected while carrying out simulations run. This can be improved and hence, corrected by including polarizable force fields.However, acquiring knowledge based on structure based drug designing and the dynamics functionalities of the macromolecules is pertinent in order to develop a deeper understanding of the main features of discovery of newer drugs with great potential. There are various reports published that have used molecular dynamics simulations to develop novel drugs against diseases such as Nasution *et al.* to analyse the stability of complexes between lead compounds and nucleoprotein of Ebola virus [69], Sulaiman *et al.* used Amber14 for simulations in order to calculate binding free energies of drugs with VP35 protein of Ebola virus [70].

Quantum/Molecular Mechanics

Quantum mechanics (QM) can be effectively used to improve the accuracy of various steps that are involved in CADD. The advantage of QM in defining chemical processes in combination with molecular mechanics (MM) provides superior information about various interactions and conformational changes. This hybrid approach is used to study enzymatic reactions that are generally the targets for drug designing. The role of QM/MM calculations is to portray the various binding events happening between the drug and the target protein. QM/MM method takes various parameters into account like polarization and charge density in order to define the interactions at each step. While other methods use the average crystal structure with static charge density which is not able to describe the pathways and various transition states involved in the processes [71].

The various application of QM/MM approach includes:

• To study the electrostatic interaction of active sites of target with the drug molecules.

• To more accurately assess the electron density of various conformations.

• To measure ligand binding free energy more precisely [72].

QM/MM method performs calculations based on charges developed during interactions between ligand and active sites of the protein which provides accurate results in their natural state. Basically, the active site of protein target is described at an appropriate QM level. Then molecular mechanics force field is applied to define the remaining region of the targeted structure which is not involved in any kind of interaction. The total energy of the system can be written as shown in equation 1:

$$E_{Total} = E_{MM} + E_{QM/MM} + E_{QM} \tag{1}$$

Where, terms E_{MM} and E_{QM} corresponds to energies related to MM and QM sections respectively. $E_{QM/MM}$ describes the interactions between the QM and MM sections. Thus, the hybrid QM/MM approach should combine both classical and quantum concepts in order to get a significant $E_{QM/MM}$ quantity. Based on the forms of $E_{QM/MM}$ quantity (electrostatic, mechanical and polarization) three different schemes for QM/MM methods have been formulated:

(i) *Electrostatic coupling* is the most popular scheme where QM Hamiltonian is directly inserted with MM charges. It is also defined as the interaction between point charges and charge density.

(ii) *Mechanical coupling*: there is no effect of point charges of MM section on QM Hamiltonian. Although, the interaction between MM and QM sections is accounted by force field.

(iii) *Polarization coupling* is the costliest one due to complexity introduced by providing flexibility to classical charges. Here, both MM and QM sections are exposed to polarization process that enhances the electrostatic forces [73].

Fragment Molecular Orbital Method

Quantum mechanical techniques are being efficiently used in order to gain theoretical accuracy to study molecular activities. However, the calculations involved in QM methods are highly complex and require huge time. So, the need of much cheaper and faster alternative for processing QM calculations accurately is met by fragment molecular orbital method (FMO) [74]. This method was first introduced by Kitaura *et al.* [75]. The basic principle was to distribute electrostatic field of complete system into fragmented calculations. The FMO approach can be applied to various substrates like organic, inorganic, nucleotides and proteins. however, it fails to provide accurate results with some systems such as metal-nanoclusters.

Here, atoms are represented as electron density distributed in a localised fashion. FMO method generally utilize two steps, in the first step QM fragment calculations are performed in the electrostatic field of active site of the protein and quantum effects are achieved at the fragment part. Then in the next step fragment pair calculations are analysed in the electrostatic field while taking repulsion and charge transfer into account.

The FMO method is being widely used in the development of novel drugs some of the major advantages of it are:

• It is considered to be less expensive as compared to other QM computational methods.

• It allows the analysis of intra-fragment interactions of various active sites of the target with the drugs.

• FMO also allows the prediction of geometry with thorough analysis of molecular interactions.

• FMO methods can be performed simultaneously with other programs on supercomputers which allow large scale screening of active compounds [76].

A BRIEF OVERVIEW OF THE ROLE OF CADD IN THE DEVELOPMENT OF DRUGS AGAINST VIRAL DISEASES

The economic, social and clinical load of the various viral infections on the population of different countries is huge. Bringing a new drug in the market takes years and also not that much effective. The treatment of most of the viral diseases is insufficient till now due to lack of effective drugs and continuous development of resistance of viruses against the medicines. Also, these viral infections are highly contagious and can affect huge number of people if proper care and measures are not taken. So, continuous efforts are required in order to improve the efficacy of drugs in which computational methods combined with quantum mechanics tools are providing desireful results [77].

An overview of last six years is given for describing the role of *in silico* approach towards the development of novel drugs against viral diseases such as Ebola, Zika, Hepatitis C and Coronavirus.

Ebola Virus

Ebola virus disease also known as Ebola haemorrhagic fever is yet very rare but fatal disease. It is generally transmitted through animals and was first discovered near Ebola River in Congo in 1976 which after time to time showed occasional outbreaks in nearby African countries. The serious outbreak in 2014-16 resulted in above 28000 cases of Ebola virus disease and near about 11000 casualties. It spreads through direct contact of body fluids or by ingestion. There are no specific symptoms at the initial stage of viral entry but after one-week host started to have symptoms similar to flu. The patients generally suffer from fever, muscle pain, sore throat, headache and fatigue. As the viral infection progresses it results in vomiting, diarrhoea and abnormality in liver functions. In severe cases, it can cause external and internal bleeding with haematological disorder which finally results in death. Diagnosis of Ebola virus disease is very difficult as its symptoms are similar to other infectious disease like typhoid and malaria. For the same reason World Health Organisation (WHO) has classified Ebola as one of the deadliest diseases on the Earth and in 2019, declared it as Public Health Emergency of International Concern (PHEIC) in Democratic Republic of the Congo in 2019. Currently there is only one emergency vaccine (rVSV-ZEBOV) available for the prevention of outbreak of Ebola virus [78].

Ebola virus is an enveloped, negative, single stranded RNA virus belonging to Filoviridae family. The genome of Ebola virus contains approximately nineteen thousand bases and there are seven proteins which have their own specific role. The seven associated proteins include membrane proteins (VP24 and VP40), Glycoproteins (GP), Phosphoprotein (VP30), Nucleoproteins (NP), Nucleocapsid protein (VP35) and RNA polymerase protein (L). These proteins are associated with certain function such as glycoprotein helps in the attachment of virus through host cell and protein interaction. After the attachment virions enter the host cell by endocytosis which are further matured with the help of VP40. While VP35 have two functions, first is viral replication and second task is to inhibit the production of interferons in the cell. Nucleoprotein on the other hand forms helical complex with the help of RNA while VP24 acts as inhibitor for signalling paths and helps in formation of nucleocapsid. Further the interaction between nucleocapsid (NP) and VP30 interfere with transcription step [79, 80].

Due to lack of efficient vaccine continuous efforts in the following direction is still going on. Computer aided drug designing is playing a vital role in the filtration of active compounds from huge databases through structure modeling for Ebola virus. Recent advancement in development of lead compounds that can target various progressing stage of Ebola virus genome and can inhibit the viral infection is given in Table **1**.

Table 1. Role of CADD in the development of drugs against Ebola virus.

Target	Database	Software	Lead Compound	Ref
Ebola virus nucleoprotein	ZINC database & RCSB Protein Databank	MOE 2014.09, FAF-Drugs3, Toxtree v2.6.13 & SwissADME	ZINC56874155 & ZINC85628951	[69]
Ebola VP35	Mcule database	AutoDock Vina & Amber14	Compound A (MCULE-1018045960-0-1)	[70]
Ebola VP30	PubChem database	AutoDock 4.2	6-Hydroxyluteolin & (-)-Arctigenin	[79]
Ebola virus glycoprotein	PubChem database & Protein Databank	Autodock 4.2.6	AC1LA8DY	[80]

(Table 1) cont.....

Target	Database	Software	Lead Compound	Ref
Ebola VP24	NANPDB databases & AfroDB	AutoDock Vina & GROMACS	ZINC000095486070, ZINC000003594643, ZINC000095486008 & NANPDB135	[81]
Ebola (VP24, VP35 & VP40)	Protein Data Bank & ZINC Database	AutoDock Vina 4.2 & iGEMDOCK	Deslanoside & Digoxin	[82]
Ebola VP24 and MTase	Drugbank & ZINC database	Autodock Vina, Audodock4, Autodock Vina, PLANTS, and Surflex	Indinavir & Sinefungin	[83]
Ebola VP35	Protein Data Bank	Vida version 4.2.1, AutoDock 3.0 & Omega version 2.4.6	Nateglinide, Telmisartan & Ticarcillin	[84]
Ebola VP40	RCSB, Traditional Chinese Medicine Database (TCMD)	AutoDock 4.2.6 & GROMACS version 4.6.4	Emodin-8-beta-D-glucoside	[85]
Ebola VP40	Protein Data Bank, UCSF Chimera & NCBI PubChem	AutoDock Vina	Gymnastatin, Sorbicillactone, Marizomib & Daryamide	[86]
Ebola (VP35 and VP40)	OTAVA chemicals, Pubchem & Analyticon Discovery	Autodock Vina & PyMOL Version 1.5.0.	-	[87]
Ebola (VP40, VP35, VP30 & VP24)	Protein Data Bank & Flavonoid library	Maestro version 9.7 & Glide module version 6.2	Gossypetin & Taxifolin	[88]
Ebola (VP24, VP30, VP35, VP40 & NP)	Protein Data Bank & ZINC database	Auto Dock tools v. 1.5.6, Chimera v. 1.8 & PyMol v.0.99r c6	Robustaflavone, Cepharanthine, Corilagin, Hypericin & Theaflavin	[89]
Ebola virus NP	NCI, ZINC, Asinex, LifeChemical, ChemBridge, MayBridge, Enamine & Specs databases	GROMACS 4.6.3 & Glide	NSC 100858, ZINC85593801, F2351-0320, AN-329/40521909, SEW06485, Z199219280, AE-848/12123126 & SYN 15609696	[90]
Heat shock protein 90 (Hsp90)	HerbalDB & RCSB-PDB database	MOE 2014.09	1-O-galloyl-6-O-luteoyl-α-D-glucose, Euphorbianin & Scutellarein 7-neohesperidoside	[91]

(Table 1) cont.....

Target	Database	Software	Lead Compound	Ref
Indoleamine 2,3-dioxygenase enzyme	Protein Data Bank, CHEMBL & Swiss Protein Database	CanSAR integrated protein annotation tool	CanSAR ID 438588 & CHEMBL 259312	[92]

Computational aided drug discovery has played a crucial role in prioritising the novel targets to inhibit the function of different Ebola virus proteins responsible for viral infection. This *in silico* methodology is adopted by several researchers such as Nasution *et al.* targeted nucleoprotein of the Ebola virus whose structure was obtained from RCSB Protein Databank to develop novel drug. They collected around 1,90,084 natural compounds from ZINC15 database which were filtered on the basis of their druglikeness and toxicity profile with the help of Data Warrior v.4.5.1. The molecular docking simulations to determine the stability of complex between hits and the Ebola virus nucleoprotein was performed using MOE 2014.09. Then pharmacokinetic features and bioavailability were predicted using FAF-Drugs3, Toxtree v2.6.13 and SwissADME softwares which helped in selecting two ligands (ZINC56874155 and ZINC85628951). Finally, Calbistrin C (ZINC56874155) was selected with the help of docking simulation due to lowest binding free energy and they predicted that it can show good potential against Ebola virus [69].

Sulaiman *et al.* targeted VP35 protein of Ebola virus through virtual screening. They screened out five active molecules out of 36 million molecules from Mcule database. The screening and docking analysis were carried out using AutoDock Vina. Molecular dynamic simulations were done to measure binding free energy of protein and drug using Amber14. Later bioavailability and druglikeness was tested by utilizing Swiss ADME tool. The toxicity studies were done by using pkCSM tool and they predicted that compounds A, C and E were non-mutagenic. All the five compounds demonstrated good potency (Ki range 0.3-1.6 μM) against the inhibition of VP35. Compound A was having the highest potency with K_i value of 0.7 μM and is a novel predicted drug as VP35 inhibitor [70].

Another work from Venkatesan *et al.*, who targeted VP30 protein of Ebola virus which is crucial for initiation of transcription process. The SMILES file of 120 lead molecules were obtained from PubChem database. Then the selected molecules were subjected to ADMET analysis and Lipinski rule approval with the help of pkCSM online server. The screened out 36 compounds were then selected for docking analysis using AutoDock 4.2. The selected compounds demonstrated non-mutagenic, non-toxic with high permeability and binding affinity towards VP30 protein. Finally, the two lead molecules (Hydroxyluteolin and (-)-

Arctigenin) were obtained that are promising novel candidate to stop the Ebola virus infection at an early stage [79].

Similarly, Shinyguruce *et al.* targeted glycoprotein of Ebola virus using two antiviral drugs Fostemsavir and Vicriviroc. The crystal structure of glycoprotein was obtained from Protein Data Bank. The docking analysis of compounds with protein structure was performed using Autodock 4.2.6. The results showed that Vicriviroc had better affinity towards glycoproteins of Ebola virus. Thus, they further focused on twenty analogue structures of Vicriviroc which were obtained from PubChem database. The five compounds out of twenty were found to have better hydrogen bonding and binding score with the glycoproteins. The ADMET properties and Lipinski's rule of five were utilized for evaluation of toxicity, metabolism and absorption by using SwissADME predictor tool. The results from ADMET analysis and binding score showed that one of the compounds (AC1LA8DY) was able to achieve best orientation with the active site of the glycoprotein. Hence, they concluded that their lead molecule has the potential to inhibit the infection of Ebola virus [80].

Kwofie *et al.* on the other hand targeted VP24 protein of Ebola virus which plays a vital role in replication and pathology cycle. AutoDock Vina was utilized to screen out 7675 natural compounds from NANPDB and AfroDB databases (catalogue of ZINC database). Then ADMET, DataWarrior and SwissADME tools were used to test pharmacokinetics and bioavailability of the top 19 hits. Finally, four compounds sarcophine, ZINC000003594643, ZINC000095486008 and ZINC000095486070 shown good inhibitory properties against VP24. The docking results revealed that the ZINC000095486070 had the highest binding affinity due to lowest binding free energy (-9.7 kcal/mol) which was much better than other standard VP24 inhibitors [81].

Zhao *et al.* focussed on Ebola's RNA polymerase activity and targeted VP24 and MTase by adopting drug repurposing approach. Homology modelling was used to build MTase structure which was further verified by Verify3D. The binding site analysis was done using SMAP for thousands of compounds. The top hits from DrugBank and ZINC database were subjected to docking analysis through four programs Audodock4, Autodock Vina, PLANTS, and Surflex. They found that the Sinfungin (antifungal drug) was able to efficiently inhibit RNA polymerase activity whereas the second candidate, Indinavir (HIV protease inhibitor) was found to be effective in disrupting interactions between human and Ebola virus [83].

Remarkable progress in the development of structures of therapeutically relevant Ebola virus protein targets has been made in order to identify potential hits against this disease. Such hit compounds can further help in forming a base for designing newer efficient drugs. Nevertheless, it is clear that both SBDD and LBDD methods have been used extensively against the targets but still there is lot of scope left in exploring new CADD based applications.

Zika Virus

Zika virus is an arbovirus (arthropod borne virus) with single stranded RNA and belongs to flavivirus genus, Flaviviridae family. The primary carrier of Zika virus is *Ades aegypti* and *Aedes albopictus* mosquitos which are generally found in tropical regions. But due to continuous climatic changes these species are spreading to other parts of the world. The viral infection is associated with fever, vomiting, abnormalities in central nervous system and Guillain-Barre syndrome. After its first detection in Uganda in 1947, it has shown increased number of cases in different countries of Africa, Asia and South America. The serious outbreak of Zika virus was observed in 2015 in Brazil which later on spread to other South America countries which resulted in over 20,000 cases. According to WHO, above 86 nations provided confirmation of Zika virus infections and in 2016, the organisation declared it as PHEIC due to its severity. Even after the serious outbreaks, there is no vaccine available for this disease [93].

The genome of Zika virus contains 10,794 bases with non structural proteins (NS) and three structural proteins (Capsid (C), envelope (E) and precursor membrane (prM)). The different non structural proteins have specific functions like NS1 forms the connection between host and the virus through glycolipid, NS3 helps in replication and capping process of Zika virus, NS5 regulates the inhibition of host immune system. Other structural proteins like NS2A, NS2B, NS4A and NS4B helps in packaging, viral replication and RNA replication. C protein plays a vital role in capsid formation for the maturation of virus. An E protein function is to help in the binding and penetration into the host cell. The last structural protein, prM gives protection to the envelope through formation of heterodimers. Thus, all the non structural and structural proteins have vital role in replication process and providing stability respectively which can be targeted to stop the infection [94].

As already discussed, there is no specific vaccine and treatments to inhibit the infection of Zika virus. So, continuous efforts are required to develop novel drugs against the virus in which computational aided drug designing is playing a vital role to cut cost, labour and time. The recent development in *in silico* approaches towards the discovery of drugs for Zika virus from various databases is provided in Table **2**.

Table 2. Role of CADD in the development of drugs against Zika virus.

Target	Database	Software	Lead Compound	Ref.
Zika virus envelope protein	ZINC database	AutoDock Vina	ZINC33683341 & ZINC49605556	[64]
Zika virus	Protein Data Bank	OEDocking 3.0.0, GNOME 2.28.2 & SZYBKI 1.5.7	-	[95]
Zika virus (MTase & RdRp) of NS5 protein	ZINC database, NCBI & RCSB protein databank	AutoDock Vina & Amber 14, Modeller (version 9.1)	ZINC64717952 & ZINC39563464	[96]
Zika virus related Genes	CanSAR Database, CheMBL & Swiss Protien Database	canSar 2.0 & GeneCards Suite	AXL & EIF2AK2	[97]
Zika virus RNA polymerases (RdRp)	-	AutoDock 4.2 software package & Python Molecule Viewer 1.4.5	Pyridobenzothiazole compound (HeE1-2Tyr)	[98]
Zika virus NS1 protein	Protein Data Bank	AMBER, AutoDock Vina & GOLD	Deoxycalyxin-A, Sanggenon-O & AC1M2ZZJ	[99]
Zika virus NS2B-NS3 protease	Protein Data Bank, ZINC & PubChem database	Glide version 6.9	ECGC-7-oa-Glucopyranoside, Isoquercetin, Rutin & ECGC--oa- Glucopyranoside	[100]
Zika NS2B-NS3 protease & NS5 methyltransferase	Protein Data Bank & NCBI Genbank	Molegro Virtual Docker (Version 6.0)	Polyphenolics	[101]
Zika virus NS2B-NS3 protein	ASINEX	GLIDE module, Desmond module & Qikprop module	BAS 19192837	[102]
Zika virus NS3-helicase	Protein Data Bank	Glide & Schrodinger Maestro 18.2	MN-9 & MN-10	[103]
Zika virus NS3 helicase	Protein Data Bank & NCI database	SYBYL-X 2.1.1 & MOLARIS version 9.15	Compound (1 & 2)	[104]
Zika virus NS3 Protein	NCBI database, ZINC database & PubChem database	AutoDock Vina, PyRx 0.8 & Phyre2	ZINC53047591	[105]

(Table 2) cont.....

Target	Database	Software	Lead Compound	Ref.
Zika virus NS5	Protein Data Bank & Life chemicals database	GOLD version 5.4.1	F3043-0013, F0922-0796, F1609-0442 & F1750-0048	[106]
Zika virus NS5 Protein	Protein Data Bank	PyMOL software	-	[107]
Zika virus NS5 Methyltransferase	DTP AIDS Antiviral Screen Database & Protein Data Bank	DOCK6/MolAr, AMBER score & DataWarrior V4.7.2	ZINC1652386	[108]
Zika virus NS5 & NS3 proteins	Swissprot database & Immune Epitope Database	PatchDock & AutoDock	VEMGEAAAI	[109]

Huge diversities of therapeutically relevant drugs against Zika Virus proteins or enzymes have been developed through computational methods. The methodology is adopted by numerous researchers to save money and time such as Ramharack and Soliman, followed *in silico* approach for targeting NS5 protein for the development of novel drug for Zika virus. Due to lack of NS5 structure, homology model was created with the help of Modeller (version 9.1) by using protein sequence template obtained from National Centre for Biotechnology Information's (NCBI) and RCSB protein databank. Then two-dimensional structure of the protein was obtained by uploading the sequence to PSIPRED V 3.3. The molecular docking analysis was performed using AutoDock Vina to screen ZINC database to reduce the hits from a total of 41 to 2 compounds (ZINC39563464 and ZINC64717952) based on binding affinities and interactions. Later molecular dynamic simulations to analyse ligand-protein interactions were done using Amber 14 for the two compounds. The complex formed with target by lead molecules (ZINC39563464 and ZINC64717952) have demonstrated high stability and have potential to inhibit the effect of Zika virus [96].

Raza *et al.* targeted the NS1 protein of Zika virus whose three-dimensional structure was obtained from Protein Data Bank which is responsible for the replication process. Docking analysis was done by following two routines with help of AutoDock Vina and GOLD. AutoDock Vina was used for blind docking to remove non-specific interaction of protein with ligands. While GOLD was used to analyse active site directed attachment to NS1 protein through genetic algorithm under optimized conditions. Finally, three compounds (AC1M2ZZJ, Deoxycalyxin-A and Sanggenon) were screened out from 90 compounds. Hydrogen bonding and other interactions between docked ligand and active site of protein was done by sing Visualizer 3.5 and Discovery Studio. ADME and molecular properties were measured by using ADMETSAR while CarcinPred-EL was used to analyse carcinogenic properties. Molecular dynamic simulations were carried out using AMBER to calculate binding free energy and all the three compounds showed excellent inhibitor properties against NS1 protein. They concluded that out of three, Deoxycalyxin-A demonstrated good binding free energy values and bonding pattern consistency [99].

Byler *et al.* targeted NS2B-NS3 protein of Zika virus which is responsible for cleaving polyprotein of the virus into individual proteins that is crucial for viral replication and NS5 methyltransferase. The structure of the target was generated with the help of homology modelling by using crystal structure templates that were obtained from Protein Data Bank and NCBI GenBank. Based on this homology structure of NS2B-NS3 protein, docking analysis was performed through Molegro Virtual Docker (version 6.0). They found that the polyphenolics gave best docking results with both the targets [101].

Similarly, Rohini *et al.* discovered novel medicinal relevant compound by targeting NS2B-NS3 proteins of Zika virus using e-pharmacophore drug design technique. They screened out potential hits from 467,802 molecules of ASINEX database. PHASE module was used for the generation of three-dimensional e-pharmacophore model after which high throughput virtual screening was used to filter hit molecules. These hit molecules were then examined for ADME properties and binding free energies with the help of Qikprop module and Prime MM-GBSA respectively. All the docking analysis was performed using GLIDE module which provide information about electrostatic, hydrophobic and hydrogen bond interactions. The stability of these selected compounds was then analysed using molecular dynamics simulation. Finally, they suggested one compound (BAS 19192837) from the hits that could show good inhibition against NS2B-NS3 proteins attributed to the presence of imidazole ring [102].

Zhu *et al.* applied *in silico* approach towards the discovery of inhibitor against NS3 protein. The crystal structure of NS3 protein was obtained from Protein Data Bank. All the docking analysis was carried out using SYBYL-X2.1.1 for the screening of potential hits. The binding free energies calculations were done through molecular dynamics simulations by using MOLARIS (version 9.15). Cytopathic inhibition assay was employed to determine potency (IC_{50}) of the hit compounds. Finally, two novel compounds (Compound 1 and Compound 2) were screened out from NCI database by carrying antiviral assays with computational analysis which showed high potency (8.5 and 15.2 μM) against NS3 protein [104].

Sahoo *et al.* focussed on targeting NS3 protein of Zika virus which is responsible for viral replication. The sequence of the NS3 protein was obtained from NCBI database which helped in predicting the structure with the help of homology modeling technique. They identified that berberine (structure obtained from PubChem database) was showing good binding affinity and done the similarity search by using ZINC database. The virtual screening was done using AutoDock Vina. The scoring results of docking analysis provide ten novel drugs, out of which ZINC53047591 (2-(benzylsulfanyl)-3-cy-clohexyl-3H-spiro[benzo[h] qui-nazoline-5,1'-cyclopentan]-4(6H)-one) was found to interact with NS3 protein with least binding energy [105].

It is clear from above discussions that the CADD is gaining lot of interest in medicinal field in identifying novel drugs against relevant Zika virus proteins. The classification of such lead compounds from various works on the basis of their structural and conformational bioactive properties can be a milestone in the development of effective vaccine for the Zika virus disease.

Hepatitis C Virus

Hepatitis C is a disease related to liver that is triggered by hepatitis C virus. The virus has the potential to affect the host both acutely and chronically, ranging from few days to lifetime illness. Hepatitis C is one of the predominant reasons for liver cancer and according to WHO approximately 4 lakh people had died by this disease in 2016. The main reason of spreading of hepatitis C infection is through blood exposure which can be due to transfusion, sexual practices and unsafe handling of medicinal equipment. The virus is a single stranded RNA, positive sense and belongs to Flaviviridae family. According to WHO there is no effective vaccine to treat Hepatitis C till now [110]. The genome of the virus contains three structural proteins (Core protein and two glycoproteins) and seven non structural proteins (NS2, NS3, NS4A, NS4B, NS5A, NS5B and p7 virions). Non structural proteins are associated with the RNA replication, viral assembly and release of hepatitis virus. While the structural proteins like core protein is responsible for the synthesis of nucleocapsid and production of infectious particles. Whereas, glycoproteins have crucial role in the life cycle of the virus along with establishing contact and entry into the host cell [111].

Since, there is no vaccine available for Hepatitis C and current therapeutic methods suffer from low potency and side effects like anaemia, depression *etc.* So, development of new drugs for inhibiting the hepatitis C virus by targeting its genome using computational methods is useful to minimise both time and cost. Recent *in silico* approaches that helped in prioritizing the lead compounds from vast molecule databases to design active drugs for Hepatitis C virus are shown in Table **3**.

Table 3. Role of CADD in the development of drugs against Hepatitis C virus.

Target	Database	Software	Lead Compound	Ref.
Hepatitis C virus	Protein Data Bank	Surflex-Dock	7e	[112]
Hepatitis C virus NS3-4A Protease-Helicase Enzyme	Protein Data Bank & Bank of Molecules of the Amazon	Molegro Virtual Docker 5.0, DOCK 6.3 & OMEGA 2.2	(-)-6-transmetoxi-khellactona & Surinamensin	[113]
Hepatitis C virus NS3-4A Protease	RCSB Protein Data Bank & PubChem database	Patchdock program, GROMACS package 4.5.3 & MOLINSPIRATION program	CID 58428446, CID 71276250 & CID 71276290	[114]
Hepatitis C virus NS3/4A protein	Protein Databank	AMBER 10 & Gaussian09 suite	4VA	[115]
Hepatitis C virus NS3 and NS5B proteins	Protein data bank, PubMed literature database & PubChem database	Auto Dock 4.2 & DRAGON	Indinavir & AT130	[116]
Hepatitis C virus NS5B protein	ZINC database & Protein Data Bank	CDOCKER Module & AMBER10	ZINC 49888724	[117]
Hepatitis C virus NS5B protein	InterBioScreen database & CHEMBL database	Glide	N1, N2, N3, N4 & N5	[118]
Hepatitis C Virus NS5A Protein	-	MOE & AMBER14	ABT-267, AZD, GSK2336805, Ledipasvir, SYN-395 & SYN-776	[119]
Hepatitis C virus NS5B Protein	Prestwick database & Gene Expression Omnibus database	GLIDE, GROMACS 5.1.0, LIGANDSCOUT 3.12 & Maestro version 9.2	Fluvastatin & Olopatadine	[120]
Hepatitis C virus NS5B protein	Protein Data Bank	AutoDock Tools 1.5.4	Sofosbuvir, IDX184, R7128 & Ribavirin	[121]

(Table 2) cont.....

Target	Database	Software	Lead Compound	Ref.
Hepatitis C virus NS5B protein	Protein data bank & CHEMBL database	STATISTICA & DRAGON	AHV-1, AHV-2, AHV-3, AHV-4, AHV-5 & AHV-6	[122]
Hepatitis C virus NS5B	Protein Data Bank	AMBER 10 package, Gaussian09 software & PyMol program	PF-00868554	[123]
Hepatitis C virus NS5B	Protein data bank	AMBER10 & Gaussian09 program	BMS-791325	[124]
Hepatitis C virus p7 protein	ChEMBL database & RCSB Protein Data Bank	AutoDock Vina & AMBER 10 package	ChEMBL ID (564216, 552486, 558654, 1956329, 1956330 & 453664)	[125]
Hepatitis C virus p7 protein	ZINC Database	Autodock version 1.5.6, MOE & LeadIT	BIT225	[126]
Hepatitis C virus gene	STITCH & Drugbank	Weka	Acyclovir (CID000002022) & Ganciclovir (CID000003454)	[127]

Remarkable varieties of medicinally relevant Hepatitis C proteins or enzymes have been characterized and their specific function has been identified. These proteins structure have provided a base for the computational discovery of drugs against Hepatitis C virus. This methodology is adopted by numerous researchers such as Pinheiro *et al.* used virtual screening method for the development of NS3-4A enzyme inhibitor by using natural products derivatives from the Amazon region. They employed 664 molecules from the extracts of vegetal species from the Amazon area and such set of molecules is usually called as Bank of Molecules of the Amazon (BMA). The two-dimensional structure of such compounds was converted in three dimensional structures with help of OMEGA 2.2. All the docking experiments were carried out using DOCK 6.3 and Molegro Virtual Docker 5.0.0 software to predict active conformations of the ligands of BMA that are docked in the pocket of NS3-4A protease whose crystal structure was obtained from Protein Data Bank. Through the docking scores 5 compounds (250, 263, 393, 394 and 396) were selected. Later three compounds were eliminated by using Lipinski's of five. The compounds 250 ((-)-6-trans-methoxy-khellactona) and 263 (surinamensin) were found to show hydrogen interactions with the amino acid residues. Also, root mean square deviation graphs depicted that compound 250 suffered mild fluctuations in structure which was stable up-to 1000 ps of simulation time while compound 263 was stable after 1000 ps of simulation time. Finally, they concluded that the selected two compounds showed good inhibition properties against NS3-4A enzyme and high stability in the active site. Small molecules derived from natural products that exhibit anti-Hepatitis C Virus activity can be a promising candidate to cut the problems of availability and hence cost [113].

Kumar *et al.* used virtual screening approach to target NS3-4A protease whose structure was retrieved from RCSB Protein Data Bank, which is responsible for cleaving polyproteins. They have done a similarity search based on an already known potent inhibitor, MK-5172 in order to develop new potent leads. The virtual screening was performed for PubChem database to identify potent compounds. The similarity search provided 32 hits which were undergone pharmacokinetic analysis with the help of MOLINSPIRATION program. After virtual screening the docking analysis of hit compounds were done using Patchdock program. Finally, three compounds (CID 58428446, CID 71276250 & CID 71276290) were shortlisted based on their docking energies and Root mean square deviations (RMSD) values [114].

Venkatesan and Dass approached direct acting antivirals (DAAs) to target the NS5B and NS3 non-structural proteins of the Hepatitis C virus. PubMed and PubChem database were used to select 82 antiviral inhibitors based on binding affinities with active sites of NS5B and NS3. STATISTICA 13.0 was used for hierarchical clustering based on the principle that similar inhibitors have similar properties. After clustering non Hepatitis C Virus inhibitors having same features as that of Hepatitis C Virus inhibitors, docking experiment was performed using Auto Dock 4.2. Protein data bank was used for docking to retrieve cartesian coordinates of NS3 and NS5B. Finally, on the basis of docking studies it was hypothesized that Indinavir and AT-130 were considered to have good characteristics to target Hepatitis C virus proteins by preventing the viral replication [116].

Another work by Nutho *et al.*, they identified NS5B polymerase inhibitor by virtual screening. The structure of NS5B polymerase of Hepatitis C virus complex was retrieved from Protein Data Bank. They used docking studies with the help of CDOCKER module to screen out 5 hits (49888724, 49780355, 49777239, 49793673, and 49054741) from top 50 ZINC compounds of the database. The selected hits were then subjected to steered molecular dynamics simulations based on the hypothesis that high binding efficiency is related high rupture force. Finally, the results demonstrated that ZINC compound 49888724 was found to have maximum rupture force showing excellent inhibitory potency and binding strength against NS5B polymerase. They predicted that the selected compound can yield double fold increase in inhibition properties than other compounds due to high Van der Waals interactions between ligand and NS5B. Also, it can serve as an excellent alternate for NS5B inhibitor or help in lead optimization as a template in future [117].

Wei *et al.* employed screening approaches based on e-pharmacophore, random forest and docking to screen out various hits from interBioScreen and CHEMBL database to target NS5B polymerase. Random forest was used to develop models to classify non-inhibitors and inhibiters for NS5B polymerase. To prepare e-pharmacophore models, six crystal structure of NS5B polymerase combined with inhibitors were used for the improvement in potency. Glide SP and XP was used with default settings to perform the docking studies. Finally, five new compounds were selected from the hits, which were obtained by applying above screening methods which were never reported as inhibitor of NS5B polymerase. The selected 5 compounds were found to have anti-Hepatitis C Virus activities (EC_{50} range, 1.61 - 21.28 μM) with good inhibition potency (IC_{50} range, 2.01 - 23.84 μM) against NS5B polymerase. Cytotoxicity studies revealed that only N2 compound displayed very low cellular cytotoxicity (CC_{50} = 51.3 μM) while other having almost no cytotoxicity (CC_{50} > 100 μM). The best activity against Hepatitis C virus was found to shown by N2 compound having selective index of 32.1 [118].

Ahmed *et al.* investigated the mechanism of action of daclatasvir (DCV) of great potency (IC_{50} ~ picomolar) as an inhibitor of NS5A protein. They used computational modeling to investigate symmetrical and asymmetrical interaction of NS5A with DCV and 6 other compounds of structure similar to DCV. They explained the definition of symmetrical ligand is related to pharmacophore symmetry and has nothing to do with symmetry in chemical composition. To support the hypothesis, they used different range of derivatives of DCV that showed excellent potency despite of variation in their chemical structure. They included three asymmetric (AZD, GSK and Ledipasvir) and three symmetric (ABT-267, SYN-395 and SYN-776) derivatives. Docking simulations were performed through pharmacophore placement method by using MOE whereas MD simulations were carried out in AMBER14. The docking studies revealed that all the DCV related compounds fits in distinctive region of space. It was observed that the bulky peptide caps on both the terminals helped in improving the potency of these inhibitors. Such enhanced potency due to bulky groups could be attributed to better interaction with lipophilic residues [119].

The above discussions clearly suggest that *in silico* techniques greatly improve the chances of getting therapeutically relevant hit compounds against the various targets of Hepatitis C virus. These recently reported efforts based on CADD will definitely help in treatment of Hepatitis C virus disease in future.

Coronavirus

Coronaviruses are generally enveloped RNA viruses that belong to the

Coronaviridae family. They all contain a genome composed of long RNA strands having crown shaped peplomers [128]. The Coronaviruses (CoV) name was first coined during 1968 because of their crown shaped morphology. The four out of six known species of CoV (229E, NL63, OC43 and HKU1) are responsible for common cold and are not so dangerous. While the other two Forms of CoV are severe acute respiratory syndrome (SARS) and Middle East Respiratory Syndrome (MERS) that can cause serious respiratory health issues. They come in the category of zoonotic pathogens as they have the ability to get transmitted in-between animals and humans. Earlier, they were not taken into account of being highly pathogenic to humans until the breakout of SARS in 2002 in the Guangdong state of China. After ten years of SARS breakout another pathogenic MERS emerged in the Middle East countries [129]. Recently in December 2019, novel Coronavirus (nCoV) or COVID-19 has emerged as another public health concern from Huanan Seafood Market in Wuhan State of China which has been declared as pandemic that has infected more than 100 countries [130].

Transmission of CoV is generally through contact of respiratory aerosols released during coughing or sneezing between two individuals. The viral infection can spread via mouth, nose, and eyes when droplets of cough or sneeze of an infected patient in atmosphere comes in contact with a healthy individual. The common symptoms of CoV include high fever, migraine, discomfort in respiration, slight respiratory problem, diarrhoea, and cough after 2 to 7 days [131]. In severe conditions patients infected with the virus can develop pneumonia. Those older than 60 years of age and have complications such as diabetes or hepatitis are generally at greater risk [132, 133]. According to WHO, there are no vaccine available for the effective treatment of CoV at present. However, various trials are going on in order to evaluate potential treatments [134].

The virion of CoV contains positive strand RNA along with the nucleocapsid protein (N) that plays a vital role in pathogenesis and transcription. The membrane of all the strains of CoV generally contains at least three proteins known as spike (S), glycoprotein and the membrane protein (M) which plays a major role in establishing contact with the host. In addition to that they contain non-structural proteins also [135].

Since, there are no specific approved drugs for the treatment of CoV. Various virtual screening methods are being employed to prioritize potential drug like compounds with least side effects to target various sites of CoV. The *in silico* development for designing novel drugs to inhibit the viral infection since the breakout of CoV is given in Table **4**.

Table 4. Role of CADD in the development of drugs against Coronavirus.

Target	Database	Software	Lead Compound	Ref
COVID-19 Protease	DrugBank, PubChem database & Protein Data Bank	Schrodinger software, Glide & AMBER18	Carfilzomib, Eravacycline, Valrubicin, Lopinavir & Elbasvir	[136]
COVID-19 RdRp	NCBI database & Protein Data Bank	AutoDock Vina	Sofosbuvir, IDX-184, Ribavirin, & Remidisvir	[137]
CoV 3CL^{pro}	Protein Data Bank, ZINC, PubChem database	Vina, Glide, GOLD & SwisParam	ZINC27332786 & ZINC09411012	[138]
CoV-Nucleocapsid protein	Zinc database, RCSB Protein Data Bank	LIBDOCK (DS 2.5) & BIA evaluation software	6-Chloro-7-(2-morpholin-4-yl-ethylamino) quinoxaline-5, 8-dione	[139]
HCoV-NL63	ZINC database, NCBI database16	Autodock	Compound N3	[140]
MERS-CoV	Coriandrum sativum L herb extract, Protein Data Bank	iGEMDOCK	Dodecanal	[141]
MERS-CoV 3C-like protease (3CLpro)	PubChem database	Schrödinger suite of software (maestro, version 11.5.011) & LigPrep	Herbacetin, Isobavachalcone, Quercetin 3-β-d-glucoside & Helichrysetin	[142]
MERS-CoV 3CLpro enzyme	NCI database & Protein Data Bank	AutoDock Vina & LigandScout	NSC159375, NSC29007, NSC337571, NSC335985 & NSC648199	[143]
MERS-CoV papain-like protease (PL^{pro})	FDA approved Prestwick, Maybridge and Chembridge libraries	GOLD5.2.2 & AMBER14 package	ZT626	[144]
MERS-CoV ORF1ab gene	GenBank, NCBI database	RNAxs tool	siRNA-1: 220.2, siRNA-2:408.3, siRNA-3:345.2, siRNA-4:322.9, siRNA-5:270.7, siRNA-6:491.6 & siRNA-7:187.3	[145]
MERS-CoV ORF1ab replicase polyprotein	NCBI database	siDirect 2.0 & RNAcofold program	JX869059.2, KF192507.1, KC667074.1, KC164505.2, KC776174.1, KF186567.1, KF186566.1, KF186565.1 & KF186564	[146]
MERS-CoV Spike protein	Antimicrobial Peptide Database version 3 (APD3), Protein Data Bank	Piper module of Schrodinger	AP00225, AP00180, AP00549, AP00744, AP00729, AP00764 & AP00223	[147]
MERS-COV Spike protein	NCBI database & Protein Data Bank	Molecular Operating Environment (MOE) tool & vaxijen v2.0	QLQMGFGITVQYGT, YKLQPLTFL & YCILEPRSG	[148]
MERS & SARS CoVs polymerases	NCBI database	SCIGRESS 3.0 software & PyMOL	IDX-184, MK-0608, Sofosbuvir & Ribavirin	[149]
MERS-CoV 3CL^{pro} & HKU4-CoV	NCI database, Protein Data Bank	LibDock & CDOCKER	Compounds 222 & 223	[150]

(Table 4) cont.....

Target	Database	Software	Lead Compound	Ref
SARS-CoV	PubChem, Chemspider, Drugbank	Material Studio program	Compounds 5, 15 & 22	[151]
SARS-CoV 3CL^{pro}	NCBI database	MODELLER (9v10), PROCHECK & SwisParam	Compound ML188 (16R)	[152]
SARS-CoV proteases (3CL^{pro} and PL^{pro})	Protein Data Bank	Autodock 3.0.5, Corina program & Chimera 1.4.1	Chalcone 6	[153]
SARS-CoV M^{pro}	-	FlexX, Sybyl 7.3	Compound 3-31 (unsymmetrical disulfides)	[154]
SARS-CoV Main Protease	NCBI database & Protein Data Bank	AutoDock vina & SeeSAR (version 9.2)	Prulifloxacin, Bictegravir, Nelfinavir & Tegobuvi	[155]
SARS-CoV-2 S Protein	ZINC15 database & Protein Data Bank	Autodock Vina & GROMACS	Cepharanthine, Ergoloid & Hypericin	[156]
SARS-CoV-2	Drug Bank Database	Patchdock	lopinavir & Remdesivir	[157]
SARS-CoV-2	Protein Data Bank, GalaXi_2019-10, KnowledgeSpace_2019-05 & REALspace_2019-12 databases	Discovery Studio software 2.5, Molegro Virtual Docker v6.0, ligandScout 4.3 & AMBER18	Losartan	[158]
SARS-CoV-2	NCBI GenBank database	SABLE server	Peptide (PELDSFKEELDKYFKNHTSPDVDLGDISGIN, FSQILPDPSKPSKRSFI, TMSLGAENSVAYSNNS, NSNNLDSKVGGNYN)	[159]
SARS-CoV-2	ZINC database & Drug Bank database	AutoDock 4.2.6 & Phyre2	Paritaprevir, Semeprevir, Grazoprevir & Velpatasvir	[160]
SARS-CoV-2	NCBI database, Drug Target Common (DTC) database & BindingDB database	AutoDock Vina (version 1.1.2) & MGLTools (version 1.5.6)	lopinavir, Ritonavir & Darunavir	[161]
SARS-CoV-2	RCSB Protein Data Bank	Autodock vina	Resveratrol	[162]
SARS-CoV-2	PubChem database, Protein Data Bank & Drug Bank	Patchdock	Atazanavir	[163]
SARS-CoV-2	PubChem database & RCSB Protein Data Bank	GROMACS & Hex Cuda 8.0.0	2-deoxy-D-glucose & 1, 3, 4, 6-Tetra-O-acet-l-2-deoxy-D-glucopyranose	[164]
SARS-CoV2 3CLpro	RCSB Protein Data Bank & ChEMBL database	Autodock4 & GROMACS-5.1.4	PI-06, PI-08, PI-11 & PI-14	[165]
SARS-CoV-2 3C-like protease (3CLpro)	SWEETLEAD database & Protein Data Bank	DOCK 6	Indinavir, Ivermectin, Cephalosporin-derivatives, Neomycin & Amprenavir	[166]
SARS-CoV-2 (3CL^{pro})	PubChem database & Protein Data Bank	Ligprep, Desmond & Schrodinger	Compound 16	[167]

(Table 4) cont.....

Target	Database	Software	Lead Compound	Ref
SARS-CoV-2 main protease (3CLpro)	ZINC15 database & Protein Data Bank	MGLTools (version 1.5.6) & AutoDock Vina (version 1.1.2)	ZINC000118795962, ZINC000003775281, ZINC000028827350, ZINC000043206238, ZINC000100472223 & ZINC000095930125	[168]
COVID-19 3CL hydrolase & protease enzymes	ZINC15 database & Protein Data Bank	Discovery Studio 4.5 & AutoDock 4.2	Talampicillin, Lurasidone, Rubitecan, Loprazolam, ZINC000000702323, ZINC000012481889, ZINC000015988935 & ZINC000103558522	[169]
SARS-CoV-2 Chymotrypsin-like protease (3CL^{pro})	RCSB Protein Data Bank	MOE & AMBER 18	Saqunavir, Remdesivir, Darunavir, Syn-16 & Nat-1	[170]
SARS-CoV-2 protease C3Lpro	Drug Bank & PubChem database	Vina & PyMol	Bictegraivir & Indinavir	[171]
SARS-CoV-2 3CL^{pro}	Protein Data Bank & PubChem Database	AutoDock vina 4.2 & Discovery Studio Visualizer version 16	10-Hydroxyusambarensine, Cryptoquindoline, 6-Oxoisoiguesterin, 22-Hydroxyhopan-3-one, Cryptospirolepine, Isoiguesterin & 20-Epibryonolic	[172]
SARS-CoV2 E protein	NCBI protein database, DrugBank 2.5 & TIPdb database	GROMACS 5.1 & AutoDock, Discovery Studio	Belachinal, Macaflavanone E & Vibsanol B	[173]
SARS-CoV-2 endoribonuclease NendoU	Phase database & Protein Data Bank	LigPrep & Schrodinger	DB00876 (Eprosartan), DB15063 (Inarigivir soproxil), DB12307 (Foretinib) & DB01813	[174]
SARS-CoV-2 main protease	ZINC database & Protein Data Bank	DOCK6, FESetup1.2 & SOMD	Aliskiren, Capreomycin & Isovuconazonium	[175]
SARS-CoV-2 M^{pro}	RCSB Protein Data Bank, NCBI GenBank database & PubChem database	AutoDock Vina & MEGA 6.0	Lopinavir-Ritonavir, Tipranavir & Raltegravir	[176]
SARS-CoV-2 M^{pro}	ZINC database & DrugBank library	AutoDock Vina & GROMACS	Amentoflavone	[177]
SARS-CoV-2 M^{pro}	PubChem database & Vietherb database	Autodock Vina & GROMACS version 5.1.5	Cannabisin A isoacteoside & Darunavir	[178]
2019-nCoV main protease	RCSB Protein Data Bank & PubChem database	AutoDock Vina	Robustone, SchizolaenoneB, Osajin, IsosilybinA & SilybinA	[179]
SARS-CoV-2 main protease	Protein Data Bank	Gaussian03 software & Discovery studio visualizer software	TMB607	[180]
COVID-19 main protease (M^{pro})	ZINC database & Phenol explorer database (version 3.6)	AutoDock Vina 1.1.2 & Pymol	Sanguiin H-6, Theaflavin 3,3'-Odigallate, Theaflavin 3-Ogallate, Protocatechuic acid 4-Oglucoside, Kaempferol 3-Oglucuronide & Punicalagin	[181]
SARS-CoV-2 M^{pro}	ZINC Database	Autodock 4.2 & Pymol 2.1	ZINC1845382, ZINC1875405, ZINC2092396, ZINC2104424, ZINC44018332, ZINC2101723, ZINC2094526, ZINC2094304, ZINC2104482, ZINC3984030 & ZINC1531664	[182]

(Table 4) cont.....

Target	Database	Software	Lead Compound	Ref
SARS-CoV-2 main protease (M^{pro})	ChEMBL database & Protein Data Bank	Glide & Standardizer v.20.8.0	Ipamorelin, Tilmicosin, Budipine, Atazanavir, Pentagastrin, Indinavir, Vinblastine, Afimoxifene, Navitoclax & Venetoclax	[183]
SARS-Cov-2 main protease	RSCB Protein Data Bank	AutoDock Vina, GOLD suite & PyMol (version 1.1)	L6, L7, L12, L15, L16, L22 & L25	[184]
SARS-CoV-2 Mpro	PubChem database & RCSB protein database	Raccoon, MGLTools-1.5.6 & Pymol-2.3.3	Carnosol, Arjunglucoside-I & Rosmanol	[185]
SARS-CoV-2 M^{pro}	PubChem database	DOCK6	Andrographolide	[186]
COVID-19 main protease	Protein Data Bank	PyMOL & AutoDock Vina	Quercetin, Hispidulin & Cirsimaritin	[187]
SARS-CoV-2 main protease	RCSB Protein Data Bank & SuperDRUG2 database	Autodock Vina & Pymol	Saquinavir & Beclabuvir	[188]
COVID-19 main protease (M^{pro})	SELLEKCHEM database & Protein Data Bank	LIGPREP version 4.2 & Maestro	Leupeptin Hemisulphate, Pepstatin A, Nelfinavir, Birinapant, Lypression & Octeotide	[189]
SARS-CoV-2 main protease	Protein Data Bank	SwissDock & Discovery Studio Visualizer	Rutin, Ritonavir, Emetine, Hesperidin, lopinavir & Indinavir	[190]
SARS-CoV-2 main protease, Nsp12 polymerase & Nsp13 helicase	ZINC database, Mcule database & Protein Data Bank	Autodock Vina & SWISSMODEL	Cmp3, Cmp12, Cmp14, Cmp17, Cmp18, Cmp1, Cmp3a, Cmp11, Cmp15, Cmp2, Cmp17a & Cmp21	[191]
2019-nCoV N-protein	RCSB database & ZINC database	QikProp & Maestro version 10.2	ZINC00003118440 & ZINC0000146942	[192]
SARS-CoV-2 2-O methyltransferase (2'OMTase)	NCBI Database & SwissProt database	Modeller v9.23, ProCheck, PyRx 0.8 & AutoDock v4.2.6	Ergotamine, Dihydroergotamine, Saquinavir, Digitoxin Chlorthalidone, Irinotecan, Teniposide, Eribulin & Zafirlukast	[193]
SARS-CoV-2 RNA dependent RNA polymerase (RdRp)	Protein Data Bank	iGEMDOCK v2.1, AutoDockVina & PyMOL	Darinaparsin	[194]
SARS-CoV2 RNA depended RNA polymerase	Protein Data Bank, NCI & ZINC database	Maestro platform	Valproic acid Co-A	[195]
nCov-2019 spike protein (S-protein)	ZINC15 database & Protein Data Bank	Autodock Vina & GROMACS	Pemirolast (ZincID: 5783214)	[196]
SARS-CoV-2 glycoprotein	RCSB Protein Data Bank & NCBI database	GROMACS 5.1.1 suite, AutoDock Vina (26) & PyMol	GR 127935 hydrochloride hydrate, GNF-5, RS504393, TNP, Eptifibatide acetate, KT203, BMS195614, KT185, RS504393 & GSK1838705A	[197]
SARS-CoV-2 SNSP15 & spike glycoprotein	PubChem database & Protein Data Bank	Schrodinger Maestro	Saikosaponin U & V	[198]

Computational aided drug discovery has attracted various researchers to develop novel drugs to inhibit the function of different CoV proteins responsible for viral infection. This *in-silico* methodologies are used by various researchers such as Rao *et al.* tried to target all the structural and non-structural proteins of the MERS-CoV whose three-dimensional images were retrieved from Protein Data Bank. Chem Sketch was used to construct the ligands that were selected from extract of *Coriandrum sativum* L herb. The docking analysis of the selected ligands was done through iGEMDOCK software. They revealed that four compounds showed affinity towards all the MERS-CoV proteins. The Dodecanal (compound A) out of four showed the best results and can be considered for inhibiting the viral activity [141].

Similarly, Jo *et al.* focused on targeting MERS-CoV 3C-like protease (3CLpro). They followed a proteolytic technique to search MERS-CoV 3CLpro inhibitory molecules with peptide labelled with EDANS-DABCYL Fluorescence resonance energy transfer (FRET) pair. With the knowledge that flavonoids have antiviral properties they screened Flavonoid library to find MERS-CoV 3CLpro inhibitors. The docking studies was done using Maestro (version 11.5.011) software to screen the compounds from PubChem database. Finally, four compounds Herbacetin, helichrysetin, isobavachalcone and quercetin 3-β-d-glucoside were found to inhibit the activity of MERS-CoV 3CLpro. They concluded that flavonoids have antiviral effects and can inhibit viral proteases functions. So, derivatives of flavonoid can be utilized to design antiviral agents against broad-spectrum of viruses [142].

Radwan and Alanazi targeted MERS-CoV 3CLpro enzyme which is essential for replication. The x-ray crystal structure of MERS-CoV 3CLpro enzyme was retrieved from Protein Data Bank. Ligand structures were obtained from National Cancer Institute (NCI) database for screening. Pharmacophore models for the targets were build using LigandScout software. All the docking studies were performed using AutoDock Vina. Finally, five compounds (NSC159375, NSC29007, NSC337571, NSC335985 and NSC648199) were selected as a potential hits against the MERS-CoV 3CLpro [143].

Lee *et al.* targeted papain like protease (PL[pro]) in order to inhibit replication of viral. The potential hits were screened from a total of 30,000 molecules from different libraries (Maybridge, FDA approved Prestwick and Chembridge libraries). The docking studies for selected hits were performed using GOLD5.2.2 while molecular dynamics simulations to measure binding affinities was done using AMBER14 program. Finally compound 6 (ZT626) was found to have potential inhibitory characteristics against the MERS-CoV target [144].

Qamar *et al.* targeted MERS-COV spike (S) protein in order to inhibit viral infection. The structure of the target was obtained from NCBI database and Protein Data Bank. The docking analysis was performed using MOE tool and later Pymol and UCSF Chimera tools were used to get images of docked complexes. The docking results depicted efficient inhibitory properties against the viral infection by predicted peptides (QLQMGFGITVQYGT, YKLQPLTFL and YCILEPRSG) [148].

The recent outbreak of COVID-19 has challenged the medical security of various nations and compelled their researchers to find vaccine against this contagious disease immediately. This resulted in availability of numerous reports suggesting class of hit compounds that are ready for further improvement and clinical trials. Here, a tabulated list of such reports containing updated efforts toward developing a novel vaccine for COVID-19 has been provided in Table **4**. Both structure and ligand based drug designing approach have played crucial role in the improvement of chances for repurposing and screening of effective drugs against this viral disease.

FUTURE PERSPECTIVE

The aim of CADD is to design and discover therapeutically relevant compounds by applying different quantum and molecular techniques. The *in silico* approach have made notable contributions in screening novel drugs against numerous disease which are either under clinical trials or being medicinally used. Even with advent of new strategies the success rate of passing clinical phases by the lead compounds is quite low. So, there is still a lot of scope left to design new methods or techniques that includes multidisciplinary concepts in a better fashion to develop novel drugs. Structure based approaches generally suffers from poor formatting of energy calculations and scoring functions. Such issues can be tackled by providing more structures of targets with good resolutions and by introducing conformational search algorithms for better results. While in ligand based approaches, the problem lies in the poor establishment of relations between activity similarity and molecular similarity. The similarity-based methods require further improvement in QSAR and pharmacophore models to get better hits. Apart from this development in ADME and toxicity predictions tools is highly anticipated which plays a crucial role in further filtering of lead compounds to avoid unnecessary efforts. Improvement in techniques related to toxicity mechanism can help in predicting toxicity profile of hit compounds more accurately. Finally, we can expect that more attention towards the neglected or less explored diseases can be paid to develop efficient vaccine timely. This is truly verified in the case of COVID 19 infectious disease. A novel virus has been annihilating the entire world since it was reported in Wuhan (China) in December 2019, for the first time. However with the knowledge and understanding of CADD, in such a short span of time, the leading scientists are ready with a beta version of its vaccination which has to be approved after the clinical testing.

CONSENT FOR PUBLICATION

Not applicable.

CONFLICT OF INTEREST

The authors confirm that they have no conflict of interest to declare for this publication.

ACKNOWLEDGEMENTS

The authors would like to thank Kirori Mal College, University of Delhi, New Delhi, India and Council of Scientific and Industrial Research (CSIR) for infrastructural and financial aid.

REFERENCES

[1]   Macalino, S.J.Y.; Gosu, V.; Hong, S.; Choi, S. Role of computer-aided drug design in modern drug discovery. *Arch. Pharm. Res.,* **2015**, *38*(9), 1686-1701.
[http://dx.doi.org/10.1007/s12272-015-0640-5] [PMID: 26208641]

[2]   Singh, B.; Mal, G.; Gautam, S.K.; Mukesh, M. Computer-Aided Drug Discovery.*Advances in Animal Biotechnology*; Springer: Cham, **2019**, pp. 471-481.
[http://dx.doi.org/10.1007/978-3-030-21309-1_44]

[3]   Prieto-Martínez, F.D.; López-López, E.; Juárez-Mercado, K.E.; Medina-Franco, J.L. Computational Drug Design Methods—Current and Future Perspectives.*In Silico Drug Design*; Academic Press, **2019**, pp. 19-44.
[http://dx.doi.org/10.1016/B978-0-12-816125-8.00002-X]

[4]   Aparoy, P.; Reddy, K.K.; Reddanna, P. Structure and ligand based drug design strategies in the development of novel 5- LOX inhibitors. *Curr. Med. Chem.,* **2012**, *19*(22), 3763-3778.
[http://dx.doi.org/10.2174/092986712801661112] [PMID: 22680930]

[5]   Bosch, F.; Rosich, L. The contributions of Paul Ehrlich to pharmacology: a tribute on the occasion of the centenary of his Nobel Prize. *Pharmacology,* **2008**, *82*(3), 171-179.
[http://dx.doi.org/10.1159/000149583] [PMID: 18679046]

[6]   Gashaw, I.; Ellinghaus, P.; Sommer, A.; Asadullah, K. What makes a good drug target? *Drug Discov. Today,* **2011**, *16*(23-24), 1037-1043.
[http://dx.doi.org/10.1016/j.drudis.2011.09.007] [PMID: 21945861]

[7]   Owens, J. **2018**. *Phenotypic versus Target-based Screening for Drug Discovery.,* https://www.tech-nologynetworks.com/drug-discovery/articles/phenotypic-versus-target-based-screening-f-r-drug-discovery-300037

[8]   Terstappen, G.C.; Schlüpen, C.; Raggiaschi, R.; Gaviraghi, G. Target deconvolution strategies in drug discovery. *Nat. Rev. Drug Discov.,* **2007**, *6*(11), 891-903.
[http://dx.doi.org/10.1038/nrd2410] [PMID: 17917669]

[9]   Moffat, J.G.; Vincent, F.; Lee, J.A.; Eder, J.; Prunotto, M. Opportunities and challenges in phenotypic drug discovery: an industry perspective. *Nat. Rev. Drug Discov.,* **2017**, *16*(8), 531-543.
[http://dx.doi.org/10.1038/nrd.2017.111] [PMID: 28685762]

[10]   Katsila, T.; Spyroulias, G.A.; Patrinos, G.P.; Matsoukas, M.T. Computational approaches in target identification and drug discovery. *Comput. Struct. Biotechnol. J.,* **2016**, *14*, 177-184.
[http://dx.doi.org/10.1016/j.csbj.2016.04.004] [PMID: 27293534]

[11]   Ehrlich, P. Über den jetzigen Stand der Chemotherapie. *Ber. Dtsch. Chem. Ges.,* **1909**, *42*(1), 17-47.
[http://dx.doi.org/10.1002/cber.19090420105]

[12]   Wermuth, C.G.; Ganellin, C.R.; Lindberg, P.; Mitscher, L.A. Glossary of terms used in medicinal chemistry (IUPAC recommendations 1998). *Pure Appl. Chem.,* **1998**, *70*, 1129-1143.
[http://dx.doi.org/10.1351/pac199870051129]

[13]   Güner, O.F., Ed. *Pharmacophore perception, development, and use in drug design*; Internat'l University Line, **2000**, Vol. 2, .

[14]   Gaurav, A.; Gautam, V.; Pereira, S.; Alvarez-Leite, J.; Vetri, F.; Choudhury, M.; Krieger, A. Structure-based three-dimensional pharmacophores as an alternative to traditional methodologies. *J. Receptor Ligand Channel Res.,* **2014**, *7*, 27-38.
[http://dx.doi.org/10.2147/JRLCR.S46845]

[15]   Barbaro, R.; Betti, L.; Botta, M.; Corelli, F.; Giannaccini, G.; Maccari, L.; Manetti, F.; Strappaghetti, G.; Corsano, S. Synthesis, biological evaluation, and pharmacophore generation of new pyridazinone derivatives with affinity toward α(1)- and α(2)-adrenoceptors. *J. Med. Chem.,* **2001**, *44*(13), 2118-2132.
[http://dx.doi.org/10.1021/jm010821u] [PMID: 11405649]

[16]   Wolber, G.; Langer, T. LigandScout: 3-D pharmacophores derived from protein-bound ligands and their use as virtual screening filters. *J. Chem. Inf. Model.,* **2005**, *45*(1), 160-169.
[http://dx.doi.org/10.1021/ci049885e] [PMID: 15667141]

[17]   Ortuso, F.; Langer, T.; Alcaro, S. GBPM: GRID-based pharmacophore model: concept and application studies to protein-protein recognition. *Bioinformatics,* **2006**, *22*(12), 1449-1455.
[http://dx.doi.org/10.1093/bioinformatics/btl115] [PMID: 16567363]

[18]   McGregor, M.J.; Muskal, S.M. Pharmacophore fingerprinting. 2. Application to primary library design. *J. Chem. Inf. Comput. Sci.,* **2000**, *40*(1), 117-125.
[http://dx.doi.org/10.1021/ci990313h] [PMID: 10661558]

[19]   Guner, O.F.; Bowen, J.P. Pharmacophore modeling for ADME. *Curr. Top. Med. Chem.,* **2013**, *13*(11), 1327-1342.
[http://dx.doi.org/10.2174/15680266113139990037] [PMID: 23675939]

[20]   Desaphy, J.; Azdimousa, K.; Kellenberger, E.; Rognan, D. *Comparison and druggability prediction of protein–ligand binding sites from pharmacophore-annotated cavity shapes.,* **2012**, *52*(8), 2287-2299.
[http://dx.doi.org/10.1021/ci300184x]

[21]   Qing, X.; Lee, X.Y.; De Raeymaecker, J.; Tame, J.R.; Zhang, K.Y.; De Maeyer, M.; Voet, A. Pharmacophore modeling: advances, limitations, and current utility in drug discovery. *J. Receptor Ligand Channel Res.,* **2014**, *7*, 81-92.

[22]   Sanders, M.P.; McGuire, R.; Roumen, L.; de Esch, I.J.; de Vlieg, J.; Klomp, J.P.; de Graaf, C. From the protein's perspective: the benefits and challenges of protein structure-based pharmacophore modeling. *MedChemComm,* **2012**, *3*(1), 28-38.
[http://dx.doi.org/10.1039/C1MD00210D]

[23]   Böhm, H.J. The computer program LUDI: a new method for the *de novo* design of enzyme inhibitors. *J. Comput. Aided Mol. Des.,* **1992**, *6*(1), 61-78.
[http://dx.doi.org/10.1007/BF00124387] [PMID: 1583540]

[24]   Barillari, C.; Marcou, G.; Rognan, D. Hot-spots-guided receptor-based pharmacophores (HS-Pharm): a knowledge-based approach to identify ligand-anchoring atoms in protein cavities and prioritize structure-based pharmacophores. *J. Chem. Inf. Model.,* **2008**, *48*(7), 1396-1410.
[http://dx.doi.org/10.1021/ci800064z] [PMID: 18570371]

[25]   Tintori, C.; Corradi, V.; Magnani, M.; Manetti, F.; Botta, M. Targets looking for drugs: a multistep computational protocol for the development of structure-based pharmacophores and their applications for hit discovery. *J. Chem. Inf. Model.,* **2008**, *48*(11), 2166-2179.
[http://dx.doi.org/10.1021/ci800105p] [PMID: 18942779]

[26]   Goodford, P.J. A computational procedure for determining energetically favorable binding sites on biologically important macromolecules. *J. Med. Chem.,* **1985**, *28*(7), 849-857.
[http://dx.doi.org/10.1021/jm00145a002] [PMID: 3892003]

[27]   Yang, S.Y. Pharmacophore modeling and applications in drug discovery: challenges and recent advances. *Drug Discov. Today,* **2010**, *15*(11-12), 444-450.
[http://dx.doi.org/10.1016/j.drudis.2010.03.013] [PMID: 20362693]

[28]   Pagadala, N.S.; Syed, K.; Tuszynski, J. Software for molecular docking: a review. *Biophys. Rev.,* **2017**, *9*(2), 91-102.
[http://dx.doi.org/10.1007/s12551-016-0247-1] [PMID: 28510083]

[29]   Sisodiya, D.; Pandey, P.; Dashora, K. Drug Designing Softwares and Their Applications in New Drug Discover. *J. Pharm. Res.,* **2012**, *5*(1), 124-126.

[30]   Bortolato, A.; Fanton, M.; Mason, J.S.; Moro, S. Molecular docking methodologies.*Biomolecular Simulations*; Humana Press: Totowa, NJ, **2013**, Vol. 924, pp. 339-360.
[http://dx.doi.org/10.1007/978-1-62703-017-5_13]

[31]   Hernández-Santoyo, A.; Tenorio-Barajas, A.Y.; Altuzar, V.; Vivanco-Cid, H.; Mendoza-Barrera, C. Protein-protein and protein-ligand docking.*Protein Eng*; InTech Rijeka, **2013**, p. 66.

[32]   Forli, S.; Huey, R.; Pique, M.E.; Sanner, M.F.; Goodsell, D.S.; Olson, A.J. Computational protein-ligand docking and virtual drug screening with the AutoDock suite. *Nat. Protoc.,* **2016**, *11*(5), 905-919.
[http://dx.doi.org/10.1038/nprot.2016.051] [PMID: 27077332]

[33]   Ghosh, S.; Nie, A.; An, J.; Huang, Z. Structure-based virtual screening of chemical libraries for drug discovery. *Curr. Opin. Chem. Biol.,* **2006**, *10*(3), 194-202.
[http://dx.doi.org/10.1016/j.cbpa.2006.04.002] [PMID: 16675286]

[34]   Sousa, S.F.; Ribeiro, A.J.; Coimbra, J.T.S.; Neves, R.P.P.; Martins, S.A.; Moorthy, N.S.H.N.; Fernandes, P.A.; Ramos, M.J. Protein-ligand docking in the new millennium--a retrospective of 10 years in the field. *Curr. Med. Chem.,* **2013**, *20*(18), 2296-2314.
[http://dx.doi.org/10.2174/09298673311320180002] [PMID: 23531220]

[35]   ZELENÝ, M. Optimizing given systems vs. designing optimal systems: The *De Novo* programming approach. . *Int. J. Gen. Syst.,* **1990**, *17*(4), 295-307.
[http://dx.doi.org/10.1080/03081079008935113]

[36]   Willock, D.J.; Lewis, D.W.; Catlow, C.R.A.; Hutchings, G.J.; Thomas, J.M. Designing templates for the synthesis of microporous solids using de novo molecular design methods. *J. Mol. Catal. Chem.,* **1997**, *119*(1-3), 415-424.
[http://dx.doi.org/10.1016/S1381-1169(96)00505-5]

[37]   Dean, P.M.; Lloyd, D.G.; Todorov, N.P. De novo drug design: integration of structure-based and ligand-based methods. *Curr. Opin. Drug Discov. Devel.,* **2004**, *7*(3), 347-353.
[PMID: 15216939]

[38]   Yuan, Y.; Pei, J.; Lai, L. LigBuilder 2: a practical de novo drug design approach. *J. Chem. Inf. Model.,* **2011**, *51*(5), 1083-1091.
[http://dx.doi.org/10.1021/ci100350u] [PMID: 21513346]

[39]   Lavecchia, A.; Di Giovanni, C. Virtual screening strategies in drug discovery: a critical review. *Curr. Med. Chem.*, **2013**, *20*(23), 2839-2860.
[http://dx.doi.org/10.2174/09298673113209990001] [PMID: 23651302]

[40]   Ekins, S.; Mestres, J.; Testa, B. In silico pharmacology for drug discovery: applications to targets and beyond. *Br. J. Pharmacol.*, **2007**, *152*(1), 21-37.
[http://dx.doi.org/10.1038/sj.bjp.0707306] [PMID: 17549046]

[41]   Lavecchia, A.; Cerchia, C. In silico methods to address polypharmacology: current status, applications and future perspectives. *Drug Discov. Today*, **2016**, *21*(2), 288-298.
[http://dx.doi.org/10.1016/j.drudis.2015.12.007] [PMID: 26743596]

[42]   Thomas, C., Ed. *Complex Systems, Sustainability and Innovation*; BoD–Books on Demand, **2016**.
[http://dx.doi.org/10.5772/62659]

[43]   Prathipati, P.; Mizuguchi, K. Integration of ligand and structure based approaches for CSAR-2014. *J. Chem. Inf. Model.*, **2016**, *56*(6), 974-987.
[http://dx.doi.org/10.1021/acs.jcim.5b00477] [PMID: 26492437]

[44]   Moro, S.; Bacilieri, M.; Deflorian, F. Combining ligand-based and structure-based drug design in the virtual screening arena. *Expert Opin. Drug Discov.*, **2007**, *2*(1), 37-49.
[http://dx.doi.org/10.1517/17460441.2.1.37] [PMID: 23496036]

[45]   Abdel-Ilah, L.; Veljović, E.; Gurbeta, L.; Badnjević, A. Applications of QSAR Study in Drug Design. *Int. J. Eng. Res. Technol.*, **2017**, *6*(06), 582-587.

[46]   Abuhammad, A.; Taha, M.O. QSAR studies in the discovery of novel type-II diabetic therapies. *Expert Opin. Drug Discov.*, **2016**, *11*(2), 197-214.
[http://dx.doi.org/10.1517/17460441.2016.1118046] [PMID: 26558613]

[47]   Wang, T.; Wu, M.B.; Lin, J.P.; Yang, L.R. Quantitative structure-activity relationship: promising advances in drug discovery platforms. *Expert Opin. Drug Discov.*, **2015**, *10*(12), 1283-1300.
[http://dx.doi.org/10.1517/17460441.2015.1083006] [PMID: 26358617]

[48]   Agrawal, R.; Jain, P.; Dikshit, S.N. Ligand-based pharmacophore detection, screening of potential gliptins and docking studies to get effective antidiabetic agents. *Comb. Chem. High Throughput Screen.*, **2012**, *15*(10), 849-876.
[http://dx.doi.org/10.2174/138620712803901090] [PMID: 23140189]

[49]   Sharma, V.; Kumar, V. Efficient way of drug designing: a comprehensive review on computational techniques. *Bull. Pharm. Res.*, **2014**, *4*, 118-123.

[50]   Seidel, T.; Ibis, G.; Bendix, F.; Wolber, G. Strategies for 3D pharmacophore-based virtual screening. *Drug Discov. Today. Technol.*, **2010**, *7*(4), e203-e270.
[http://dx.doi.org/10.1016/j.ddtec.2010.11.004] [PMID: 24103798]

[51]   van Drie, J.H. Pharmacophore discovery--lessons learned. *Curr. Pharm. Des.*, **2003**, *9*(20), 1649-1664.
[http://dx.doi.org/10.2174/1381612033454568] [PMID: 12871063]

[52]   Leach, A.R.; Gillet, V.J. *An introduction to chemoinformatics*; Springer Science & Business Media, **2007**.
[http://dx.doi.org/10.1007/978-1-4020-6291-9]

[53]   Hamza, A.; Wei, N.N.; Zhan, C.G. Ligand-based virtual screening approach using a new scoring function. *J. Chem. Inf. Model.*, **2012**, *52*(4), 963-974.
[http://dx.doi.org/10.1021/ci200617d] [PMID: 22486340]

[54]   Melagraki, G.; Afantitis, A. Ligand and structure based virtual screening strategies for hit-finding and optimization of hepatitis C virus (HCV) inhibitors. *Curr. Med. Chem.*, **2011**, *18*(17), 2612-2619.
[http://dx.doi.org/10.2174/092986711795933759] [PMID: 21568888]

[55]   Renner, S.; Schneider, G. Scaffold-hopping potential of ligand-based similarity concepts. *ChemMedChem*, **2006**, *1*(2), 181-185.

[http://dx.doi.org/10.1002/cmdc.200500005] [PMID: 16892349]

[56]   Sun, H.; Tawa, G.; Wallqvist, A. Classification of scaffold-hopping approaches. *Drug Discov. Today,* **2012**, *17*(7-8), 310-324.
        [http://dx.doi.org/10.1016/j.drudis.2011.10.024] [PMID: 22056715]

[57]   Hessler, G.; Baringhaus, K.H. The scaffold hopping potential of pharmacophores. *Drug Discov. Today. Technol.,* **2010**, *7*(4), e203-e270.
        [http://dx.doi.org/10.1016/j.ddtec.2010.09.001] [PMID: 24103802]

[58]   Lu, J.J.; Pan, W.; Hu, Y.J.; Wang, Y.T. Multi-target drugs: the trend of drug research and development. *PLoS One,* **2012**, *7*(6), e40262.
        [http://dx.doi.org/10.1371/journal.pone.0040262] [PMID: 22768266]

[59]   Gupta, N.; Pandya, P.; Verma, S. Computational predictions for multi-target drug design.*Multi-Target Drug Design Using Chem-Bioinformatic Approaches*; Humana Press: New York, NY, **2018**, pp. 27-50.
        [http://dx.doi.org/10.1007/7653_2018_26]

[60]   Vincetti, P.; Caporuscio, F.; Kaptein, S.; Gioiello, A.; Mancino, V.; Suzuki, Y.; Rastelli, G. Discovery of multitarget antivirals acting on both the Dengue virus NS5-NS3 interaction and the host Src/Fyn kinases. *J. Med. Chem.,* **2015**, *58*(12), 4964-4975.
        [http://dx.doi.org/10.1021/acs.jmedchem.5b00108] [PMID: 26039671]

[61]   Yang, S.; Xu, M.; Lee, E.M.; Gorshkov, K.; Shiryaev, S.A.; He, S.; Lu, B. Emetine inhibits Zika and Ebola virus infections through two molecular mechanisms: inhibiting viral replication and decreasing viral entry. *Cell Discov.,* **2018**, *4*(1), 31.
        [http://dx.doi.org/10.1038/s41421-018-0034-1] [PMID: 29872540]

[62]   Muhammed, M.T.; Aki-Yalcin, E. Homology modeling in drug discovery: Overview, current applications, and future perspectives. *Chem. Biol. Drug Des.,* **2019**, *93*(1), 12-20.
        [http://dx.doi.org/10.1111/cbdd.13388] [PMID: 30187647]

[63]   Cavasotto, C.N.; Phatak, S.S. Homology modeling in drug discovery: current trends and applications. *Drug Discov. Today,* **2009**, *14*(13-14), 676-683.
        [http://dx.doi.org/10.1016/j.drudis.2009.04.006] [PMID: 19422931]

[64]   Fernando, S.; Fernando, T.; Stefanik, M.; Eyer, L.; Ruzek, D. An approach for Zika virus inhibition using homology structure of the envelope protein. *Mol. Biotechnol.,* **2016**, *58*(12), 801-806.
        [http://dx.doi.org/10.1007/s12033-016-9979-1] [PMID: 27683255]

[65]   Ekins, S.; Liebler, J.; Neves, B.J.; Lewis, W.G.; Coffee, M.; Bienstock, R.; Southan, C.; Andrade, C.H. Illustrating and homology modeling the proteins of the Zika virus. *F1000 Res.,* **2016**, *5*(5), 275.
        [http://dx.doi.org/10.12688/f1000research.8213.1] [PMID: 27746901]

[66]   McCammon, J.A.; Gelin, B.R.; Karplus, M. Dynamics of folded proteins. *Nature,* **1977**, *267*(5612), 585-590.
        [http://dx.doi.org/10.1038/267585a0] [PMID: 301613]

[67]   Salomon-Ferrer, R.; Götz, A.W.; Poole, D.; Le Grand, S.; Walker, R.C. Routine microsecond molecular dynamics simulations with AMBER on GPUs. 2. Explicit solvent particle mesh Ewald. *J. Chem. Theory Comput.,* **2013**, *9*(9), 3878-3888.
        [http://dx.doi.org/10.1021/ct400314y] [PMID: 26592383]

[68]   Pissurlenkar, R.R.; Shaikh, M.S.; Iyer, R.P.; Coutinho, E.C. Molecular mechanics force fields and their applications in drug design. *Antiinfect. Agents Med. Chem.,* **2009**, *8*(2), 128-150.
        [http://dx.doi.org/10.2174/187152109787846088]

[69]   Nasution, M.A.F.; Toepak, E.P.; Alkaff, A.H.; Tambunan, U.S.F. Flexible docking-based molecular dynamics simulation of natural product compounds and Ebola virus Nucleocapsid (EBOV NP): a computational approach to discover new drug for combating Ebola. *BMC Bioinformatics,* **2018**, *19*(14) Suppl. 14, 419.

[http://dx.doi.org/10.1186/s12859-018-2387-8] [PMID: 30453886]

[70] Sulaiman, K.O.; Kolapo, T.U.; Onawole, A.T.; Islam, M.A.; Adegoke, R.O.; Badmus, S.O. Molecular dynamics and combined docking studies for the identification of Zaire ebola virus inhibitors. *J. Biomol. Struct. Dyn.,* **2019**, *37*(12), 3029-3040.
[http://dx.doi.org/10.1080/07391102.2018.1506362] [PMID: 30058446]

[71] Gleeson, M.P.; Gleeson, D. QM/MM calculations in drug discovery: a useful method for studying binding phenomena? *J. Chem. Inf. Model.,* **2009**, *49*(3), 670-677.
[http://dx.doi.org/10.1021/ci800419j] [PMID: 19434900]

[72] Olsson, M.A.; Ryde, U. Comparison of QM/MM methods to obtain ligand-binding free energies. *J. Chem. Theory Comput.,* **2017**, *13*(5), 2245-2253.
[http://dx.doi.org/10.1021/acs.jctc.6b01217] [PMID: 28355487]

[73] Lodola, A.; De Vivo, M. The increasing role of QM/MM in drug discovery.*Advances in protein chemistry and structural biology*; Academic Press, **2012**, Vol. 87, pp. 337-362.

[74] Fedorov, D.G.; Kitaura, K. Extending the power of quantum chemistry to large systems with the fragment molecular orbital method. *J. Phys. Chem. A,* **2007**, *111*(30), 6904-6914.
[http://dx.doi.org/10.1021/jp0716740] [PMID: 17511437]

[75] Starikov, E.; Lewis, J.P.; Tanaka, S. *Modern methods for theoretical physical chemistry of biopolymers*; Elsevier, **2012**.

[76] Mazanetz, M.P.; Chudyk, E.; Fedorov, D.G.; Alexeev, Y. Applications of the fragment molecular orbital method to drug research.*Computer-Aided Drug Discovery*; Humana Press: New York, **2015**, pp. 217-255.
[http://dx.doi.org/10.1007/7653_2015_59]

[77] Mercorelli, B.; Palù, G.; Loregian, A. Drug repurposing for viral infectious diseases: how far are we? *Trends Microbiol.,* **2018**, *26*(10), 865-876.
[http://dx.doi.org/10.1016/j.tim.2018.04.004] [PMID: 29759926]

[78] Ebola virus disease, World Health Organisation. https://www.who.int/news-room/fact-sheets/detail/ebola-virus-disease

[79] Venkatesan, A.; Ravichandran, L.; Dass, J.F.P. Computational Drug Design against Ebola Virus Targeting Viral Matrix Protein VP30. *Borneo Journal of Pharmacy,* **2019**, *2*(2), 71-81.
[http://dx.doi.org/10.33084/bjop.v2i2.836]

[80] Shinyguruce, A.; Sathya, R.; Devayani, S.; Keerthiga, K.; Pavithra, P.; Vinitha, M.; Vaidheeswari, R.; Eswara, P.B. *In silico* antiviral drug screening and molecular docking studies against ebola virus glycoprotein. *J. App. Sci. Comp.,* **2018**, *5*(9), 211-216.

[81] Kwofie, S.K.; Broni, E.; Teye, J.; Quansah, E.; Issah, I.; Wilson, M.D.; Miller, W.A., III; Tiburu, E.K.; Bonney, J.H.K. Pharmacoinformatics-based identification of potential bioactive compounds against Ebola virus protein VP24. *Comput. Biol. Med.,* **2019**, *113*, 103414.
[http://dx.doi.org/10.1016/j.compbiomed.2019.103414] [PMID: 31536833]

[82] Shah, R.; Panda, P.K.; Patel, P.; Panchal, H. Pharmacophore based virtual screening and molecular docking studies of inherited compounds against Ebola virus receptor proteins. *World J. Pharm. Pharm. Sci.,* **2015**, *4*(5), 1268-1282.

[83] Zhao, Z.; Martin, C.; Fan, R.; Bourne, P.E.; Xie, L. Drug repurposing to target Ebola virus replication and virulence using structural systems pharmacology. *BMC Bioinformatics,* **2016**, *17*(1), 90.
[http://dx.doi.org/10.1186/s12859-016-0941-9] [PMID: 26887654]

[84] Kharkar, P.S.; Ramasami, P.; Choong, Y.S.; Rhyman, L.; Warrier, S. Discovery of anti-Ebola drugs: a computational drug repositioning case study. *RSC Advances,* **2016**, *6*(31), 26329-26340.
[http://dx.doi.org/10.1039/C6RA01704E]

[85] Karthick, V.; Nagasundaram, N.; Doss, C.G.P.; Chakraborty, C.; Siva, R.; Lu, A.; Zhang, G.; Zhu, H.

Virtual screening of the inhibitors targeting at the viral protein 40 of Ebola virus. *Infect. Dis. Poverty,* **2016**, *5*(1), 12.
[http://dx.doi.org/10.1186/s40249-016-0105-1] [PMID: 26888469]

[86]   Skariyachan, S.; Acharya, A.B.; Subramaniyan, S.; Babu, S.; Kulkarni, S.; Narayanappa, R. Secondary metabolites extracted from marine sponge associated Comamonas testosteroni and Citrobacter freundii as potential antimicrobials against MDR pathogens and hypothetical leads for VP40 matrix protein of Ebola virus: an *in vitro* and in silico investigation. *J. Biomol. Struct. Dyn.,* **2016**, *34*(9), 1865-1883.
[http://dx.doi.org/10.1080/07391102.2015.1094412] [PMID: 26577929]

[87]   Mirza, M.U.; Ikram, N. Integrated computational approach for virtual hit identification against ebola viral proteins VP35 and VP40. *Int. J. Mol. Sci.,* **2016**, *17*(11), 1748.
[http://dx.doi.org/10.3390/ijms17111748] [PMID: 27792169]

[88]   Raj, U.; Varadwaj, P.K. Flavonoids as multi-target inhibitors for proteins associated with Ebola virus: In silico discovery using virtual screening and molecular docking studies. *Interdiscip. Sci.,* **2016**, *8*(2), 132-141.
[http://dx.doi.org/10.1007/s12539-015-0109-8] [PMID: 26286008]

[89]   Uzochukwu, I.C.; Olubiyi, O.O.; Ezebuo, F.C.; Obinwa, I.C.; Ajaegbu, E.E.; Eze, P.M.; Ikegbunam, M.N. Ending the Ebola virus scourge: a case for natural products. *Pharm. Res.,* **2016**, *1*(1), 000105.

[90]   Muthumanickam, S.; Langeswaran, K.; Sangavi, P.; Boomi, P. Screening of inhibitors as potential remedial against Ebolavirus infection: pharmacophore-based approach. *J. Biomol. Struct. Dyn.,* **2020**, 1-14.
[http://dx.doi.org/10.1080/07391102.2020.1715260] [PMID: 31928158]

[91]   Haikal, M.C.; Nasution, M.A.F.; Saputro, L.; Tambunan, U.S.F. *In silico* identification of potent inhibitors of heat shock protein 90 (Hsp90) from Indonesian natural product compounds as a novel approach to treat Ebola virus disease. In IOP Conference Series; Materials Science and Engineering; IOP Publishing, 2019, Vol. 509, No. 1, pp. 012082.

[92]   Narayanan, R. Druggableness of the ebola associated genes in the human genome: Chemoinformatics approach. *MOJ Proteomics Bioinform,* **2015**, *2*(2), 00038-00044.
[http://dx.doi.org/10.15406/mojpb.2015.02.00038]

[93]   Zika virus, World Health Organisation. https://www.who.int/news-room/fact-sheets/detail/zika-virus

[94]   Panwar, U.; Singh, S.K. An overview on Zika Virus and the importance of computational drug discovery. *J. Expl. Res. Pharm.,* **2018**, *3*(2), 43-51.
[http://dx.doi.org/10.14218/JERP.2017.00025]

[95]   Mascini, M.; Dikici, E.; Robles Mañueco, M.; Perez-Erviti, J.A.; Deo, S.K.; Compagnone, D.; Wang, J.; Pingarrón, J.M.; Daunert, S. Computationally Designed Peptides for Zika Virus Detection: An Incremental Construction Approach. *Biomolecules,* **2019**, *9*(9), 498.
[http://dx.doi.org/10.3390/biom9090498] [PMID: 31533374]

[96]   Ramharack, P.; Soliman, M.E.S. Zika virus NS5 protein potential inhibitors: an enhanced in silico approach in drug discovery. *J. Biomol. Struct. Dyn.,* **2018**, *36*(5), 1118-1133.
[http://dx.doi.org/10.1080/07391102.2017.1313175] [PMID: 28351337]

[97]   Narayanan, R. Zika virus therapeutic lead compounds discovery using chemoinformatics approaches. *MOJ Proteomics Bioinform,* **2016**, *3*(3), 00084.
[http://dx.doi.org/10.15406/mojpb.2016.03.00088]

[98]   Tarantino, D.; Cannalire, R.; Mastrangelo, E.; Croci, R.; Querat, G.; Barreca, M.L.; Bolognesi, M.; Manfroni, G.; Cecchetti, V.; Milani, M. Targeting flavivirus RNA dependent RNA polymerase through a pyridobenzothiazole inhibitor. *Antiviral Res.,* **2016**, *134*, 226-235.
[http://dx.doi.org/10.1016/j.antiviral.2016.09.007] [PMID: 27649989]

[99]   Raza, S.; Abbas, G.; Azam, S. S. Screening pipeline for Flavivirus based inhibitors for Zika virus NS1. *IEEE/ACM Trans. Comput. Biol,* **2019**.

[100] Yadav, R.; Selvaraj, C.; Aarthy, M.; Kumar, P.; Kumar, A.; Singh, S.K.; Giri, R. Investigating into the molecular interactions of flavonoids targeting NS2B-NS3 protease from ZIKA virus through *in-silico* approaches. *J. Biomol. Struct. Dyn.*, **2020**, 1-13.
[http://dx.doi.org/10.1080/07391102.2019.1709546] [PMID: 31920173]

[101] Byler, K.G.; Ogungbe, I.V.; Setzer, W.N. In-silico screening for anti-Zika virus phytochemicals. *J. Mol. Graph. Model.*, **2016**, *69*, 78-91.
[http://dx.doi.org/10.1016/j.jmgm.2016.08.011] [PMID: 27588363]

[102] Rohini, K.; Agarwal, P.; Preethi, B.; Shanthi, V.; Ramanathan, K. Exploring the lead compounds for Zika Virus NS2B-NS3 protein: an e-pharmacophore-based approach. *Appl. Biochem. Biotechnol.*, **2019**, *187*(1), 194-210.
[http://dx.doi.org/10.1007/s12010-018-2814-3] [PMID: 29911269]

[103] Kumar, D.; Kaur, N.; Giri, R.; Singh, N. Biscoumarin scaffold as an efficient anti-Zika virus lead with NS3-Helicase inhibitory potential: in-vitro and in-silico Investigation. *New J. Chem.*, **2020**, *44*, 1872-1880.
[http://dx.doi.org/10.1039/C9NJ05225A]

[104] Zhu, S.; Zhang, C.; Huang, L.S.; Zhang, X.Q.; Xu, Y.; Fang, X.; Zhou, J.; Wu, M.; Schooley, R.T.; Huang, Z.; An, J. Discovery and computational analyses of novel small molecule Zika virus inhibitors. *Molecules*, **2019**, *24*(8), 1465.
[http://dx.doi.org/10.3390/molecules24081465] [PMID: 31013906]

[105] Sahoo, M.; Lingaraja Jena, S.D.; Kumar, S. Virtual screening for potential inhibitors of NS3 protein of Zika virus. *Genomics Inform.*, **2016**, *14*(3), 104-111.
[http://dx.doi.org/10.5808/GI.2016.14.3.104] [PMID: 27729840]

[106] Stephen, P.; Baz, M.; Boivin, G.; Lin, S.X. Structural insight into NS5 of Zika virus leading to the discovery of MTase inhibitors. *J. Am. Chem. Soc.*, **2016**, *138*(50), 16212-16215.
[http://dx.doi.org/10.1021/jacs.6b10399] [PMID: 27998085]

[107] Alshiraihi, I.M.A.; Bennett, J.; Hamrick, M.; Good, M.; Henderson, K.; Sobolewski, C.; Brown, M.A. Targeting the NS5 Protein of Zika Virus. *J. Multidiscip. Eng. Sci. Stud.*, **2016**, *2*(12), 1237-1240.

[108] Santos, F.R.S.; Lima, W.G.; Maia, E.H.; Assis, L.C.; Davyt, D.; Taranto, A.G.; Ferreira, J.M.S. Identification of a Potential Zika Virus Inhibitor Targeting NS5 Methyltransferase Using Virtual Screening and Molecular Dynamics Simulations. *J. Chem. Inf. Model.*, **2020**, *60*(2), 562-568.
[http://dx.doi.org/10.1021/acs.jcim.9b00809] [PMID: 31985225]

[109] Sharma, P.; Kaur, R.; Upadhyay, A.K.; Kaushik, V. In-Silico Prediction of Peptide Based Vaccine Against Zika Virus. *Int. J. Pept. Res. Ther.*, **2020**, *26*(1), 85-91.
[http://dx.doi.org/10.1007/s10989-019-09818-2]

[110] Hepatitis, C. *World Health Organisation.*, https://www.who.int/news-room/fact-sheets/detail/hepatit-s-c

[111] Moradpour, D.; Penin, F. Hepatitis C virus proteins: from structure to function.*Hepatitis C virus: from molecular virology to antiviral therapy*; Springer: Berlin, Heidelberg, **2013**, Vol. 369, pp. 42-67.

[112] Han, J.; Lee, H.W.; Jin, Y.; Khadka, D.B.; Yang, S.; Li, X.; Kim, M.; Cho, W.J. Molecular design, synthesis, and biological evaluation of bisamide derivatives as cyclophilin A inhibitors for HCV treatment. *Eur. J. Med. Chem.*, **2020**, *188*, 112031.
[http://dx.doi.org/10.1016/j.ejmech.2019.112031] [PMID: 31923861]

[113] Pinheiro, A.S.; Duarte, J.B.C.; Alves, C.N.; de Molfetta, F.A. Virtual screening and molecular dynamics simulations from a bank of molecules of the Amazon region against functional NS3-4A protease-helicase enzyme of hepatitis C virus. *Appl. Biochem. Biotechnol.*, **2015**, *176*(6), 1709-1721.
[http://dx.doi.org/10.1007/s12010-015-1672-5] [PMID: 26009474]

[114] Kumar, A.; Gupta, R.; Verma, K.; Iyer, K.; Shanthi, V.; Ramanathan, K. Identification of Novel Hepatitis C virus NS3-4A protease inhibitors by virtual screening approach. *J. Microb. Biochem.*

Technol., **2014**, *6*(4), 1-7.

[115]   Xue, W.; Yang, Y.; Wang, X.; Liu, H.; Yao, X. Computational study on the inhibitor binding mode and allosteric regulation mechanism in hepatitis C virus NS3/4A protein. *PLoS One*, **2014**, *9*(2), e87077.
[http://dx.doi.org/10.1371/journal.pone.0087077] [PMID: 24586263]

[116]   Venkatesan, A.; Febin Prabhu Dass, J. Deciphering molecular properties and docking studies of hepatitis C and non-hepatitis C antiviral inhibitors - A computational approach. *Life Sci.*, **2017**, *174*, 8-14.
[http://dx.doi.org/10.1016/j.lfs.2017.02.014] [PMID: 28259653]

[117]   Nutho, B.; Meeprasert, A.; Chulapa, M.; Kungwan, N.; Rungrotmongkol, T. Screening of hepatitis C NS5B polymerase inhibitors containing benzothiadiazine core: a steered molecular dynamics. *J. Biomol. Struct. Dyn.*, **2017**, *35*(8), 1743-1757.
[http://dx.doi.org/10.1080/07391102.2016.1193444] [PMID: 27236925]

[118]   Wei, Y.; Li, J.; Qing, J.; Huang, M.; Wu, M.; Gao, F.; Li, D.; Hong, Z.; Kong, L.; Huang, W.; Lin, J. Discovery of novel hepatitis C virus NS5B polymerase inhibitors by combining random forest, multiple e-pharmacophore modeling and docking. *PLoS One*, **2016**, *11*(2), e0148181.
[http://dx.doi.org/10.1371/journal.pone.0148181] [PMID: 26845440]

[119]   Ahmed, M.; Pal, A.; Houghton, M.; Barakat, K. A comprehensive computational analysis for the binding modes of hepatitis C virus NS5A inhibitors: the question of symmetry. *ACS Infect. Dis.*, **2016**, *2*(11), 872-881.
[http://dx.doi.org/10.1021/acsinfecdis.6b00113] [PMID: 27933783]

[120]   Polamreddy, P.; Vishwakarma, V.; Saxena, P. Identification of potential anti-hepatitis C virus agents targeting non structural protein 5B using computational techniques. *J. Cell. Biochem.*, **2018**, *119*(10), 8574-8587.
[http://dx.doi.org/10.1002/jcb.27071] [PMID: 30058078]

[121]   Elfiky, A.A.; Gawad, W.A.; Elshemey, W.M. *Hepatitis C viral polymerase inhibition using directly acting antivirals, a computational approach*; Software and Techniques for Bio-Molecular Modeling, **2016**, pp. 197-209.

[122]   Speck-Planche, A.; Dias Soeiro Cordeiro, M.N. Speeding up early drug discovery in antiviral research: A fragment-based in silico approach for the design of virtual anti-hepatitis C leads. *ACS Comb. Sci.*, **2017**, *19*(8), 501-512.
[http://dx.doi.org/10.1021/acscombsci.7b00039] [PMID: 28437091]

[123]   Jiao, P.; Xue, W.; Shen, Y.; Jin, N.; Liu, H. Understanding the drug resistance mechanism of hepatitis C virus NS5B to PF-00868554 due to mutations of the 423 site: a computational study. *Mol. Biosyst.*, **2014**, *10*(4), 767-777.
[http://dx.doi.org/10.1039/c3mb70498j] [PMID: 24452008]

[124]   Pan, D.; Niu, Y.; Xue, W.; Bai, Q.; Liu, H.; Yao, X. Computational study on the drug resistance mechanism of hepatitis C virus NS5B RNA-dependent RNA polymerase mutants to BMS-791325 by molecular dynamics simulation and binding free energy calculations. *Chemom. Intell. Lab. Syst.*, **2016**, *154*, 185-193.
[http://dx.doi.org/10.1016/j.chemolab.2016.03.015]

[125]   Ying, B.; Pang, S.; Yang, J.; Zhong, Y.; Wang, J. Computational study of HCV p7 channel: insight into a new strategy for HCV inhibitor design. *Interdiscip. Sci.*, **2019**, *11*(2), 292-299.
[http://dx.doi.org/10.1007/s12539-018-0306-3] [PMID: 30194627]

[126]   Dahl, S.L.; Kalita, M.M.; Fischer, W.B. Interaction of antivirals with a heptameric bundle model of the p7 protein of hepatitis C virus. *Chem. Biol. Drug Des.*, **2018**, *91*(4), 942-950.
[http://dx.doi.org/10.1111/cbdd.13162] [PMID: 29251816]

[127]   Chen, L.; Lu, J.; Huang, T.; Yin, J.; Wei, L.; Cai, Y.D. Finding candidate drugs for hepatitis C based on chemical-chemical and chemical-protein interactions. *PLoS One*, **2014**, *9*(9), e107767.

[http://dx.doi.org/10.1371/journal.pone.0107767] [PMID: 25225900]

[128]   Sahin, A.R.; Erdogan, A.; Agaoglu, P.M.; Dineri, Y.; Cakirci, A.Y.; Senel, M.E.; Tasdogan, A.M. 2019 Novel Coronavirus (COVID-19) Outbreak: A Review of the Current Literature. *Eurasian J. Med. Oncol.,* **2020**, *4*(1), 1-7.
[http://dx.doi.org/10.14744/ejmo.2020.12220]

[129]   Zhu, N.; Zhang, D.; Wang, W.; Li, X.; Yang, B.; Song, J.; Zhao, X.; Huang, B.; Shi, W.; Lu, R.; Niu, P.; Zhan, F.; Ma, X.; Wang, D.; Xu, W.; Wu, G.; Gao, G.F.; Tan, W. A novel coronavirus from patients with pneumonia in China, 2019. *N. Engl. J. Med.,* **2020**, *382*(8), 727-733.
[http://dx.doi.org/10.1056/NEJMoa2001017] [PMID: 31978945]

[130]   World Health Organization. Coronavirus disease 2019 (COVID-19): situation report. **2020**, 51.

[131]   Leung, C.W.; Chiu, W.K. Clinical picture, diagnosis, treatment and outcome of severe acute respiratory syndrome (SARS) in children. *Paediatr. Respir. Rev.,* **2004**, *5*(4), 275-288.
[http://dx.doi.org/10.1016/j.prrv.2004.07.010] [PMID: 15531251]

[132]   Koren, G.; King, S.; Knowles, S.; Phillips, E. Ribavirin in the treatment of SARS: A new trick for an old drug? *CMAJ,* **2003**, *168*(10), 1289-1292.
[PMID: 12743076]

[133]   Müller, M.A.; Raj, V.S.; Muth, D.; Meyer, B.; Kallies, S.; Smits, S.L.; Wollny, R.; Bestebroer, T.M.; Specht, S.; Suliman, T.; Zimmermann, K.; Binger, T.; Eckerle, I.; Tschapka, M.; Zaki, A.M.; Osterhaus, A.D.; Fouchier, R.A.; Haagmans, B.L.; Drosten, C. Human coronavirus EMC does not require the SARS-coronavirus receptor and maintains broad replicative capability in mammalian cell lines. *MBio,* **2012**, *3*(6), e00515-e12.
[http://dx.doi.org/10.1128/mBio.00515-12] [PMID: 23232719]

[134]   Coronavirus, World Health Organisation. https://www.who.int/health-topics/coronavirus#tab=tab_3

[135]   Weiss, S.R.; Navas-Martin, S. Coronavirus pathogenesis and the emerging pathogen severe acute respiratory syndrome coronavirus. *Microbiol. Mol. Biol. Rev.,* **2005**, *69*(4), 635-664.
[http://dx.doi.org/10.1128/MMBR.69.4.635-664.2005] [PMID: 16339739]

[136]   Wang, J. Fast Identification of Possible Drug Treatment of Coronavirus Disease -19 (COVID-19) Through Computational Drug Repurposing Study. *ChemRxiv,* **2020**.
[PMID: 32510523]

[137]   Elfiky, A.A. Anti-HCV, nucleotide inhibitors, repurposing against COVID-19. *Life Sci.,* **2020**, *248*, 117477.
[http://dx.doi.org/10.1016/j.lfs.2020.117477] [PMID: 32119961]

[138]   Berry, M.; Fielding, B.C.; Gamieldien, J. Potential broad spectrum inhibitors of the coronavirus 3CLpro: a virtual screening and structure-based drug design study. *Viruses,* **2015**, *7*(12), 6642-6660.
[http://dx.doi.org/10.3390/v7122963] [PMID: 26694449]

[139]   Chang, C.K.; Jeyachandran, S.; Hu, N.J.; Liu, C.L.; Lin, S.Y.; Wang, Y.S.; Chang, Y.M.; Hou, M.H. Structure-based virtual screening and experimental validation of the discovery of inhibitors targeted towards the human coronavirus nucleocapsid protein. *Mol. Biosyst.,* **2016**, *12*(1), 59-66.
[http://dx.doi.org/10.1039/C5MB00582E] [PMID: 26542199]

[140]   Wang, F.; Chen, C.; Tan, W.; Yang, K.; Yang, H. Structure of main protease from human coronavirus NL63: insights for wide spectrum anti-coronavirus drug design. *Sci. Rep.,* **2016**, *6*, 22677.
[http://dx.doi.org/10.1038/srep22677] [PMID: 26948040]

[141]   Rao, S.V.; Tulasi, D.P.; Pavithra, K.; Nisha, R.; Taj, R. In silico studies on dengue and mers coronavirus proteins with selected coriandrum sativum l. herb constituents. *World J. Pharm. Pharm. Sci.,* **2018**, *7*(12), 971-989.

[142]   Jo, S.; Kim, H.; Kim, S.; Shin, D.H.; Kim, M.S. Characteristics of flavonoids as potent MERS-CoV 3C-like protease inhibitors. *Chem. Biol. Drug Des.,* **2019**, *94*(6), 2023-2030.
[http://dx.doi.org/10.1111/cbdd.13604] [PMID: 31436895]

[143] Radwan, A.A.; Alanazi, F.K. *In silico* studies on novel inhibitors of MERS-CoV: Structure-based pharmacophore modeling, database screening and molecular docking. *Trop. J. Pharm. Res.,* **2018**, *17*(3), 513-517.
[http://dx.doi.org/10.4314/tjpr.v17i3.18]

[144] Lee, H.; Ren, J.; Pesavento, R.P.; Ojeda, I.; Rice, A.J.; Lv, H.; Kwon, Y.; Johnson, M.E. Identification and design of novel small molecule inhibitors against MERS-CoV papain-like protease via high-throughput screening and molecular modeling. *Bioorg. Med. Chem.,* **2019**, *27*(10), 1981-1989.
[http://dx.doi.org/10.1016/j.bmc.2019.03.050] [PMID: 30940566]

[145] Sohrab, S. S.; El-Kafrawy, S. A.; Mirza, Z.; Azhar, E. I. *In silico Prediction and Cytotoxicity Evaluation of siRNAs Targeting Conserved Regions of MERS-CoV in Cell Culture,* **2019**.

[146] Nur, S.M.; Hasan, M.A.; Amin, M.A.; Hossain, M.; Sharmin, T. Design of potential RNAi (miRNA and siRNA) molecules for Middle East respiratory syndrome coronavirus (MERS-CoV) gene silencing by computational method. *Interdiscip. Sci.,* **2015**, *7*(3), 257-265.
[http://dx.doi.org/10.1007/s12539-015-0266-9] [PMID: 26223545]

[147] Mustafa, S.; Balkhy, H.; Gabere, M. Peptide-protein interaction studies of antimicrobial peptides targeting middle east respiratory syndrome coronavirus spike protein: an in silico approach. *Adv. Bioinforma.,* **2019**, *1*, 1-16.
[http://dx.doi.org/10.1155/2019/6815105] [PMID: 31354813]

[148] Tahir Ul Qamar, M.; Saleem, S.; Ashfaq, U.A.; Bari, A.; Anwar, F.; Alqahtani, S. Epitope-based peptide vaccine design and target site depiction against Middle East Respiratory Syndrome Coronavirus: an immune-informatics study. *J. Transl. Med.,* **2019**, *17*(1), 362.
[http://dx.doi.org/10.1186/s12967-019-2116-8] [PMID: 31703698]

[149] Mahdy, S.M.; EL-Fiky, A.A.; EL-Shemey, W.M. MERS and SARS CoV Inhibition Using Anti-HCV Drugs: A Computational Biophysical Approach. *Med. J. Cairo Univ.,* **2017**, *85*(2), 481-485.

[150] Abuhammad, A.; Al-Aqtash, R.A.A.; Anson, B.J.; Mesecar, A.D.; Taha, M.O. Computational modeling of the bat HKU4 coronavirus 3CL^{pro} inhibitors as a tool for the development of antivirals against the emerging Middle East respiratory syndrome (MERS) coronavirus. *J. Mol. Recognit.,* **2017**, *30*(11), e2644.
[http://dx.doi.org/10.1002/jmr.2644] [PMID: 28608547]

[151] Abdallah, H. Theoretical study for the inhibition ability of some bioactive imidazole derivatives against the Middle-East respiratory syndrome corona virus (MERS-Co). *ZANCO Journal of Pure and Applied Sciences,* **2019**, *31*(2), 71-78.

[152] Berry, M.; Fielding, B.; Gamieldien, J. Human coronavirus OC43 3CL protease and the potential of ML188 as a broad-spectrum lead compound: homology modelling and molecular dynamic studies. *BMC Struct. Biol.,* **2015**, *15*(1), 8.
[http://dx.doi.org/10.1186/s12900-015-0035-3] [PMID: 25928480]

[153] Park, J.Y.; Ko, J.A.; Kim, D.W.; Kim, Y.M.; Kwon, H.J.; Jeong, H.J.; Kim, C.Y.; Park, K.H.; Lee, W.S.; Ryu, Y.B. Chalcones isolated from Angelica keiskei inhibit cysteine proteases of SARS-CoV. *J. Enzyme Inhib. Med. Chem.,* **2016**, *31*(1), 23-30.
[http://dx.doi.org/10.3109/14756366.2014.1003215] [PMID: 25683083]

[154] Wang, L.; Bao, B.B.; Song, G.Q.; Chen, C.; Zhang, X.M.; Lu, W.; Wang, Z.; Cai, Y.; Li, S.; Fu, S.; Song, F.H.; Yang, H.; Wang, J.G. Discovery of unsymmetrical aromatic disulfides as novel inhibitors of SARS-CoV main protease: Chemical synthesis, biological evaluation, molecular docking and 3D-QSAR study. *Eur. J. Med. Chem.,* **2017**, *137*, 450-461.
[http://dx.doi.org/10.1016/j.ejmech.2017.05.045] [PMID: 28624700]

[155] Li, Y.; Zhang, J.; Wang, N.; Li, H.; Shi, Y.; Guo, G.; Zou, Q. Therapeutic Drugs Targeting 2019-nCoV Main Protease by High-Throughput Screening. *bioRxiv,* **2020**.

[156] Smith, M.; Smith, J.C. Repurposing therapeutics for covid-19: Supercomputer-based docking to the

sars-cov-2 viral spike protein and viral spike protein-human ace2 interface. *ChemRxiv*, **2020**.

[157] Parikesit, A.A.; Nurdiansyah, R. Drug Repurposing Option for COVID-19 with Structural Bioinformatics of Chemical Interactions Approach. *Cermin Dunia Kedokteran*, **2020**, *47*(3), 222-226.

[158] Salari-Jazi, A.; Mahnam, K.; Hejazi, S.H. Predicting and repurposing of drug and drug like compounds for inhibition of the covid-19 and its cytokine storm by computational methods. *Preprints*, **2020**, 2020030457.https://www.preprints.org/manuscript/202003.0457/v2

[159] Biswas, S.; Chatterjee, S.; Dey, T. *In Silico* Approach for Peptide Vaccine Design for CoVID 19: MOL2NET, International Conference Series on Multidisciplinary Sciences, 6 th Edition; USINEWS-04: US-IN-EU Worldwide Science Workshop Series, UMN, Duluth, USA, **2020**.

[160] Kadioglu, O.; Saeed, M.; Johannes Greten, H. Identification of novel compounds against three targets of SARS CoV-2 coronavirus by combined virtual screening and supervised machine learning.. *Bull. World Health Organ.*, **2020**. Available from: http://dx.doi.org/10.2471/BLT.20.255943

[161] Beck, B.R.; Shin, B.; Choi, Y.; Park, S.; Kang, K. Predicting commercially available antiviral drugs that may act on the novel coronavirus (SARS-CoV-2) through a drug-target interaction deep learning model. *Comput. Struct. Biotechnol. J.*, **2020**, *18*, 784-790.
[http://dx.doi.org/10.1016/j.csbj.2020.03.025] [PMID: 32280433]

[162] Wahedi, H.M.; Ahmad, S.; Abbasi, S.W. Stilbene-based natural compounds as promising drug candidates against COVID-19. *J. Biomol. Struct. Dyn.*, **2020**, 1-10.
[http://dx.doi.org/10.1080/07391102.2020.1762743] [PMID: 32345140]

[163] Mahapatra, S.; Nath, P.; Chatterjee, M. Repurposing Therapeutics for COVID-19: Rapid Prediction of Commercially available drugs through Machine Learning and Docking. medRxiv. **2020**. Available from: https://doi.org/10.1101/2020.04.05.20054254

[164] Balkrishna, A.; Thakur, P.; Singh, S. Glucose antimetabolite 2-Deoxy-D-Glucose and its derivative as promising candidates for tackling COVID-19: Insights derived from in silico docking and molecular simulations. **2020**. Available from: https://10.22541/au.158567174.40895611

[165] Namsani, S.; Pramanik, D.; Khan, M.A.; Roy, S.; Singh, J.K. Potential drug candidates for SARS-CoV-2 using computational screening and enhanced sampling methods. *ChemRxiv*, **2020**.
[http://dx.doi.org/10.26434/chemrxiv.12254135.v1]

[166] Koulgi, S.; Jani, V.; Uppuladinne, M.; Sonavane, U.; Nath, A.K.; Darbari, H.; Darbari, H. Drug repurposing studies targeting SARS-nCoV2: An ensemble docking approach on drug target 3C-like protease (3CLpro). *ChemRxiv*, **2020**.
[http://dx.doi.org/10.26434/chemrxiv.12228831.v1]

[167] Pena, L. Discovery of New Hydroxyethylamine Analogs Against 3CLpro Protein Target of SARS-CoV-2: Molecular Docking, Molecular Dynamics Simulation and Structure-Activity Relationship Studies. *ChemRxiv*, **2020**.
[http://dx.doi.org/10.26434/chemrxiv.12083004.v1]

[168] Liu, S.; Zheng, Q.; Wang, Z. Potential covalent drugs targeting the main protease of the SARS-CoV-2 coronavirus. *Bioinformatics*, **2020**, *36*(11), 3295-3298.
[http://dx.doi.org/10.1093/bioinformatics/btaa224] [PMID: 32239142]

[169] Elmezayen, A.D.; Al-Obaidi, A.; Şahin, A.T.; Yelekçi, K. Drug repurposing for coronavirus (COVID-19): *in silico* screening of known drugs against coronavirus 3CL hydrolase and protease enzymes. *J. Biomol. Struct. Dyn.*, **2020**, 1-13.
[http://dx.doi.org/10.1080/07391102.2020.1758791] [PMID: 32306862]

[170] Khan, S.A.; Zia, K.; Ashraf, S.; Uddin, R.; Ul-Haq, Z. Identification of chymotrypsin-like protease inhibitors of SARS-CoV-2 *via* integrated computational approach. *J. Biomol. Struct. Dyn.*, **2020**, 1-10.
[http://dx.doi.org/10.1080/07391102.2020.1751298] [PMID: 32238094]

[171] Peterson, L. *In Silico Molecular Dynamics Docking of Drugs to the Inhibitory Active Site of SARS-CoV-2 Protease and Their Predicted Toxicology and ADME.*, **2020**.https://ssrn.com/abstract=3580951

[172] Gyebi, G.A.; Ogunro, O.B.; Adegunloye, A.P.; Ogunyemi, O.M.; Afolabi, S.O. Potential inhibitors of coronavirus 3-chymotrypsin-like protease (3CL^{pro}): an *in silico* screening of alkaloids and terpenoids from African medicinal plants. *J. Biomol. Struct. Dyn.,* **2020**, 1-19.
[http://dx.doi.org/10.1080/07391102.2020.1764868] [PMID: 32367767]

[173] Gupta, M.K.; Vemula, S.; Donde, R.; Gouda, G.; Behera, L.; Vadde, R. *In-silico* approaches to detect inhibitors of the human severe acute respiratory syndrome coronavirus envelope protein ion channel. *J. Biomol. Struct. Dyn.,* **2020**, 1-11.
[http://dx.doi.org/10.1080/07391102.2020.1751300] [PMID: 32238078]

[174] Joshi, G.; Poduri, R. Virtual screening enabled selection of antiviral agents against Covid-19 disease targeting coronavirus endoribonuclease NendoU: Plausible mechanistic interventions in the treatment of new virus strain. *ChemRxiv,* **2020**.
[http://dx.doi.org/10.26434/chemrxiv.12198966.v1]

[175] Mendoza-Martinez, C.; Rodriguez-Lezama, A. Identification of Potential Inhibitors of SARS-CoV-2 Main Protease via a Rapid In-Silico Drug Repurposing Approach. *ChemRxiv,* **2020**.
[http://dx.doi.org/10.26434/chemrxiv.12085083.v1]

[176] Kumar, Y.; Singh, H. In silico identification and docking-based drug repurposing against the main protease of SARS-CoV-2, causative agent of COVID-19. *ChemRxiv,* **2020**.

[177] Strodel, B.; Olubiyi, O.; Olagunju, M. High Throughput Virtual Screening to Discover Inhibitors of the Main Protease of the Coronavirus SARS-CoV-2. *Preprints.org,* **2020**.
[http://dx.doi.org/10.20944/preprints202004.0161.v1]

[178] Ngo, S.T.; Quynh Anh Pham, N.; Thi Le, L.; Pham, D.H.; Vu, V.V. Computational Determination of Potential Inhibitors of SARS-CoV-2 Main Protease. *J. Chem. Inf. Model.,* **2020**.
[http://dx.doi.org/10.1021/acs.jcim.0c00491] [PMID: 32530282]

[179] Rasool, N.; Akhtar, A.; Hussain, W. Insights into the inhibitory potential of selective phytochemicals against Mpro of 2019-nCoV: a computer-aided study. *Struct. Chem.,* **2020**, 1.
[http://dx.doi.org/10.1007/s11224-020-01536-6] [PMID: 32362735]

[180] Iruthayaraj, A.; Magudeeswaran, S.; Poomani, K. Possibility of HIV-1 Protease Inhibitors-Clinical Trial Drugs as Repurposed Drugs for SARSCoV-2 Main Protease: A Molecular Docking, Molecular Dynamics and Binding Free Energy Simulation Study. *ChemRxiv,* **2020**.
[http://dx.doi.org/10.26434/chemrxiv.12204668.v1]

[181] Giri, S.; Lal, A.F.; Singh, S. Battle against Coronavirus: Repurposing old friends (Food borne polyphenols) for new enemy (COVID-19). *ChemRxiv,* **2020**.
[http://dx.doi.org/10.26434/chemrxiv.12108546.v1]

[182] Andrade, B.; Ghosh, P.; Barth, D. Computational screening for potential drug candidates against SARS-CoV-2 main protease. *Preprints,* **2020**.https://www.preprints.org/manuscript/202004.0003/v2

[183] Bobrowski, T.; Alves, V.; Melo-Filho, C.C.; Korn, D.; Auerbach, S.S.; Schmitt, C.; Muratov, E.; Tropsha, A. Computational Models Identify Several FDA Approved or Experimental Drugs as Putative Agents Against SARS-CoV-2. *ChemRxiv,* **2020**.
[http://dx.doi.org/10.26434/chemrxiv.12153594.v1] [PMID: 32511287]

[184] Islam, R.; Parves, M.R.; Paul, A.S.; Uddin, N.; Rahman, M.S.; Mamun, A.A.; Hossain, M.N.; Ali, M.A.; Halim, M.A. A molecular modeling approach to identify effective antiviral phytochemicals against the main protease of SARS-CoV-2. *J. Biomol. Struct. Dyn.,* **2020**, 1-12.
[http://dx.doi.org/10.1080/07391102.2020.1761883] [PMID: 32340562]

[185] Umesh, K.; Kundu, D.; Selvaraj, C.; Singh, S.K.; Dubey, V.K. Identification of new anti-nCoV drug chemical compounds from Indian spices exploiting SARS-CoV-2 main protease as target. *J. Biomol. Struct. Dyn.,* **2020**, 1-9.
[http://dx.doi.org/10.1080/07391102.2020.1763202] [PMID: 32362243]

[186] Enmozhi, S.K.; Raja, K.; Sebastine, I.; Joseph, J. Andrographolide as a potential inhibitor of SARS-

CoV-2 main protease: an in silico approach. *J. Biomol. Struct. Dyn.,* **2020**, 1-7.
[http://dx.doi.org/10.1080/07391102.2020.1760136] [PMID: 32329419]

[187] Omar, S.; Bouziane, I.; Bouslama, Z.; Djemel, A. *In-Silico* Identification of Potent Inhibitors of COVID-19 Main Protease (Mpro) and Angiotensin Converting Enzyme 2 (ACE2) from Natural Products: Quercetin, Hispidulin, and Cirsimaritin Exhibited Better Potential Inhibition than Hydroxy-Chloroquine Against COVID-19 Main Protease Active Site and ACE2. *ChemRxiv,* **2020**.
[http://dx.doi.org/10.26434/chemrxiv.12181404.v1]

[188] Sekhar, T. Virtual Screening based prediction of potential drugs for COVID-19. *Preprints,* **2020**.https://www.preprints.org/manuscript/202002.0418/v2

[189] Mittal, L.; Kumari, A.; Srivastava, M.; Singh, M.; Asthana, S. Identification of potential molecules against COVID-19 main protease through structure-guided virtual screening approach. *J. Biomol. Struct. Dyn.,* **2020**, 1-26.
[PMID: 32396769]

[190] Das, S.; Sarmah, S.; Lyndem, S.; Singha Roy, A. An investigation into the identification of potential inhibitors of SARS-CoV-2 main protease using molecular docking study. *J. Biomol. Struct. Dyn.,* **2020**, 1-18.
[http://dx.doi.org/10.1080/07391102.2020.1763201] [PMID: 32362245]

[191] Mirza, M.U.; Froeyen, M. Structural elucidation of SARS-CoV-2 vital proteins: Computational methods reveal potential drug candidates against main protease, Nsp12 polymerase and Nsp13 helicase. *J. Pharm. Anal.,* **2020**.
[http://dx.doi.org/10.1016/j.jpha.2020.04.008] [PMID: 32346490]

[192] Sarma, P.; Shekhar, N.; Prajapat, M.; Avti, P.; Kaur, H.; Kumar, S.; Singh, S.; Kumar, H.; Prakash, A.; Dhibar, D.P.; Medhi, B. In-silico homology assisted identification of inhibitor of RNA binding against 2019-nCoV N-protein (N terminal domain). *J. Biomol. Struct. Dyn.,* **2020**, 1-9.
[http://dx.doi.org/10.1080/07391102.2020.1753580] [PMID: 32266867]

[193] Sharma, K.; Morla, S.; Goyal, A.; Kumar, S. Computational Guided Drug Repurposing for Targeting 2'-O-Ribose Methyltransferase of SARS-CoV-2. *ChemRxiv,* **2020**.
[http://dx.doi.org/10.26434/chemrxiv.12111138.v1]

[194] Chowdhury, T.; Roymahapatra, G.; Mandal, S.M. In Silico Identification of a Potent Arsenic Based Approved Drug Darinaparsin against SARS-CoV-2: Inhibitor of RNA dependent RNA polymerase (RdRp) and Necessary Proteases. *ChemRxiv,* **2020**.

[195] Bhavesh, N.S.; Patra, A. *Virtual Screening and Molecular Dynamics Simulation Suggest Valproic Acid Co-A could Bind to SARS-CoV2 RNA Depended RNA Polymerase*; Preprints, **2020**, p. 2020030393.

[196] Smith, M.; Smith, J.C. Repurposing Therapeutics for the Wuhan Coronavirus nCov-2019: Supercomputer-Based Docking to the Viral S Protein and Human ACE2 Interface. *ChemRxiv,* **2020**.

[197] Choudhary, S.; Malik, Y.S.; Tomar, S.; Tomar, S. Identification of SARS-CoV-2 cell entry inhibitors by drug repurposing using in silico structure-based virtual screening approach. *ChemRxiv,* **2020**.
[http://dx.doi.org/10.26434/chemrxiv.12005988.v2]

[198] Sinha, S.K.; Shakya, A.; Prasad, S.K.; Singh, S.; Gurav, N.S.; Prasad, R.S.; Gurav, S.S. An *in-silico* evaluation of different Saikosaponins for their potency against SARS-CoV-2 using NSP15 and fusion spike glycoprotein as targets. *J. Biomol. Struct. Dyn.,* **2020**, 1-12.
[PMID: 32345124]

Molecular Modeling Applied to Design of Cysteine Protease Inhibitors – A Powerful Tool for the Identification of Hit Compounds Against Neglected Tropical Diseases

Igor José dos Santos Nascimento[1], Thiago Mendonça de Aquino[1], Paulo Fernando da Silva Santos-Júnior[1], João Xavier de Araújo-Júnior[2] and Edeildo Ferreira da Silva-Júnior[1,2,*]

[1] *Chemistry and Biotechnology Institute, Federal University of Alagoas, Maceió, Brazil*

[2] *Laboratory of Medicinal Chemistry, Pharmaceutical Sciences Institute, Federal University of Alagoas, Maceió, Brazil*

Abstract: Cysteine proteases play numerous and extremely important roles in the life cycle of parasitic organisms with medicinal importance. From general catabolic functions and protein processing, cysteine proteases may be key to parasite immunoevasion, excystment/encystment, and cell and tissue invasion. Parasite cysteine proteases are unusually immunogenic and have been exploited as serodiagnostic markers and vaccine targets. The research focused on the development of new drugs actives toward this macromolecular target is an important task, where the rational design is considered as a critical step on it. The discovery of new drugs is a complex and multidisciplinary process, which includes an in-depth knowledge of organic chemistry, pharmacology, biochemistry, computer sciences, and others. This process involves high costs and several scientific fields, leading to the necessity to develop new processes that involve optimization of molecular modeling applied to the identification of bioactive molecules. These techniques could increase the probability of obtaining a rational-designed compound, with high activity and safety, which could be considered as a potential drug in the future. Thus, the use of computational techniques has become increasingly common in medical chemistry laboratories due to their low costs and high correlation with experimental results from assays. A broadly used technique in the rational design of active compounds is molecular docking of small ligand at the active site from the biological targets. In this chapter, we will demonstrate in detail different molecular modeling techniques applied to the development of new inhibitors against cruzain (*Trypanosoma cruzi*); falcipain (*Plasmodium falciparum*); SmHDAC8 (*Schistosoma mansoni*); nsP2 (Chikungunya virus) enzymes; and others, such as cathepsin family; caspase family, 3C^{pro} (Enterovirus 71) and 3CL^{pro} (Coronavirus).

* **Corresponding author Edeildo Ferreira da Silva-Júnior:** Chemistry and Biotechnology Institute, Federal University of Alagoas, Maceió, Brazil; Tel: (+55)879-9610-8311; E-mail: edeildo.junior@esenfar.ufal.br

Finally, studies have revealed that the application of molecular modeling is a powerful tool for predicting new active and productive molecules against infectious diseases.

Keywords: Cysteine Proteases, Drug Discovery, Molecular Modeling, Neglected Tropical Diseases, Virtual Screening.

INTRODUCTION

According to the World Health Organization (WHO), Neglected Tropical Diseases (NTDs) refer to a set of 20 bacterial, parasitic, or viral diseases involving vectors or hosts with complex life cycles. In 2018, NTDs affected about 1.5 billion individuals that are living in social exclusion and poverty in 149 different countries, representing 11% of all global diseases [1 - 3].

From 2012 to 2018, only 3.1% of all 256 approved new drugs were addressed to the NTDs' treatment [2, 4]. Among these, six approved drugs were obtained from drug repurposing based on *in silico* studies), new formulations or biological studies, evidencing a profound lack of interest in research addressed to the discovery of new drugs and treatments for such diseases [5 - 7]. Thus, the major reason for this lack of interest by pharmaceutical companies is related to the high financial costs for drug designing and development. Recently, it has been estimated that US$ 3.4 million are necessary for the preclinical phase, increasing up to 8.6 and 21.4 million dollars in the II and III clinical phases, respectively. Finally, the low financial return generated by this type of pharmacotherapeutic agents is also a complex factor [8 - 11].

Considering this, the current trend of the drug discovery process remains on the combination of Computer-Assisted Drug Design (CADD) tools (*in silico* methods) and combinatorial chemistry and/or high-throughput screening (HTS) studies to improve the ability to discover promising agents or scaffolds [12, 13]. This trend reduces costs by decreasing the number of molecules to be tested, minimizes the possibility of failure due to physicochemical properties, and allows to comprehend the pharmacological mechanism at a molecular level [14 - 16].

The CADD strategy is subdivided into two large groups: Ligand-Based Drug Design (LBDD), in which all information about ligand structures and their biological activities are considered (*e.g.* Quantitative Structure-Activity Relationship (QSAR) and similarity modeling). In contrast, Structure-Based Drug Design (SBDD) considers information about the 3D-structure of biological targets and also ligands' structures (*e.g.* molecular docking and dynamics simulations) [13 - 15]. Moreover, the homology modeling (HM) is based on the production of an atomic model (as an SBDD method), built from the amino acid sequence of a biomacromolecular target, as well as homologous proteins [17, 18]. In this context, CADD has become an essential and powerful tool for drug designing of new agents against NTDs. From this, several drugs have already been discovered by this approach, such as Saquinavir (antiviral), Zanamivir (antiviral), Norfloxacin (antimicrobial), Dorzolamide (antiglaucoma), and Oxymorphone (narcotic analgesic) (in clinical trials) [13].

CA-clan papain-like cysteine proteases (CP) represent a group of targets present in several parasites associated with NTDs that have been exhaustively explored in various virtual screening studies. These enzymes are proteases that irreversibly hydrolyze nucleophilic attack-mediated peptide bonds that generate a tetrahedral intermediate. Furthermore, cysteine proteases are present in different life forms of various NTD-mediating parasites, participating in diverse biological processes [19 - 21]. Thus, they constitute potential targets for drug design and development of antiparasitic compounds, such as the promising vinyl sulfone K777 (Fig. **1**), in the preclinical stage against Chagas disease, for example [19, 22, 23].

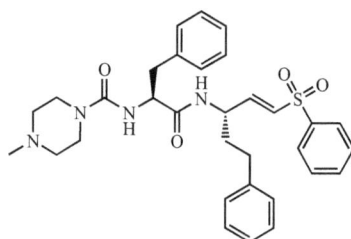

Fig. (1). Most promising vinyl sulfone (K777) found in the literature.

This chapter summarizes a compilation of studies involving the application of molecular modeling approaches for developing new cysteine protease inhibitors. It will focus on the development of inhibitors against cruzain (*Trypanosoma cruzi*), falcipain (*Plasmodium falciparum*), SmHDAC8 (*Schistosoma mansoni*), nsP2 (Chikungunya virus), cathepsin-S, -K, -B, and -L (human), 3C^{pro} (Enterovirus 71), 3CL^{pro} (Coronavirus), and caspase-1 and 3 (human) by using virtual screening techniques.

VIRTUAL SCREENING IN DRUG DEVELOPMENT

The discovery of new drugs is a complex and time-consuming process that requires a high degree of knowledge in different areas of science and has faced many changes and challenges in recent decades [24]. Recently, the development of drug has included virtual screening methods in which molecules do not physically exist. Still, they are deposited or obtained in virtual databases and can be purchased or synthesized [25]. As previously discussed, this technique allows discovering new *lead* compounds through two approaches: induced fitting, which requires prior knowledge of the interaction site at the target protein (SBDD); and the other, is based only on ligands' structures and their activities (LBDD), in which the target protein is unknown. Thus, molecules are obtained from virtual databases according to the chosen criteria [26].

Experimental studies performed in large compound libraries, such as high throughput screening (HTS) have been developed since the 1980s, aiding to find new compounds more safely. However, due to the high financial cost employed in the experiments, the virtual screening (VS) method was developed, a technique similar to the HTS, which uses only *in silico* experiments, reducing process-related financial costs [27]. Thus, some authors cite VS as a complementary method to HTS, which plays a fundamental role in drug discovery, where millions of compounds are docked in a specific target. Thus, scoring and ranking are performed by using affinity energy values obtained from the complex ligand-target, which will be used to select the most promising compounds, reducing the number of compounds to be synthesized and tested. [28, 29] Finally, in Fig. (**2**), a basic VS procedure proposed by Ghosh *et al.* (2006) is shown [30].

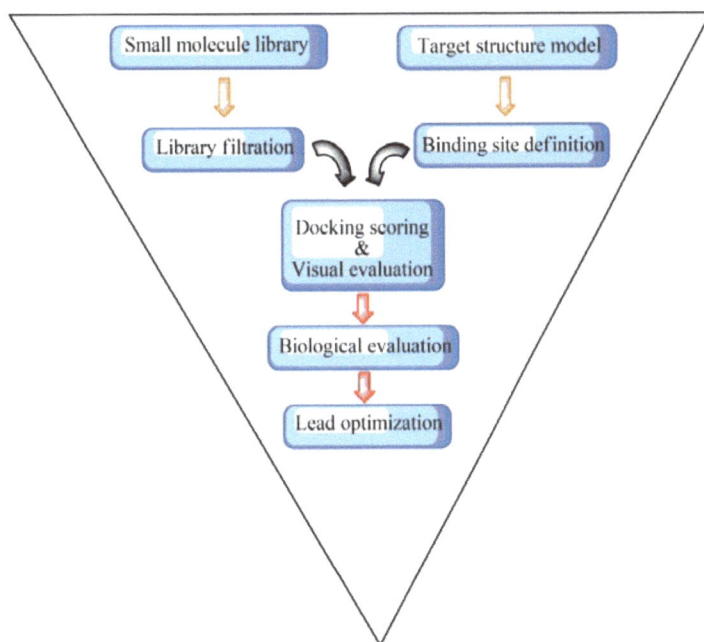

Fig. (2). A systematic overview of a virtual screening protocol.

A research model using VS proposed by Cerqueira *et al.* (2015) [31]. Fig. (**3**) shows a complete approach, where the workflow includes the treatment of the target and the selection of the filters applied to the compounds, as well as the protocol validation.

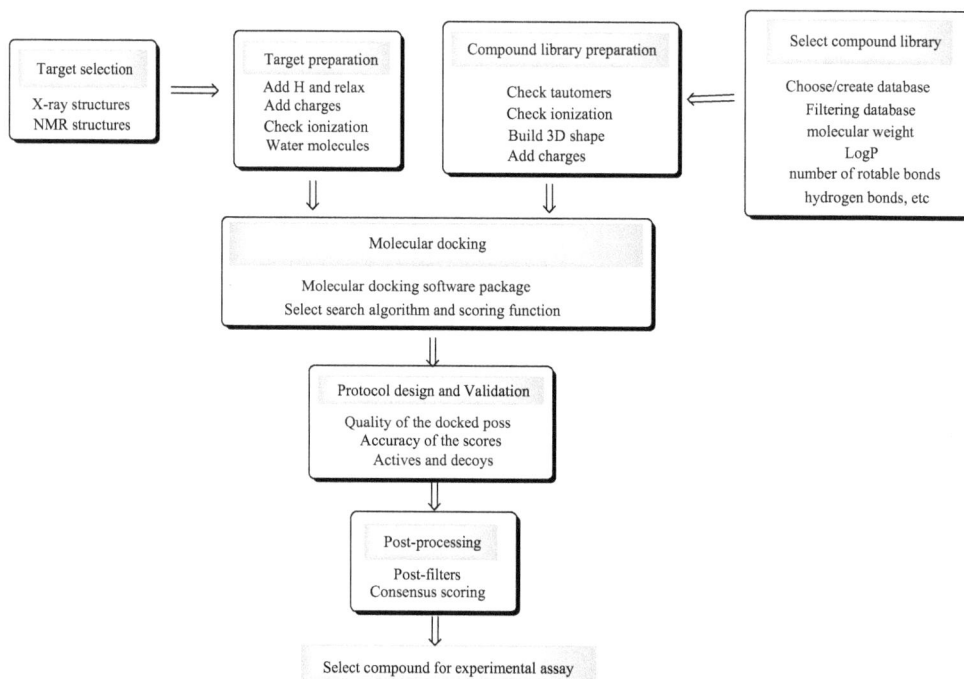

Fig. (3). Virtual screening workflow by Cerqueira *et al.*, 2015.

CYSTEINE PROTEASES AND THEIR CATALYTIC MECHANISM

Proteolytic enzymes are involved in a significant number of biological processes in ranging from the less complex (virus, bacteria, and protozoa) to the more complex organisms (mammals) [32, 33]. In general, proteases cleave proteins into smaller fragments by catalyzed-peptide bonds hydrolysis. They can be classified according to their catalytic site, and categorized into four major classes: serine, aspartic, cysteine, and metalloproteases [34]. Papain family comprises enzymes that are responsible for many physiological functions, such as antigen presentation, matrix turnover, digestion, processing surface proteins, hemoglobin hydrolysis, immune invasion, and parasite invasion and egress [35].

Cysteine proteases are synthesized as zymogens and contain a prodomain (with regulatory function) and a mature domain (with catalytic function). Moreover, this prodomain can act as an endogenous inhibitor of the mature enzyme. In order to completely activate the mature enzyme, all prodomain should be removed, in which this step could be achieved by different modes. Zymogen conversion may be accomplished by accessory molecules, such as the conversion of trypsinogen into trypsin in the presence of Ca^{2+}. Also, the pro-mature domain interaction can be classified as protein-protein interactions (PPI), and it could be targeted in a large number of diseases [36]. More information can be found in a review manuscript by Verma *et al.* (2016) [37], in which different modes of activation in cysteine proteases are highlighted and also their future aspects.

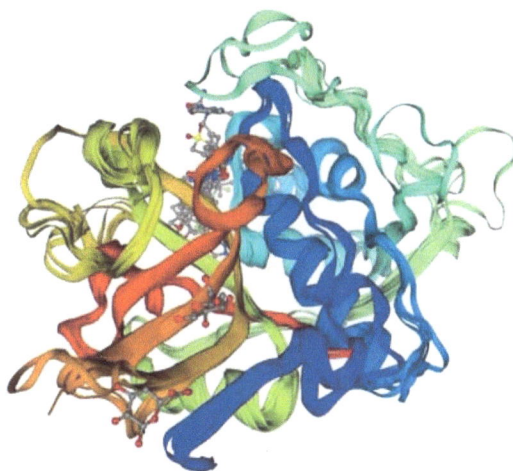

Fig. (4). Molecular alignment of some important cysteine proteases associated with neglected tropical diseases or human organism. Cruzain (PDB: 3I06), rhodesain (PDB: 6EX8), cathepsin B (PDB: 6AY2), cathepsin K (PDB: 3H7D), cathepsin L (PDB: 2NQD), cathepsin L1 (PDB: 3HWN), and cathepsin S (PDB: 2FUD). Models were produced by using the Swiss-Model server (https://swissmodel.expasy.org/).

For NTD's, cysteine proteases from viral and parasitic organisms represent promising targets for a wide diversity of diseases, such as malaria [37], Chagas disease [23], African sleeping sickness [23], Chikungunya [38], and leishmaniasis [39]. Although, the main challenge in the design and development of cysteine protease inhibitors remains on the selectivity, once that this protease group comprises several closely-related enzymes and also their isoforms from different organism species. In order to illustrate that, in Fig. (4) shows the molecular alignment of several cysteine proteases (cruzain, rhodesain, and cathepsins –B, -K, -L, -L1, and -S) of medicinal interest obtained by using molecular homology technique (SBDD approach). Finally, it is possible to verify that these enzymes have high structural similarity (identity index), once most of the fragments of them are overlapped.

Normally, cysteine proteases have a Cys-His-Asn triad at the active site, named catalytic triad. The catalytic mechanism of cysteine proteases (Fig. **5**) depends on two residues at the active site, which typically is composed of Cys and His amino acids. It is known that the physiological environment favors that the imidazole group from the His residue acts as a proton donor and polarizes the thiol group from the Cys residue, leading to the highly nucleophilic Cys(S⁻)/His(NH⁺) ion pair. Subsequently, the nucleophilic Cys[25] attacks the carbon of the reactive peptide bond, leading to three different possible mechanisms of proteolysis, involving acylation and deacylation reactions. In the mechanism I, a tetrahedral thioester intermediate is formed (acylation), in which this intermediate is stabilized by hydrogen bonding between the substrate oxyanion and a highly conserved glutamine residue (Asn). Posteriorly, amine and carboxylic acid fragments of the substrate are released in a deacylation reaction catalyzed by a water molecule. In this pathway, His[159] generates a strong nucleophile by proton abstraction from the water molecule (at the deacylation step) that hydrolyzes the thioester remaining substrate fragment. In mechanism II, the thiohemiketal intermediate also produces amine and carboxylic acid fragments a water-catalyzed deacylation reaction. Finally, in mechanism III, an alcohol intermediate is produced, and an intramolecular proton abstraction occurs, leading to thioester and amine fragments. After the hydroxyl anion attack, the tetrahedral intermediate is stabilized by a hydrogen interaction with the amine fragment, favoring the carbonyl double bond replacement and releasing of a carboxylic acid fragment [37, 40, 41]. Various computational studies involving density functional theory (DFT) have been developed to understand the high reactivity of these proteases toward electrophilic groups at substrate structures, supporting several studies focused on the covalent inhibitors' development [41 - 44].

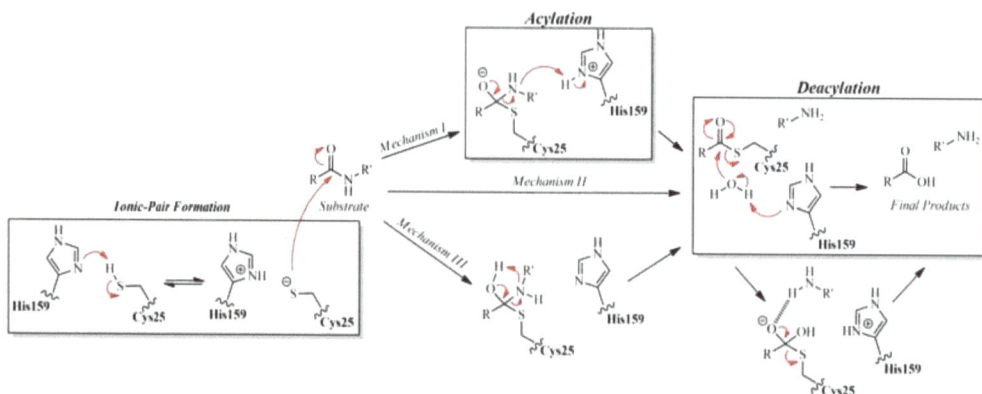

Fig. (5). Schematic overview of the catalytic mechanism of cysteine proteases.

MOLECULAR MODELING APPLIED TO THE DESIGN OF CYSTEINE PROTEASE INHIBITORS

Trypanosoma cruzi (Cruzain)

Trypanosoma cruzi is the protozoan responsible for causing the Chagas disease, which is associated with several cases worldwide, mainly in Latin America. There are an estimated 8 million infected around the world, leading to approximately 10,000 death cases per year, and still, 25 million people are living in risk zones [45]. Although it is a fairly incident disease, the pharmacological treatment is limited to two old drugs, nifurtimox **(1)** and benznidazole **(2)** (Fig. **6**), both compounds have demonstrated effectiveness in cases of acute infections. However, they have some limitations in patients, such as cutaneous adverse effects (benznidazole) and gastrointestinal tract disorders (nifurtimox), which lead to treatment discontinuation. Additionally, there are few compounds in clinical II phase trials, such as fexinidazole **(3)** (used to treat sleep sickness), AN4169 **(4)**, and GNF6702 **(5)** (Fig. **6**) [46].

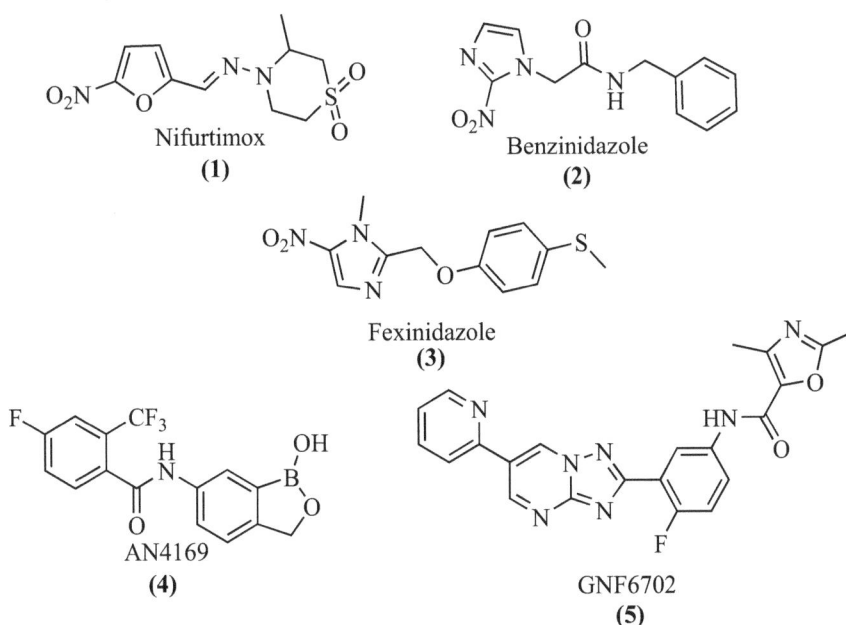

Fig. (6). Approved (and in clinical trials) compounds addressed to the treatment of Chagas disease.

Among *T. cruzi*-related molecular targets, one of the main targets is cruzain, a cysteine protease expressed at all stages of the life cycle of this parasite. It is related to the replication and differentiation of epimastigotes in metacyclic infectious forms, as well as immune evasion and parasite interactions-host [47, 48]. Then, cruzain is the most studied target in trypanocidal drug design.

Compounds are designed in order to interact with amino acid residues responsible for the catalytic activity of cruzain, which are Cys^{25}, Gly^{23} and Gly^{65}, as well as filling the subsites S1, S1', S2 e S3 (Fig. **7**) [48].

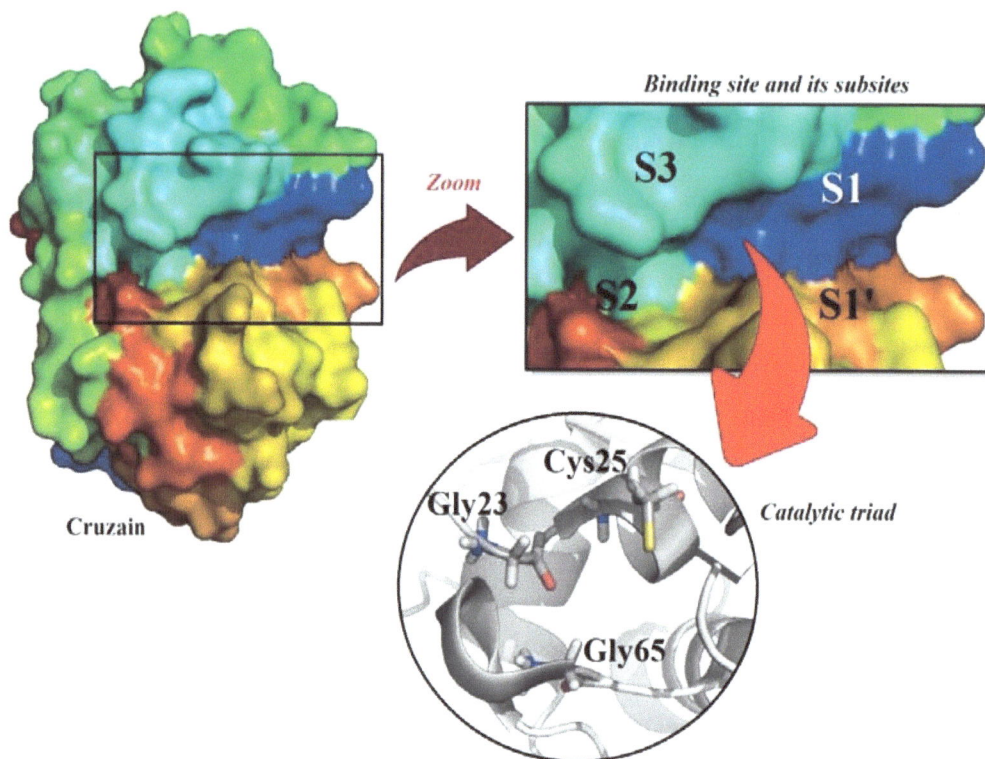

Fig. (7). Catalytic triad and subsites from the cruzain enzyme.

One of the first VS studies conducted for developing new inhibitors of cruzain from *T. cruzi* was performed by Choe *et al*. (2005) [49]. In this study, the authors emphasized the importance of using virtual libraries of compounds in VS to identify and optimize new compounds. Considering that it was a pioneering study, the authors described that the lead compound AQ665183 **(6)** was selected from a library of peptides containing an α-ketoamide moiety, leading to potential covalent inhibitors. The compounds were optimized at the positions P1' e P1, generating 25 compounds **(7-31)** (Fig. **8**) with inhibitory activity in the nanomolar range.

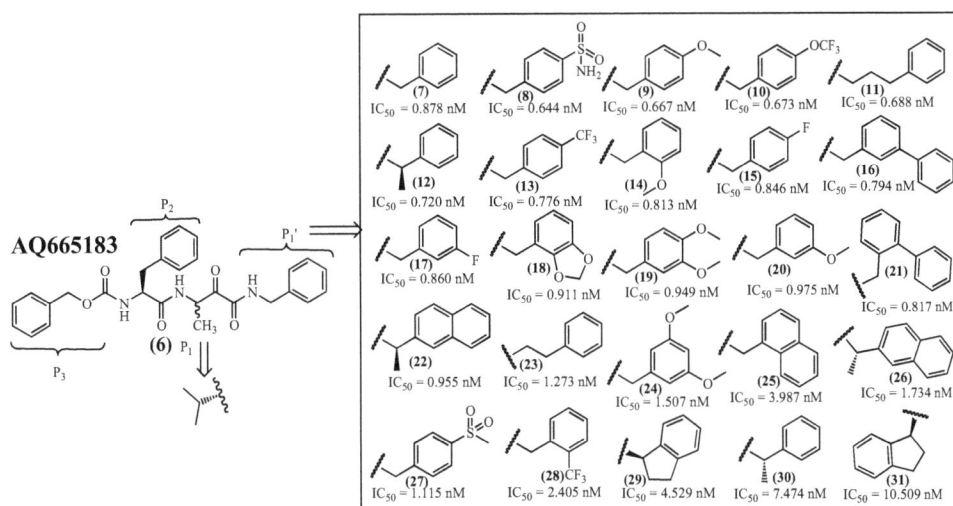

Fig. (8). Optimization of compound AQ665183 and its potent analogs.

Computational studies have become a routine increasingly in drug development research groups, with progressively more specific and accurate approaches, by using well-defined validated protocols by LBDD or SBDD methods. Freitas *et al.* (2008) [50] performed an LBDD study employing Hologram Quantitative Structure-Activity Relationship (HQSAR) and 2D-Quantitative Structure-Activity Relationship (2D-QSAR) approaches in a series of 41 cruzain inhibitors. It was found that the combination of both techniques achieved a statistically significant predictive model by HQSAR ($r^2 = 0.9$ and $q^2 = 0.77$). The authors emphasized that the selectivity is a crucial factor for the development of cruzain inhibitors, due to this enzyme has high similarity to cathepsin-L, a protease present in mammals. Thus, it was necessary to use the Rapid Overlay of Chemical Structures (ROCS) method, obtaining an area under the curve (AUC) value of 0.96, showing that the method was successful in discriminating active compounds from inactive. This information was confirmed by the fact that the three most selective inhibitors **(32-34)** (Fig. **9**) for cruzain were identified among the 37 top-hit compounds previously selected.

Another protocol that was successfully used for discovering new hits against cruzain was developed by Malvezzi *et al.* (2008) [51], using an SBDD approach. In this study, compounds were obtained from the ZINC database library (3,294,741 structures) and various physicochemical filters associated with Lipinski's rule of five were applied, resulting in 296,700 selected compounds. For these, a 3D-pharmacophore model was used to improve the selection of promising molecules. The authors observed that the selected compounds presented common characteristics, such as aromatic ring (S2 pocket occupation); LH donor domain (interaction with Asp[158]); LH acceptor domain (interaction with the Gly[66]); and electrophilic domain (potential site for attacking cysteine nucleophile). In total, these chemical characteristics allowed to select 308 compounds. Subsequently, molecular docking was employed to select the 50 compounds with best score values, after pose ranking and visual inspection. Finally, the six most promising compounds (**35-40**) (Fig. **10**) were selected for biological testing, where they displayed IC_{50} values ranging from 0.67 to 44.91 μM.

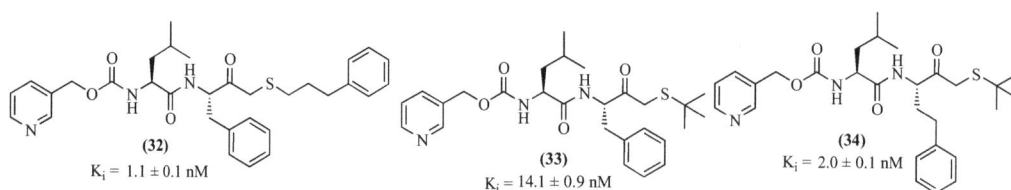

Fig. (9). Selective inhibitors of the cruzain enzyme discovered by Freitas *et al.*, 2008.

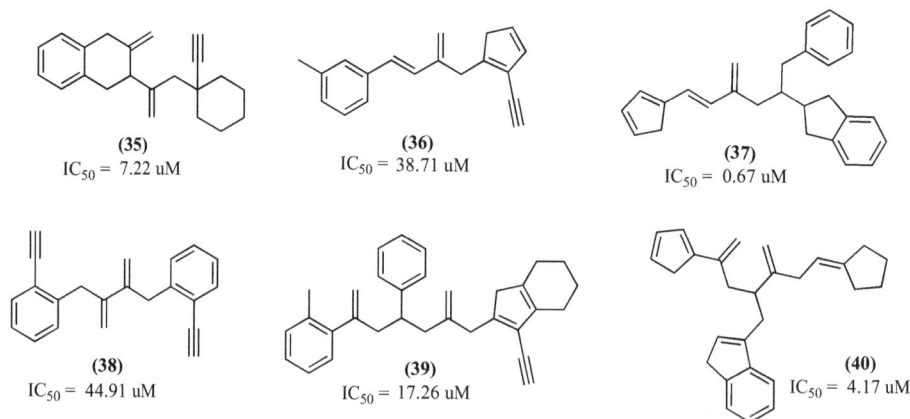

Fig. (10). Cruzain inhibitors discovered by Malvezzi *et al.*, 2008.

Due to the success of both LBDD and SBDD protocols, some authors have sought a combination of both approaches that could be used in the search for new cruzain inhibitors. Then, Wiggers *et al.* (2011) [52] obtained an excellent performance by using an SBDD approach through molecular docking of reactive groups toward the catalytic cysteine (Cys[25]), resulting in an area under the ROC curve similar to LBDD method. It was also observed that the combination of LBDD and SBDD techniques by consensus methods result in high effectiveness. Through this study, the application of the SBDD approach was provided frameworks for the robust and predictive HQSAR model with the ability to discriminate between active and inactive compounds, with better performance than ROCS or EON procedures.

Two years later, Wiggers *et al.* (2013) [53] both approaches again in order to identify new cruzain inhibitors. The compounds were selected from the ZINC database, applying physicochemical filters which resulted in a library of 3.5 million compounds. FRED and ROCS ligand-based VS methods were applied as a faster screening methodology. These methods are based on Tanimoto metric similarity and docking score, allowing that the compounds could be analyzed from predefined threshold values. The FRED program was used to discriminate between active and inactive compounds using SHAPE-GAUSS score function in order to decrease the number of molecules in the library by the selection of the best binding energy value for every compound. A consensus score strategy was also applied, calculating pKi values by the HQSAR model and Glide extra precision score function (Glide XP). Finally, some compounds were considered as most promising based on the following criteria: occupation at S1, S2, and S3 subsites; and hydrogen bonding with Gly[66] residue. 23 compounds were selected for *in vitro* testing, where 6 molecules (**41-46**) showed IC$_{50}$ values ranging from 5.6 to 73.9 μM (Fig. **11**).

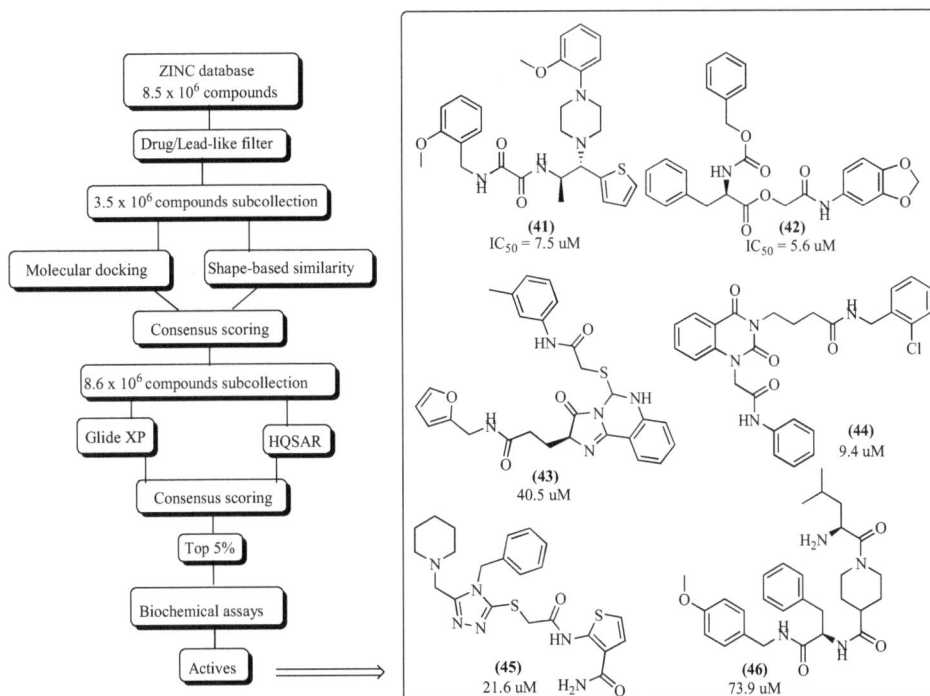

Fig. (11). Virtual screening protocol developed by Wiggers *et al.*, 2013.

Another interesting model was developed by Bellera *et al.* (2013) [54], where a 2D-classification model followed by the use of the VS technique, it was found that amiodarone **(47)** and bromocriptine **(48)** (antiarrhythmics and antiparkinson drugs, respectively) may be promising as reversible inhibitors of cruzain. In this study, 82 reversible inhibitors and 81 non-inhibitors previously described in the literature were initially selected. A cluster analysis was performed by using the LibraryMCS v0.7 (ChemAxon) hierarchical clustering combined com k-means clustering, followed by an analysis of LDA. Then, the PDD and ROC curves were used to select the limit value to be used in the VS. Starting from 6,684 molecules obtained in the Drugbank, 256 structures were found to be active. However, 54 of these compounds were commercially available drugs. Four compounds (Fig. **12**) were selected due to their easy obtainments, such as **(47)**, **(48)**, colchicines **(49)**, and escitalopram **(50)**. These compounds exhibited promising results in cruzain inhibition assays, especially amiodarone and bromocriptine, with IC_{50} values of 219.8 and 82.4 µM, respectively.

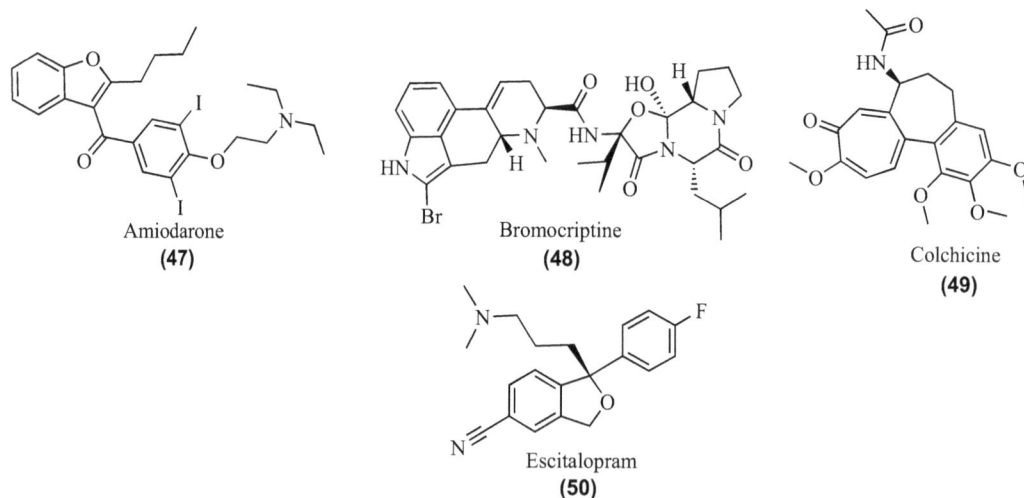

Amiodarone
(47)

Bromocriptine
(48)

Colchicine
(49)

Escitalopram
(50)

Fig. (12). Promising cruzain inhibitors identified by Bellera *et al.,* 2013.

Encouraged by these previous results, Bellera *et al.* (2015) [55] developed a similar study, leading to the identification of four other drugs on the market as potential anti-*T. cruzi* agents, namely clofazimine (anti-leprosy agent), benidipine (antihypertensive) and saquinavir (anti-HIV). In this study, 147 inhibitors and non-inhibitors were selected and submitted to the hierarchical clustering approach LibraryMCS v0.7 in combination with the k-means cluster implementing Statistica 10 Cluster Analysis module. This model was optimized using the ROC curve, thus increasing the specificity and decreasing the false positives. This model was applied in a VS against cruzain using 5,570 molecules obtained from the Merck Index 12th edition database. Four compounds (51-54) (Fig. 13) were selected due to their easy obtainment and tested against epimastigote forms. As a result, clofazimine (52) and benidipine (53) presented *in vitro* LC$_{50}$ values of 10.6 and 19.5 μM. Additionally, both of these compounds displayed similar results to benznidazole (2) in *in vivo* assays. Through molecular docking studies, it was shown that activity of the most active compound (clofazimine) is mainly due to the interaction of chloro-phenyl moiety with S2 pocket by the interaction with Leu^{67}, Ala^{133}, and Leu^{157} residues. The authors also mentioned that no strong interactions with the catalytic triad residues (Cys^{25}, His^{159}, and Asn^{175}) were observed, and it was proposed that the stability of the ligand-macromolecule complexes is associated with interactions close to the catalytic cleft.

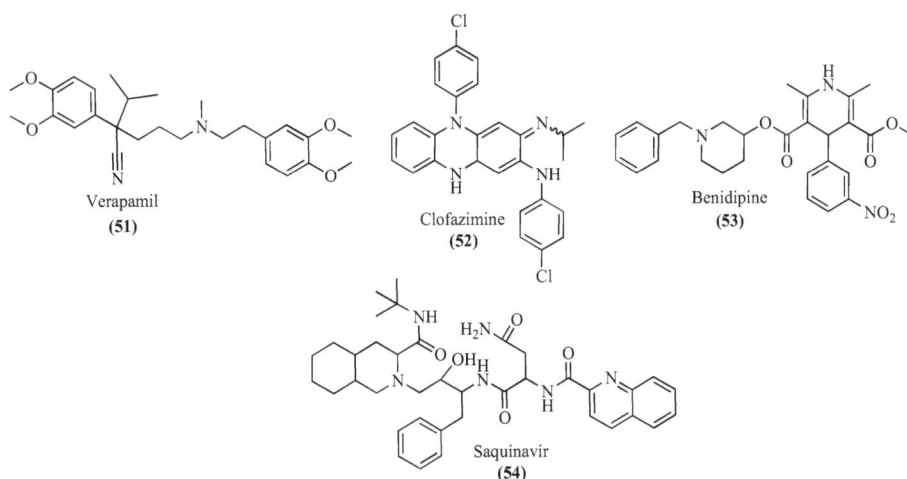

Fig. (13). Promising cruzain inhibitors identified by Bellera *et al.*, 2015.

In a more recent study by Mayorga *et al.* (2019) [56], the VS model was applied for the discovery of new *N*-acylhydrazone compounds from the ZINC database. The authors stated that this chemical class of molecules was chosen because they are starting points for designing various compounds in medicinal chemistry. The selected molecules were those that presented the general *N*-acyl hydrazone moiety (C=NNC(C)=O). Also, chemical structures above 50% similarity and respecting the Lipinski rule were included in this study. Thus, 2,221 ligands were selected for molecular docking. Then, the most promising compounds (55-58) with the lowest binding energies (less than -8.0 Kcal/mol) were selected (Fig. 14) for *in vitro* assays. Compound (56) showed similar activity to the nifurtimox (1) against epimastigote forms (LC$_{50}$ = 36.26 μM (56) and 38.36 μM(1)). Finally, (56) also presented an IC$_{50}$ value of 56.26 μM against cruzain.

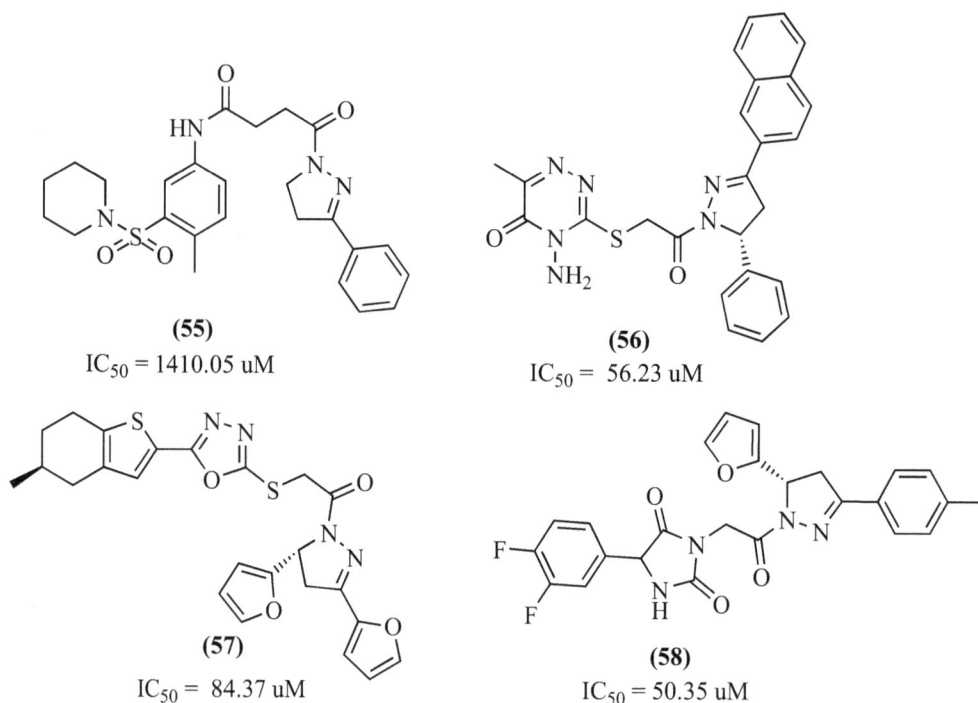

Fig. (14). Compounds identified by Mayorga *et al.,* 2019.

A disadvantage of the SBDD approach is that either false positives or false negatives could be identified as promising compounds, which can lead to the selection of compounds with unsatisfactory biological results, or even an active molecule to be excluded of biological assays [57, 58]. Thus, Ferreira *et al.* (2010) [58] performed a study in order to identify new cruzain inhibitors. The authors used molecular docking and HTS methods toward the MLSMR database in a library of compounds (in total, 197,861), in which 99% of them were eliminated as false positives. As a result, 921 compounds were considered as actives, where 34 compounds were selected based on their scores and from these, 19 molecules could be purchased. The most promising compounds **(59-69)** were categorized as promising reversible-competitive inhibitors of cruzain, with IC_{50} values ranging from 0.3 to 38 μM (Fig. **15**). The authors emphasized that the HTS could be used to identify false positives and false negatives safely.

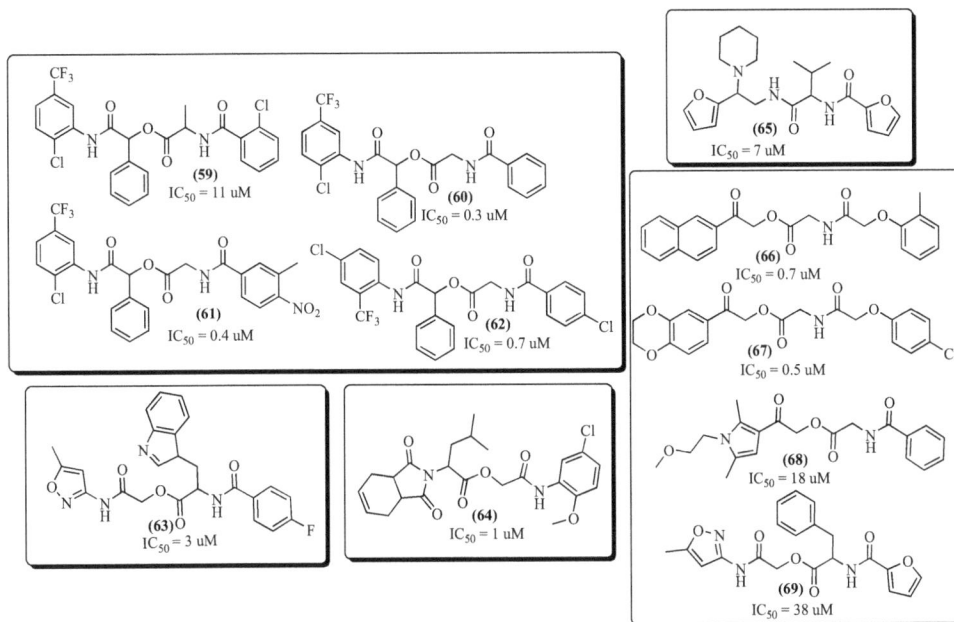

Fig. (15). Cruzain inhibitors identified by Ferreira *et al.*, 2010.

Plasmodium falciparum (Falcipain)

Malaria is a parasitic disease caused by mosquitoes from the *Anopheles* genus that affects several people around the world. It is globally estimated at about 216 million cases, mainly in African countries with high mortality rates [59]. The etiological agent of this disease belongs to the *Plasmodium* genus, which comprehends four main species, such as *P. vivax, P. ovale, P. malariae,* and *P. falciparum* [60].

In general, antimalarial drugs are divided into three categories that are related to their mechanism of action and chemical structures, being category A (aryl aminoalcohol compounds, **70-76**), category B (antifolate compounds, **77-79**), and category C (artemisinin compounds, **80-83**), as shown in Fig. (**16**) [61]. These substances have been developed for decades and are widely used in clinical practice, which led to the emergence and re-emergence of parasite resistance to drugs, is responsible for the need of new and more effective agents [61].

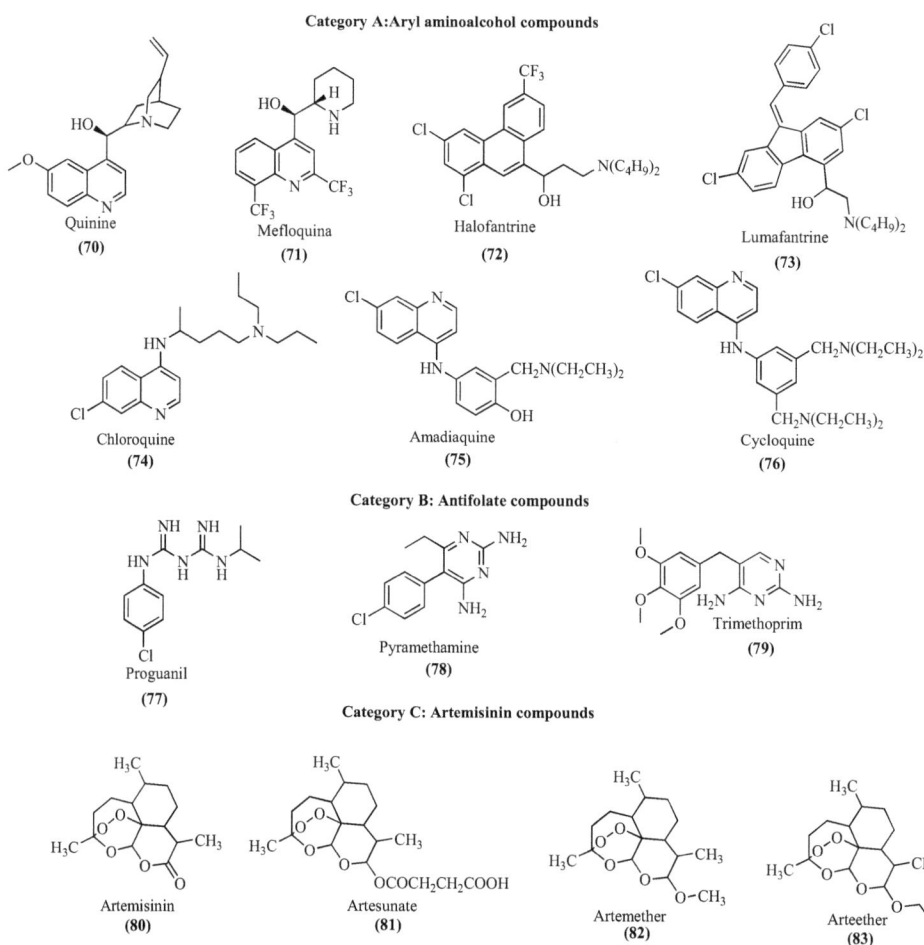

Fig. (16). Category of compounds employed in the treatment of malaria.

According to literature, the discovery of most of the antimalarials used in clinical practice was not made by using rational design focused on inhibition of a specific target, once that the targets for most of them remains uncertain [61]. With the advances in medicinal chemistry and computational chemistry fields, new prototypes were developed targeting specific macromolecules or structures from the parasite, such as cytosolic medium, plasmatic membrane, transporters, food vacuole, apicoplast, and mitochondria [61].

The food vacuole represents an interesting target for the development of new antimalarial compounds since hemoglobin is hydrolyzed in the food vacuole into small peptides, which subsequently will be the source of amino acids for the parasite [61]. The enzyme that controls the proteolysis process is named falcipain, which is a cysteine protease that presents the triad Cys-His-Asn at the active site for peptide hydrolysis. This macrobiological target have been extensively studied in the development of new antimalarial inhibitors [61, 62].

One of the first studies using VS for searching new falcipain inhibitors was proposed by Desai *et al.* (2004) [63], in which a VS was conducted using the ChemBridge database and a library of over 241,000 compounds. Specific filters were initially applied, such as metal-containing molecules, molecules with good ADME properties and those that filled the Lipinski's rule of five. Then, 60,000 compounds were selected and docked by using the GOLD program and falcipain-2 and -3, as targets. The top 200 hits were visually inspected following some criteria, such as reasonable link mode geometry, the proximity of the electrophilic center to the cysteine amino acid, spatial occupation in hydrophilic and hydrophobic regions. These steps allowed 100 compounds to reach the biological assays. So, 22 compounds with IC_{50} values ranging from 1.0 to 63 μM were obtained, being the compound (**84**) (Fig. **17**), the most active analog. Finally, the authors considered that the activity is associated with van der Waals interactions (π-π interactions) between the ligand and Trp^{206} residue at the pocket S1, and also a hydrogen interaction with Ile [85] amino acid residue at the pocket S2.

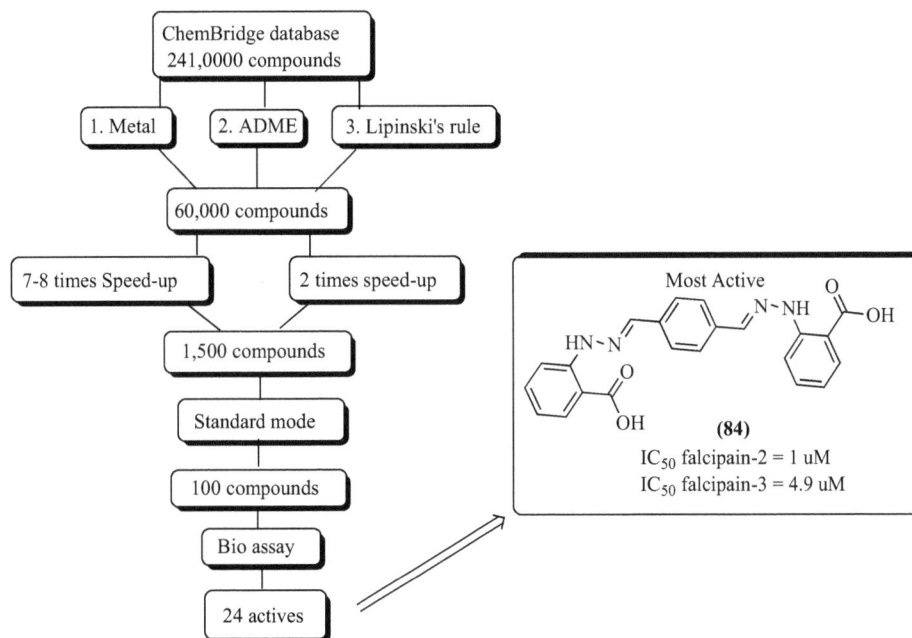

Fig. (17). The most promising falcipain inhibitor identified by Desai *et al.* 2004.

Considering these previous results, Desai and coworkers (2006) [64] conducted a second study using the Available Chemical Directory (ACD), a database of about 355,000 compounds and using exactly the same VS protocol applied in their previous study. Finally, the authors obtained 18 compounds with inhibitory activity towards falcipain-2 with IC_{50} values ranging from 1.4 to 53.6 µM, resulting in seven new leads compounds (**85-91**) (Fig. **18**), which could be optimized for designing new cysteine protease inhibitors.

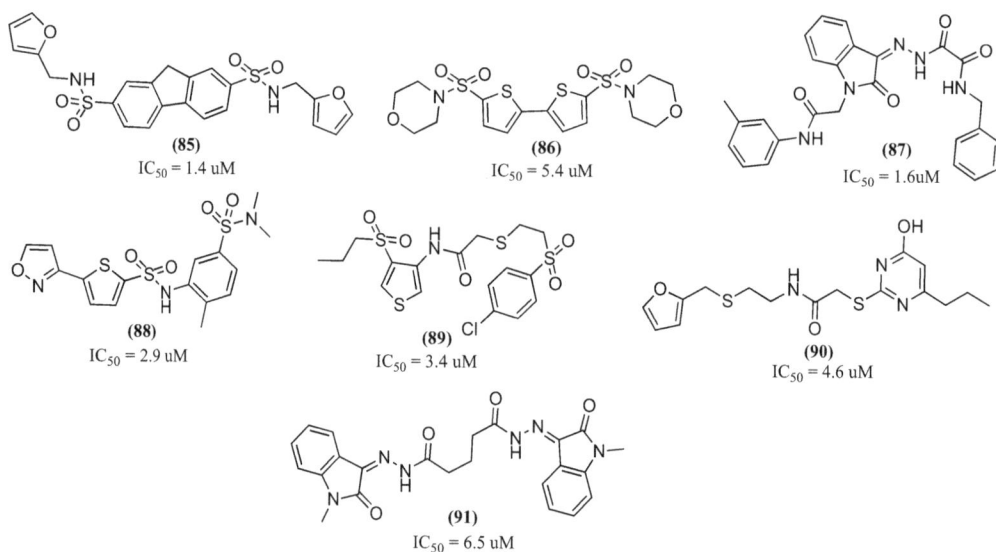

Fig. (18). Falcipain inhibitors identified by Desai *et al.*, 2006.

These studies performed by Desai and collaborators were fundamental for *in silico* prediction of *leads* against the falcipain-2 enzyme. Considering both of them, Li and colleagues (2009) [65] then applied a similar VS approach to that from Desai's studies, but using the Glide and GAsDock programs in the SPECS organic molecule database, containing approximately 287,000 compounds. The authors applied the druglike filter developed by Zheng *et al.*, reducing this dataset to 80,000 compounds. Then, these molecules were docked and ranked in the Glide and GAsDock programs by two different modes, as shown in Fig. (**19**). By using Glide XP's, 200 best compounds were subjected to visual inspection, in which specific criteria were applied, such as hydrophobic interaction with S2 pocket, hydrogen interaction with Cys[42] amino acid residue. In contrast, GAsDock compounds were selected by energy score values. Finally, 81 compounds were selected for the falcipain-2 assays, where eight molecules (**92-99**) showed IC_{50} values ranging from 2.4 to 54.2 µM.

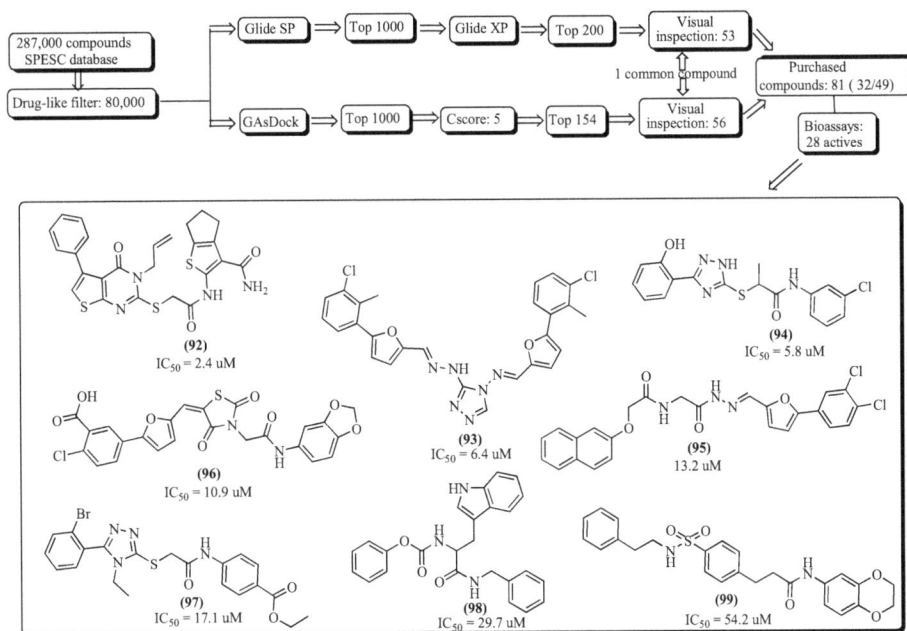

Fig. (19). Virtual screening protocol and the most promising falcipain-2 inhibitors identified from the SPECS database.

While previous studies were performed on only one database, the study of the Shah *et al.* (2011) [66] was conducted using several of them. Initially, it was building the Focused Cysteine Protease Inhibitor (FCPI) Library by using various databases of organic molecules (Asinex platinum, ChemBridge, SPECS, Enamine, IBScreen, and Aurora Fine Chemicals Ltd.) to select molecules contained in their structures soft electrophiles moieties. After removing all identical molecules, structures containing problematic groups (such as metals, aldehydes, chloroamides, peroxides, and *N*-oxides) and that comply with the Lipinski's rule of five resulted in approximately 65,000 molecules. These compounds were then submitted to the VS protocol, in which approximately 30,000 compounds containing vinyl sulfone groups were selected. These compounds were docked using the previously validated Glide XP protocol, resulting in 3,000 compounds, which were then ranked by the Emodel score and selected for the model eMBrAce minimization calculation (Macromodel). Thus, these steps resulted in 500 molecules. Moreover, compounds were visually inspected following some criteria, such as present a soft electrophile near Cys[42], hydrophobic interaction with Leu[172] and Leu[84] amino acid residues, or hydrogen interaction with Gln [36] and Gly [83] amino acid residues, coherent geometry at the binding site; occupation of S2 subsite by a hydrophobic group, and commercial availability. 200 molecules were selected and analyzed using the Dissimilarity module of Sybyl 8.1 software, resulting in 50 molecules that were screened for their biological activities. Finally, 20 molecules were found to be promising as falcipain inhibitors, where it was verified that six compounds **(100-105)** (Fig. **20**) showed dual activity toward W2 infected cells and falcipain-2 (FP-2) and -3 (FP-3) enzymes, with IC_{50} values ranging from 1.57 to 49.18 µM.

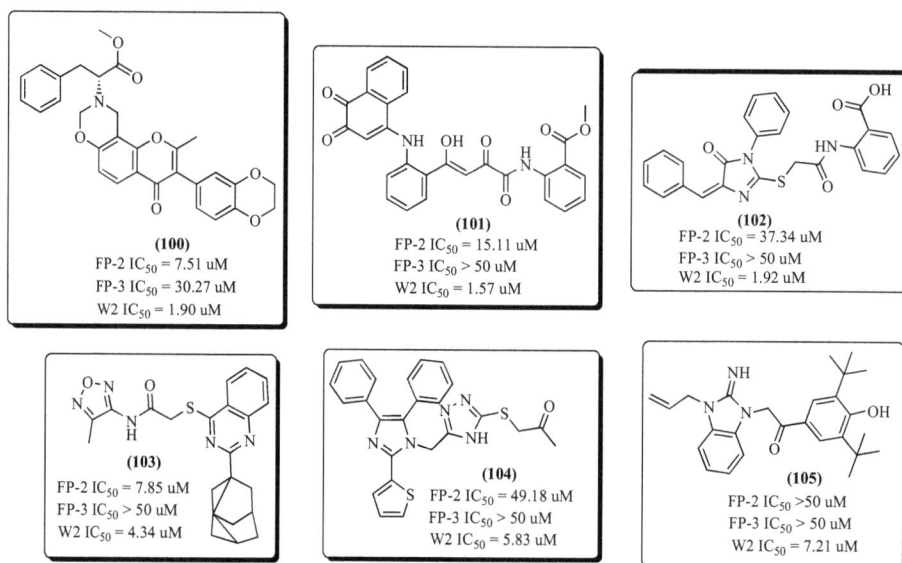

Fig. (20). Inhibitors identified by Sha and colleagues, 2011.

These promising results from previous work inspired Sha *et al.* (2012) [67] to continue the study, using a similar VS protocol in the Chembridge and Asnex databases, based on three *leads* substructures discovered in their previous study. The application of this VS protocol led to the discovery of 69 molecules that were screened by biological tests of falcipain inhibition. 28 analogs showed IC_{50} values ranging from 5.0 to 48 µM against FP-2. Also, some of them demonstrated IC_{50} less than 10 µM **(106-117)** (Fig. **21**) against infected cells with malaria.

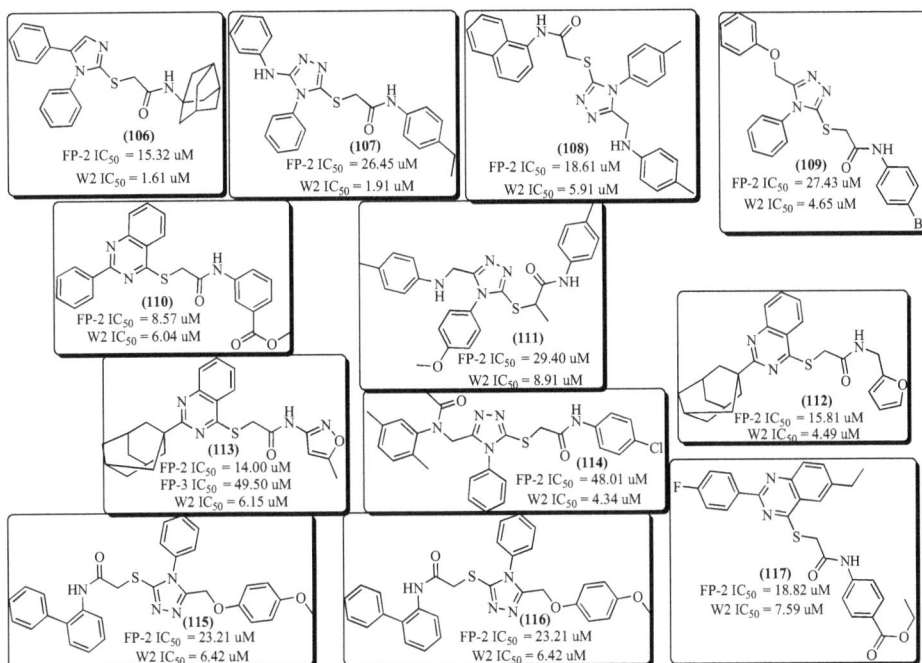

Fig. (21). Compounds identified by Sha *et al., 2012.*

VS studies are typically performed from libraries of organic molecules, with approximately one million molecules recorded in different libraries [68]. In VS procedures, the choice of library is a crucial step because although each library has a high number of molecules, these molecules must be commercially available or easily synthesizable [69]. Some studies are performed utilizing an in-house library of compounds because the process for obtaining compounds is easier [70]. Based on this information, Wang *et al.,* (2014) [71] performed a study involving the discovery of new FP-2 inhibitors through a SBDD approach by using an in-house library (Natural Product Database - NPD), containing about 4,000 compounds. The compounds were docked, and the top 500 was visually inspected. Moreover, 50 compounds were selected to be tested against FP-2, where it was verified that the 10 most promising compounds **(118-127)** showed IC_{50} values ranging from 3.18 to 68.19 μM. Also, the compound **(122)** was found to be the most promising from this series of analogs, due to its micromolar activity against chloroquine sensitive (3D7) and resistant (Dd2) strains, yielding a new compound that can be used in the optimization of antimalarial compounds (Fig. **22**).

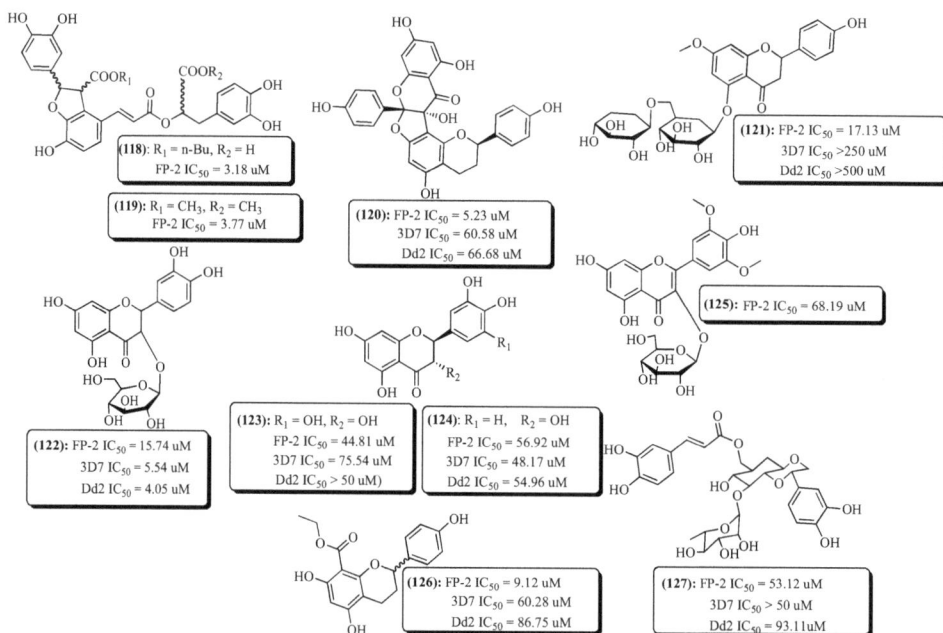

Fig. (22). Falcipain-2 inhibitors Identified by Wang *et al.*, 2014.

An interesting study was developed by Chakka *et al.* (2015) [72], which aimed the discovery of a new falcipain-2 inhibitor from a VS in the UNITY database, using about 250,000 compounds. As a result, seven compounds were selected, where one of these compounds **(128)** was found to be more promising, with a K_i value of 27 ± 6 µM. Docking studies of this compound and two other falcipain-2 inhibitors (leupeptin and K11017) (Fig. **23**) were then performed. In general, new compounds were designed based on the best filling of the S1' and S3 pockets. Then, a total of 23 compounds were synthesized in which the most potent derivatives **(129-131)** showed submicromolar falcipain-2 inhibitory activity.

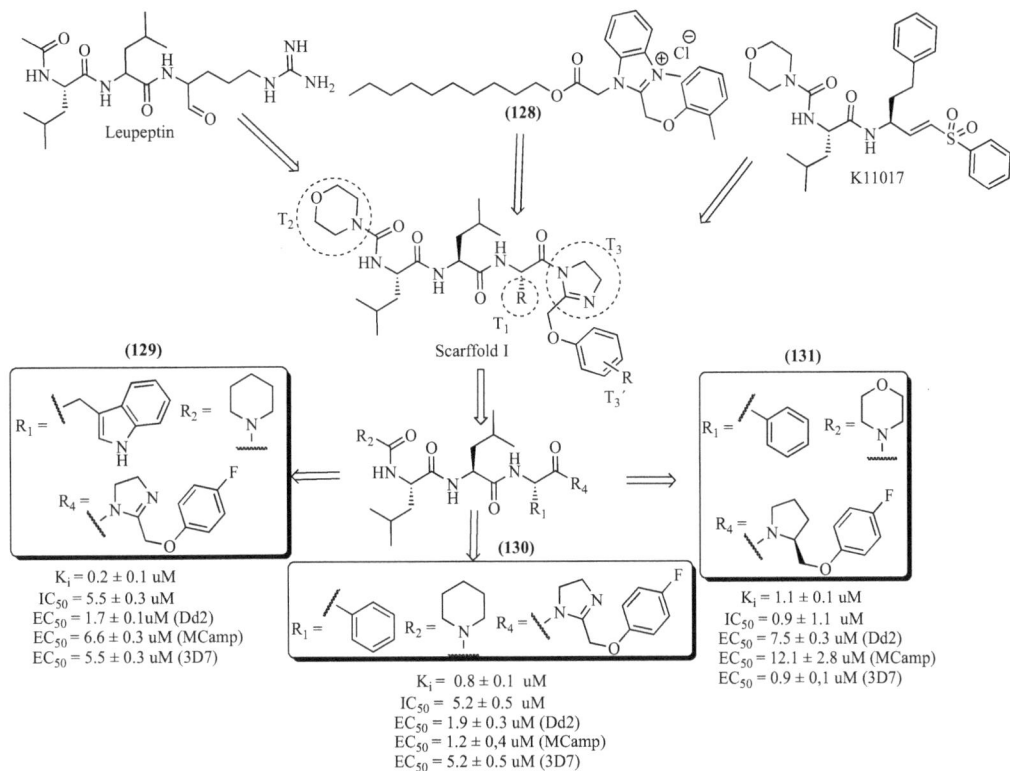

Fig. (23). Falcipain-2 Inhibitors developed by Chakka *et al.,* 2015.

There are some cases in which the target is not well understood, and some comparative models with known proteins are used that have similar sequences with the protein to be studied [14]. Thus, the homology modeling comes as a useful tool in drug design and development, also known as comparative modeling, in which the protein structure is predicted by comparing the conformation and the presence of similar amino acids of another protein. It is based on the fact that the *in natura* conformational structure is highly conserved and that small sequence changes do not drastically alter the 3D-structure [14, 73, 74]. The study by Saddala *et al.* (2018) [75] used homology modeling for discovering new inhibitors of vivapain-3 (VP-3) enzyme. It is an enzyme present in *P. vivax* that is similar to falcipain-3. After validating the models by homology and pharmacophore modeling, only the commercially accessible compounds were then searched in the PubChem database, resulting in 14 hits that were submitted to molecular docking simulations. From this step, five compounds were selected with better affinity than leupeptin (standard compound). Molecular dynamics simulations were also performed, proving the stability of compounds at the active site in physiological conditions. Compounds **(132-134)** (Fig. **24**) showed promising results against *P.*

vivax enzyme, with IC_{50} values better than the reference compounds (1.5 ± 0.24; 1.0 ± 0.35; and 2.1 ± 0.45 µM, respectively). The authors also mentioned that the compounds filled Lipinski's rule of five, leading to a good pharmacokinetic profile and that the approach used may contribute to the discovery of new antimalarial compounds.

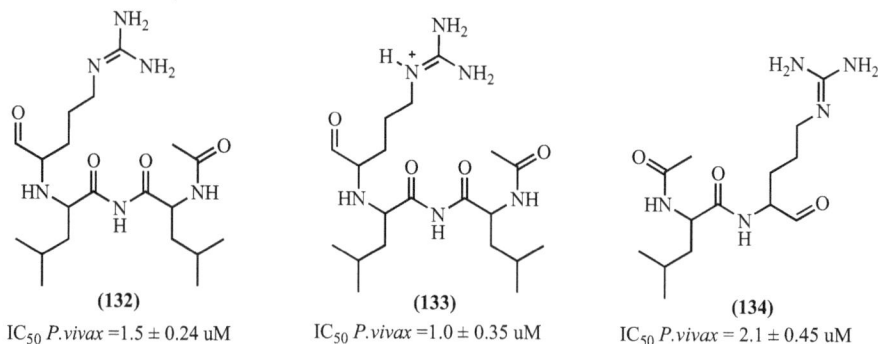

IC_{50} *P.vivax* =1.5 ± 0.24 uM IC_{50} *P.vivax* =1.0 ± 0.35 uM IC_{50} *P.vivax* = 2.1 ± 0.45 uM

Fig. (24). Compounds identified by Saddala *et al.*, 2018.

The main problem to work with cysteine protease inhibitors is the selectivity toward human proteases [23]. A study by Hernádez-González *et al.* (2018) [76] was performed employing an SBDD approach and the Maybridge HitFinder database to discover new selective falcipain-2 and -3 inhibitors with lower affinity to human cathepsin K (hCatK). Through a structure alignment study, it was observed that hCatK was the enzyme with greater similarity to falcipains 2 and 3 were chosen to evaluate the selectivity of the compounds from SBDD analysis. The SBDD procedure was performed at the Maybridge HitFinder library against falcipains 2 and 3 and hCatK through molecular docking studies. As a result, nine compounds **(135-143)** (Fig. **25**) presented IC_{50} values ranging from 7.4 to 326 µM. The compounds **(142)** and **(143)** were highlighted, as they presented excellent values of IC_{50} as well as a better selectivity index (46 and 10, respectively), better antiplasmodial activity (2.91 and 34 µM) and lower toxicity in normal cells (133 and 350 µM).

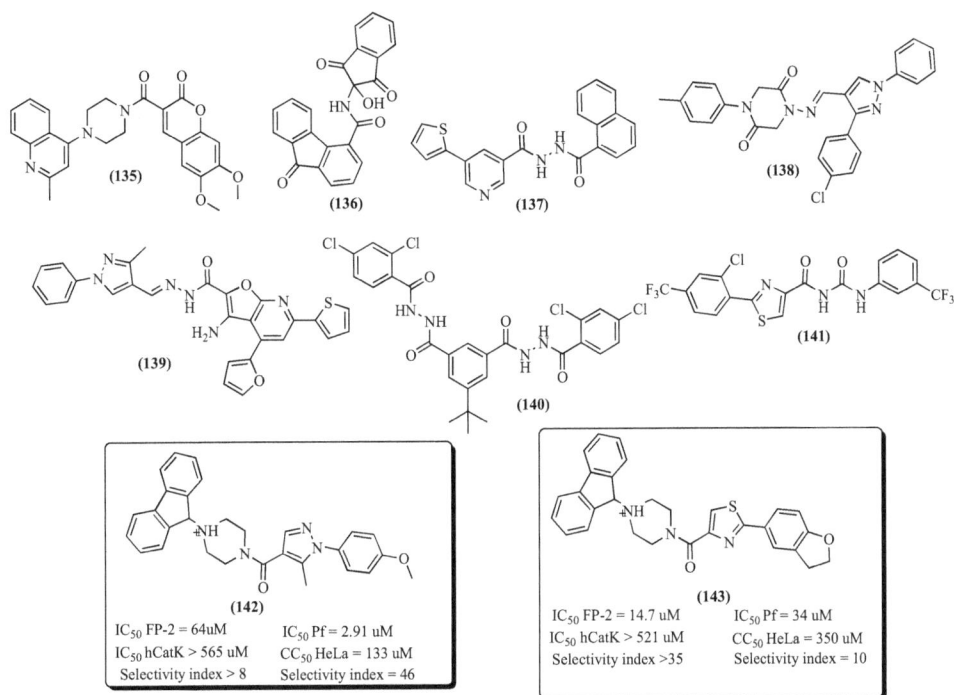

Fig. (25). Compounds identified by Hernádez-González *et al.*, 2018.

Chikungunya (nsP2)

Chikungunya fever (CHIKF) is a viral disease transmitted mainly by *Aedes aegypti* and *Ae. albopictus* mosquitoes, which affect millions of people in over 100 countries, mainly in Asian, American, and African continents [77 - 79]. In general, symptoms are characterized by fever and severe joint pain. The joint pain can persist for months or years even after the patient's recovery, but in the elderly, patients can lead to death. There are no pharmacological agents and no vaccines. The treatment is based on relieving symptoms with antipyretics and analgesics [78].

The genome of the Chikungunya virus (CHIKV) consists of structural proteins (C, E1, E2, E3, and 6K) and nonstructural proteins (nsP1, nsP2, nsP3, and nsP4), which are of fundamental importance for viral propagation and replication 80. Among nonstructural proteins, nsP2 has been reported as an attractive target for drug design of new anti-CHIKV agents, because this is the largest nonstructural protein responsible for some enzymatic functions, such as mediator of the processing of nonstructural polyproteins P123 and P1234; activator of helicase on viral RNA synthesis; NTPase activity; and function and RNA phosphatase of G and SG RNAs in buffering reactions [80 - 82].

Targeting the nsP2 enzyme, Bassetto *et al.* (2013) 38 used molecular modeling in combination with the classical approach of medicinal chemistry to develop more selective antiviral compounds. Through homology modeling and application of the virtual screening protocol resulted in the discovery of compound **(144)** (Fig. **26**), the most active candidate with EC_{50}= 5.0 μM; EC_{90}= 6.4 μM; CC_{50}= 72 μM, and selectivity index (SI) of 14 against Chikungunya virus-induced cytopathic effects (CPE) formation in Vero cells. The authors emphasize that the presence of aliphatic groups in the benzylidene position *para* is important for antiviral activity. This compound was optimized using classical medicinal chemistry strategies resulting in the synthesis of compounds **(145)** and **(146)** (Fig. **26**). The docking simulations toward nsP2 were able to show that the compounds **(145)** and **(146)** occupy the binding site, similarly to compound 1; however, the compound **(145)** was more active, with an EC_{50} value of 3.2 μM. This study was used for the design of the new compounds proposed by Tardugno *et al.* (2018) [83]. As a result, compounds **(147)** and **(148)** (Fig. **26**) were found to be the most promising of series, with better selectivity indexes. These studies also assisted the work of Giancotti *et al.* (2018) [84], that performed the synthesis of compounds **(149-154)** (Fig. **26**). In this work, the maximum inhibition effect was reached by these new derivatives, showing that this strategy may be of fundamental importance for the discovery of new compounds for the treatment of CHIKV.

The computational study performed by Nguyen *et al.* (2015) [81] combined VS, molecular docking, and dynamics simulations for developing new nsP2 inhibitors. By using VS and docking, four *hits* compounds **(155-158)** (Fig. **27**) were identified, and these were evaluated in molecular dynamics studies to determine the preferential binding modes. The pocket 4 was identified as a potential active site for ligands. Also, it was verified that the interaction of ligands with Cys^{1013} and His^{1083} amino acid residues is essential for anti-CHIKV activity. The authors claimed that these methods are useful for finding new nsP2 inhibitors and that experimental studies are needed to prove their hypothesis.

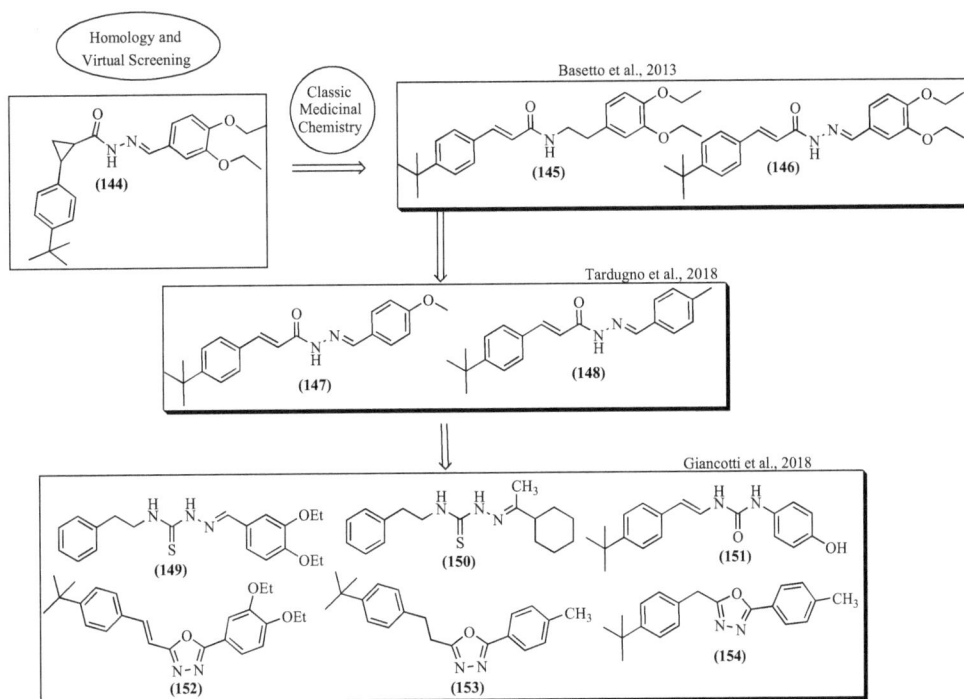

Fig. (26). Compounds discovered by Bassetto *et al.* 2013; Tardugno *et al.*; and Giancotti *et al.* 2018.

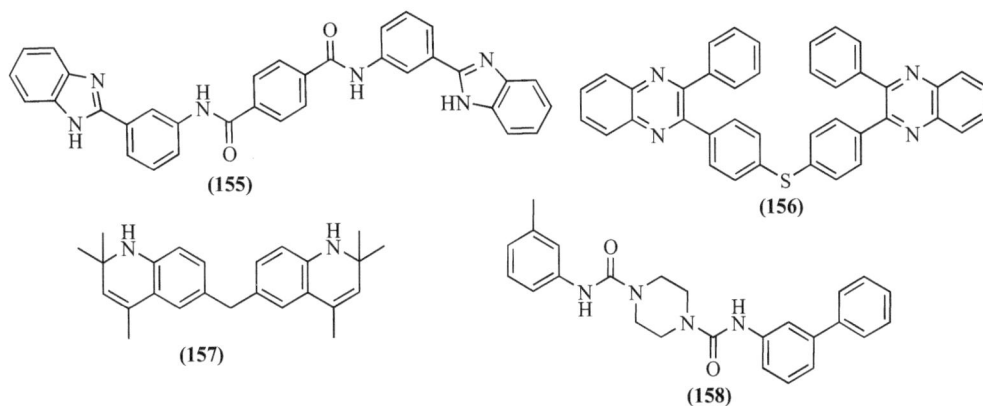

Fig. (27). Anti-CHIKV (nsP2) potential inhibitors identified by Nguyen *et al.*, 2015.

Schistosoma mansoni (SmHDAC8)

Schistosomiasis is a parasitic disease caused by worms from the *Schistosoma* genus, being *S. haematobium, S. japonicum,* and *S. mansoni* the responsible for this neglected disease. It is an NTD reported in at least 78 countries, with an estimated at least 220.8 million people needing preventive treatment [85, 86]. The main compound used in its treatment is praziquantel **(159)** (Fig. **28**), discovered in the '70s and which still has an unknown mechanism of action [87]. Another alternative is oxamniquine **(160)** (Fig. **28**), used only in cases of treatment failure with praziquantel, which although effective, both have low activity against adult worms, as well as tolerance and resistance, justifying the investment in research for new molecules [88].

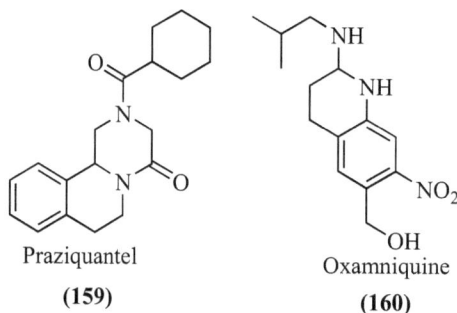

Praziquantel
(159)

Oxamniquine
(160)

Fig. (28). Chemical Structures of Praziquantel and Oxamniquine drugs.

Regarding drug development, approximately 100 proteins are reported as potential pharmacological to become targets. The most studied are thioredoxin glutathione reductase (TGR), 3-hydroxy-3-methyl-glutaryl-coenzyme A reductase, and histone deacetylases (HDACs) [87]. Some experiments indicate that HDACs (main cysteine proteases present in the parasite) are related to the death of larvae and adult worms, showing to be a possible strategy in the development of new compounds against this disease [89].

Considering this important enzyme, Kannan *et al.* (2014) [90] studied the combination of homology modeling and VS towards the smHDAC8 enzyme. A replacement of the Met[274] residue present in hmHDAC8 with a Hist[292] in smHDAC8 has been revealed. However, the authors stated that replacing Met[274] with a more polar residue assists in the development of more specific inhibitors. After the application of the VS protocol, 75 molecules were found to be promising. From this number, seven compounds **(161-166)** demonstrated promising activity (Fig. **29**), with IC_{50} values ranging from 1.22 to 5.56 µM.

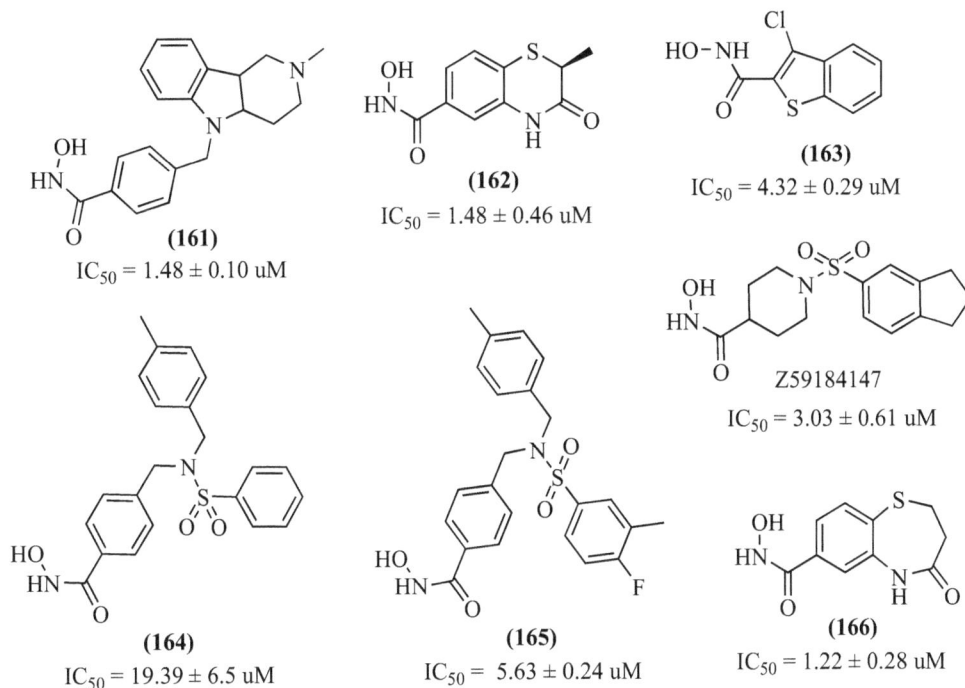

Fig. (29). smHDAC8 inhibitors identified by Kannan *et al.*, 2014.

Based on the study cited above, Heimburg *et al.* (2019) [91] used hit compounds as scaffolds for the design of new and more selective smHDAC8 inhibitors, with less selectivity for the isoform of the enzyme present in humans (hHDAC8). Several molecules **(167-178)** (Table **1**) showed higher selectivity for smHDAC8 when compared to hHDAC8 present in humans. After docking simulations, the authors proposed that the selectivity for human isoforms of the enzyme (hHDAC8 and hHDAC6) is associated with the hydrogen bond between the ligand and a conserved water molecule that leads to interaction with the His[180] amino acid residue.

Other Cysteine Proteases

Cathepsin Family

Cathepsins are lysosomal proteases of cysteines that are part of the large family of hydrolases responsible for the hydrolysis of acidic peptide bonds [92, 93]. These enzymes are responsible for regulating various signaling pathways and related immune responses, so that unregulated activity of these enzymes is related to some types of diseases, such as cancer, osteoporosis, cystic fibrosis, arthritis, cardiovascular disease, and neurodegenerative disease [92 - 94]. Studies are conducted on this class of enzymes (focusing on cathepsins -S, -K, -B, and -L) considered targets of interest in medicinal chemistry for the discovery of new compounds, mainly related to osteoporosis and cancer [19, 95].

Table 1. Inhibitors synthesized by Heimburg *et al.*, 2019.

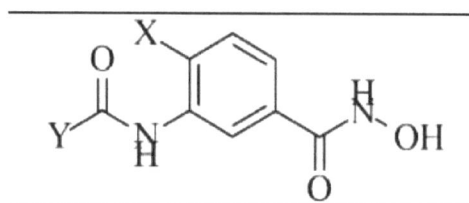

Entry	X	Y	IC$_{50}$ (µM)		
			smHDAC8	hHDAC8	hHDAC6
(167)	F	Ar	177.6 ± 8.1	317.8 ± 54.2	0.5 ± 0.01
(168)	Cl	Ar	67 ± 10.2	120 ± 36.7	0.12 ± 0.02
(169)	Br	Ar	150.4 ± 8.5	191.4 ± 26	0.15 ± 0.01
(170)	CF$_3$	Ar	139.6 ± 8.3	342.2 ± 76.1	0.14 ± 0.02
(171)	OCH$_3$	Ar	129.3 ± 7.6	171.5 ± 15.6	1.3 ± 0.1
(172)	OCH$_3$	2,4-dichlorophenyl	121.6 ± 18.7	548.3 ± 93.9	2.3 ± 0.4
(173)	OCH$_3$	4-ethoxyphenyl	305 ± 35.0	438.4 ± 48	1 ± 0.1
(174)	OCH$_3$	Bz	182.7 ± 39.3	512.2 ± 29.8	5.1 ± 0.7
(175)	Cl	4-methoxyphenyl	147.1 ± 4.8	235.6 ± 49.5	0.13 ± 0.01
(176)	Cl	3-phenoxyphenyl	396.4 ± 43.3	448.6 ± 100.4	0.3 ± 0.1
(177)	Cl	2,4-dichlorophenyl	191.4 ± 16.7	1184 ± 45.1	0.8 ± 0.1
(178)	OCH$_3$	4-biphenyl	92 ± 26	148.7 ± 22.7	0.6 ± 0.1

The study performed by Barrett *et al.* (2005) [96] focused on cathepsin-K inhibitors for osteoporosis treatment, employed the VS in the ACD database for molecules with structural similarity to compound **(179)**, a cathepsin K inhibitor previously discovered by his research group IC$_{50}$ = 1200 nM). After the application of filters, 25 compounds were selected for virtual inspection against cathepsin-K, being chosen for synthesis only seven compounds **(180-186)** (Fig. 30) that showed binding energy similar to compound **(179)**.

(**179**: R$_1$ = OTBu

IC$_{50}$ = 1200 nM

| IC$_{50}$ = 350 nM | IC$_{50}$ = 190 nM | IC$_{50}$ = 32 nM | IC$_{50}$ = 26 nM | IC$_{50}$ = 5.7 nM | IC$_{50}$ = 7.2 nM | IC$_{50}$ = 3.0 nM |
| (**180**) | (**181**) | (**182**) | (**183**) | (**184**) | (**185**) | (**186**) |

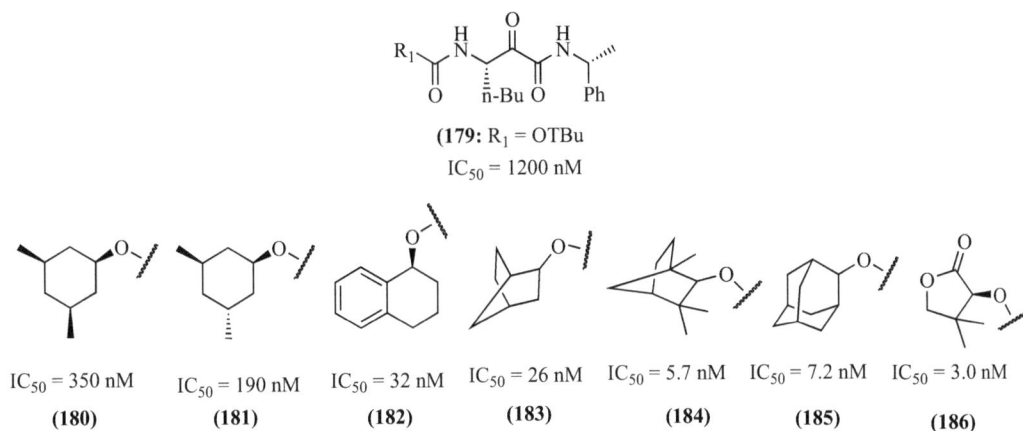

Fig. (30). Inhibitors synthesized by Barrett *et al.*, 2005.

Schröder *et al.* (2013) [97] focused on building a library of commercially available compounds that featured electrophilic chemical groups to obtain new cathepsin-K inhibitors. So, a library of 1,247 compounds was constructed in which the docking simulations were performed using the GOLD software in covalent mode by using two scoring functions, GoldScore and ChemScore. As a result, three active compounds (**187-189**) (Fig. **31**) showed promising activities, with IC$_{50}$ values ranging between 0.1 and 0.4 µM and K$_i$ values in the nanomolar range, suggesting that the methods employed in this study constitute an adequate approach to prioritize new ligands for initial experimental tests.

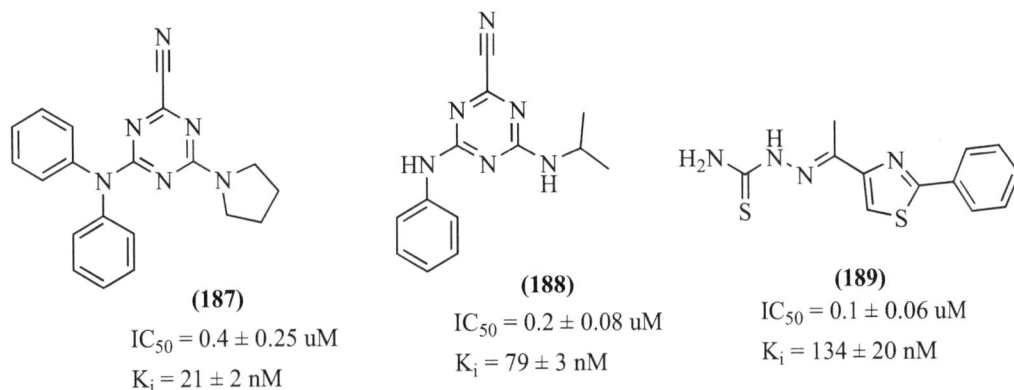

(**187**)

IC$_{50}$ = 0.4 ± 0.25 uM

K$_i$ = 21 ± 2 nM

(**188**)

IC$_{50}$ = 0.2 ± 0.08 uM

K$_i$ = 79 ± 3 nM

(**189**)

IC$_{50}$ = 0.1 ± 0.06 uM

K$_i$ = 134 ± 20 nM

Fig. (31). Compounds identified by Schröder *et al.*, 2013.

Markt *et al.* (2008) [98] focused on the discovery of cathepsin-S inhibitors, a possible target for the treatment of atherosclerosis. In their model, it was decided to search for covalent and non-covalent inhibitors in 18 databases, in a total of

2,664,754 compounds. After the application of several filters, 15 compounds **(190-204)** (Fig. **32**) were selected for visual inspection. Finally, 13 compounds presented micromolar Ki values.

Pinitglang *et al.* (2012) [99] investigated the active site of cathepsin-S towards inhibitors from the ZINC database. They obtained 10 compounds in which the main amino acid residues responsible for the inhibitory activity of the molecules were studied, and when comparing three best compounds **(205-207)** (Fig. **33**). The authors observed that the formation of a hydrogen bond is essential between residues of Cys [25], His[164], Gln [19], Trp[186], and Phe[146]. The authors also conducted a molecular dynamics study for the best compound, detecting that hydrophobic interactions are essential between inhibitor and cathepsin-S. The authors conclude that this study is of fundamental importance for the discovery of new molecules that can help cancer therapies, as well as immune disorders (multiple sclerosis), rheumatoid arthritis, allergic asthma, and pain.

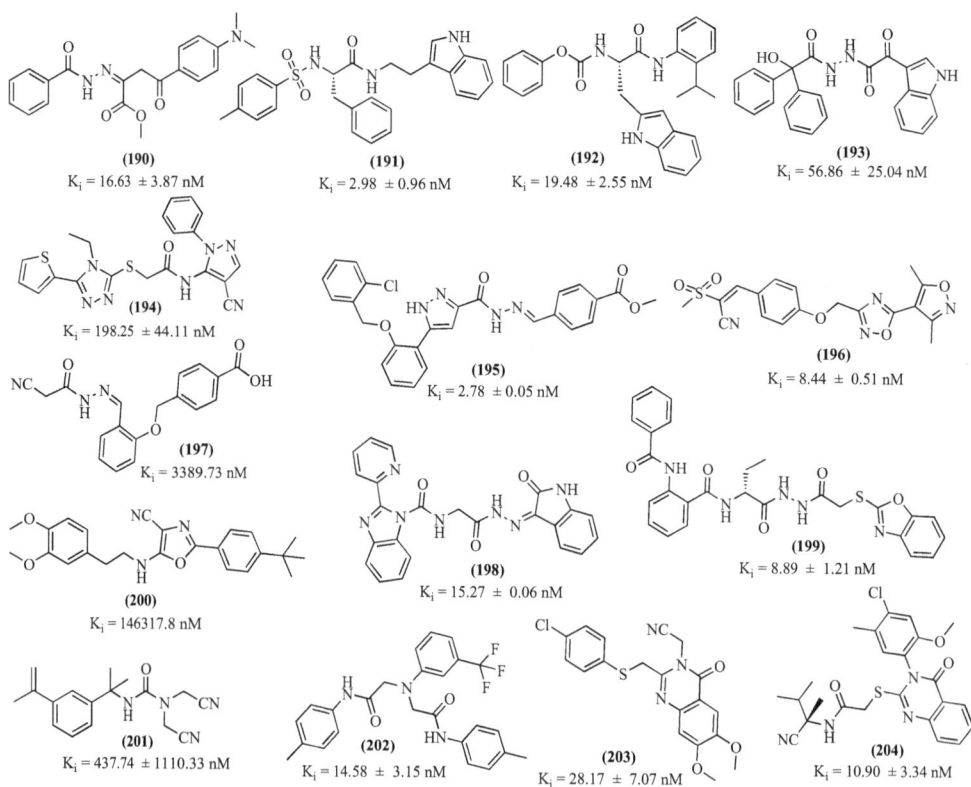

Fig. (32). Compounds identified by Markt *et al.*, 2008.

Chowdhury *et al.* (2019) [100] performed work to discover new covalent inhibitors of cathepsin-L cysteine proteases in a library of 1,648 vinyl sulfone-derived compounds. The authors found the compound **(208)** (Fig. **34**) as a covalent inhibitor of cathepsin-L, with a K_i value of 59 ± 10 μM. Also, it was demonstrated that this covalent docking method was very promising, as the discovered compound showed selectivity for cathepsin L, without inhibitory activity against other important enzymes, such as cathepsin-B, -I, and USP8.

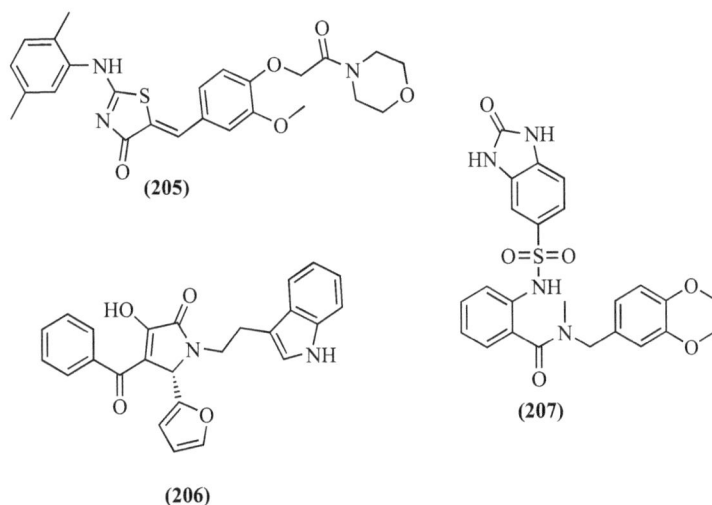

Fig. (33). Inhibitors identified by Pinitglang *et al.* 2012.

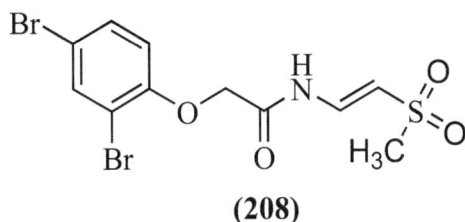

Fig. (34). Cathepsin-L inhibitor identified by Chowdhury *et al.*, 2019.

Enterovirus 71 (3C^{pro}) and Coronavirus (3CL^{pro})

Enterovirus 71 (EV71) is a species of human enterovirus from the *Picornaviridae* family. It has a positive genome of RNA and a length of approximately 7.5kb [101, 102]. EV71-associated diseases include aseptic meningitis, brainstem encephalitis, and acute flaccid paralysis associated with polio, which may result in neurological damage [102]. The initial step for translating this pinacovirus is the production of approximately 2,000 amino acid polyprotein (250 kDa), which is rapidly broken down by viral proteases 2A and 3C to form mature and functional viral proteins essential for virus survival and replication [103]. The EV71 3C[pro] is a cysteine protease of fundamental importance in this process, responsible for the virion maturation, besides presenting with different host proteins aimed at interrupting the synthesis of host proteins and inducing apoptosis [101 - 103]. The essential role of 3C[pro] in virus replication makes it an interesting target in the discovery of new antiviral therapy compounds [102, 103].

The study proposed by Ma *et al.* (2016) [104] focused on the discovery of new EV71 3C[pro] enzyme inhibitors. The VS protocol was then applied to the SPECS database, generating 50 compounds that were purchased and tested. Compound **(209)** showed an IC$_{50}$ = 21.72 ± 0.27 μM, besides being a competitive inhibitor with K$_i$ = 23.29 ±1.07μM. The VS protocol performed by Zhang *et al.* (2017) [105] was focused in an in-house library containing approximately 10,000 compounds, which resulted in 20 compounds that had interaction with catalytic triad-related amino acids. The compound **(210)** (Fig. **35**) was presented as the most active in the series, with more than 80% inhibition at 100 μM.

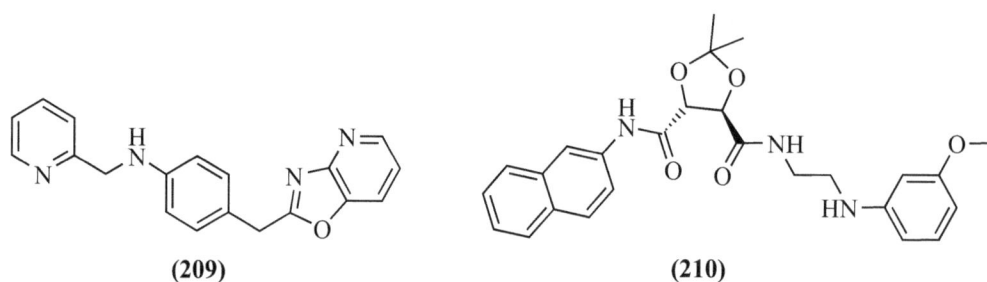

(209)   **(210)**

Fig. (35). Compounds identified by Zang *et al.*, 2017.

Another attractive protease targeting antiviral drug development is 3CL[pro], a cysteine protease responsible for the proteolytic processing of coronavirus replicase polyproteins. It is a positive single-stranded RNA virus responsible for generating upper respiratory tract infections in humans [106 - 108]. The study performed by Lee *et al.* (2014) [109] focused on developing a VS model for the discovery of new non-covalent and non-peptide inhibitors of 3CL[pro]. In his study, a low computational VS protocol was employed taking into account the protein flexibility and combining with HTS in different libraries thus identifying compound **(211)** (Fig. **36**) with an IC$_{50}$ value of 13.9 μM and K$_i$ of 11.1 μM, representing a new scaffold that can be used in other studies targeting more potent compounds.

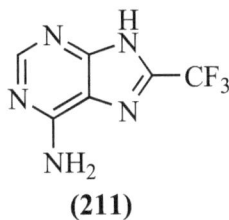

(211)

Fig. (36). The compound identified by Lee *et al.,* 2014.

Caspase Family

The caspase family are cysteine proteases responsible for the aspartate-specific hydrolysis encoded by 12 genes in humans and ten genes in rats [110, 111]. Nowadays, 13 caspase types have been identified in humans, and 10 in rats, whose primary function is programmed cell death (apoptosis), and also other functions are related to inflammation regulation and immunity [111, 112]. Caspase-1 was the first identified in mammals, responsible for the conversion of pro-interleukin (IL)-1 into its active cytokine-related inflammatory process. From this, some experiments showed that caspase-1, -4, -5, and -12 are mainly responsible for the inflammatory process and cytokine production [111, 112]. Caspase-3 is considered one of the main performers of apoptosis due to its catalytic efficiency in breaking substrates and with the ability to differentiate various cell types with neuronal functions responsible for protection against stress [113]. Thus, Caspases-1 and -3 are interesting targets in medicinal chemistry, where the inhibition of these targets plays a role in neuroinflammation and neurodegeneration, decreasing inflammation and preventing the accumulation of beta-amyloid, and it could be a target of interest for the development of drugs for the treatment of Alzheimer's disease [114].

Aiming at the development of new caspase inhibitors, Patel *et al.* (2018) [115] used only computational approaches upon caspases-1, -3, -7, and -8. Combining integrated VS strategies with molecular docking and pharmacophore mapping, the authors identified 15 compounds **(212-226)** (Fig. **37**) with optimal affinity and estimated high ADME activity and properties. Through molecular dynamics studies, the stability of the compounds at the active site was demonstrated. Moreover, this study was important because through docking, it was possible to reveal specific interactions for caspase-1 inhibitors, through interactions in pocket P4 with Trp^{340}, His^{342}, and Val^{348} residues, showing to be a study that helps in the discovery of new caspase-1 inhibitors as a strategy for the development of new anti-inflammatory drugs.

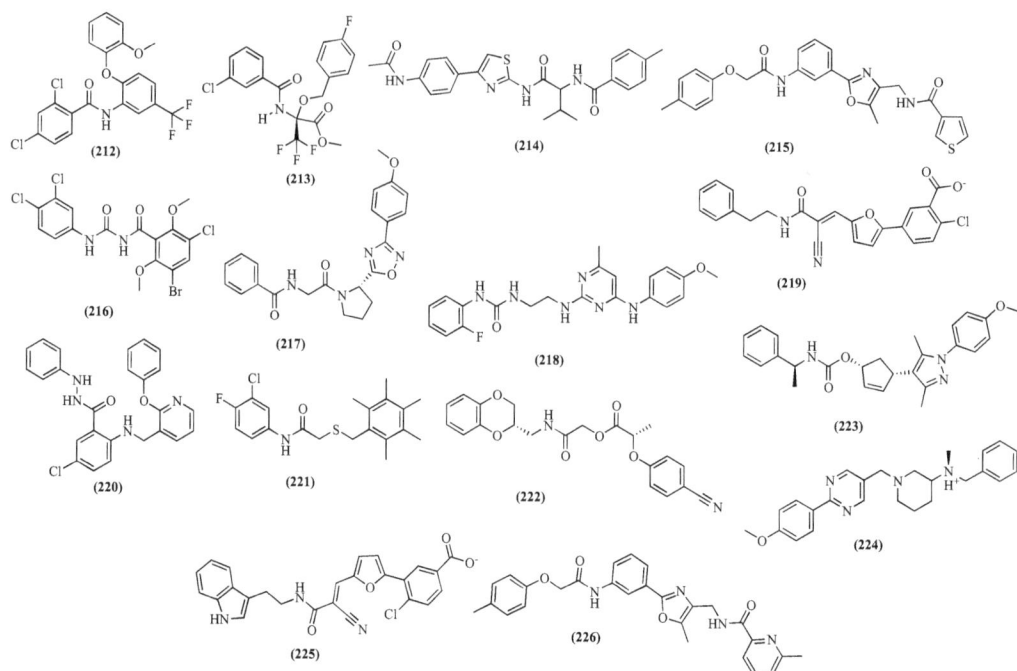

Fig. (37). Compounds identified by Patel *et al.,* 2018.

The study developed by Lakshmi *et al.* (2009) [116] proposed the VS approach to the discovery of new caspase-3 inhibitors. The authors developed a pharmacophore model based on 25 compounds, showing that for the molecule to have inhibitory activity a hydrogen bond donor atom, an acceptor, and two hydrophobic groups are required. This model was applied in the VS protocol using the Maybridge database, the filters were applied, leading to the selection of compound **(227)** (Fig. **38**) for biological assays, showing maximum fluorescence at concentrations of 10, 50 and 100µM, being identified as a non-peptide inhibitor against caspase-3 and which may be promising in the development of caspase-3 inhibitors.

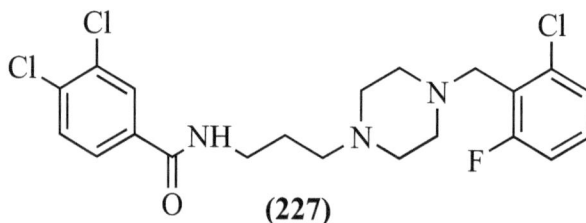

Fig. (38). The compound identified by Lakshmi *et al.,* 2009.

CRITICAL STEPS AND LIMITATIONS OF VIRTUAL SCREENING CAMPAIGNS

Despite being a very effective method, VS has some restrictions, which, if not carefully observed and treated, can generate false-positives and even false results, harming the VS campaign [29, 30, 70].

One of the factors to be considered is the protonation state of amino acid residues at the active site, considering which the dielectric conditions of this site can interfere in the pKa of functional groups during the interaction with ligands. This interference can generate alterations in the functional groups, for example, in both hydrogen bond donors and acceptors groups. Thus, this pKa alterations can directly influence the results from a VS protocol [117].

The conformational states and changes of a protein can be a problem in VS procedures, due to its dynamic nature. Thus, protein flexibility is essential for the VS results and it should be considered as a crucial factor in a virtual protocol. The protein should be used in the most stable conformation to increase the affinity between ligands and active site, reproducing conditions closer to real events. Molecular dynamic simulations are fundamentally important to reduce such limitation, considering that this makes it possible to obtain more reliable results [24, 69].

The selection of the scoring function is a critical step that deserves more attention in the VS protocols. Thus, the selection of an algorithm to be used is an extremely important step, considering that it will be employed to estimate interactions between ligands and targets, ranking all poses. Additionally, the classification of different poses should be closer to the "real", in which the highest score leads to the best experimental result, in biological assays, for example. When several ligands are present in a VS campaign, they should be distinguished between those that bind and those that do not bind at the active site, and their energy values should be precisely classified. Finally, the scoring function should be quick in order to be applied for a large number of compounds in a VS study [25, 31].

In studies of ligand-target interactions, it is important to consider that the crystal 3D-structures of biological targets usually have water molecules at the binding site, which can lead to important interactions between the ligand-target. In contrast, in many cases, they are not important and may negatively affect the VS campaign, since they can interfere in ligand-target interactions. Thus, it is difficult to define if they are essential or not and should be carefully analyzed, being necessary to study several solvation models for the target with different number of water molecules at the active site [31].

The library selection is also a very important point in a VS campaign, considering that, in some cases, the molecules obtained in the VS procedure may not be synthetically viable, so the most of studies select libraries that contain compounds commercially available. Also, the filtration of molecules should be performed in order to avoid/reduce false-positive compounds with problematic properties, such as chemical reactivity, toxicity, and promiscuity, in which their optimization would be time-consuming. Finally, part of the campaigns is conducted seeking approved compounds as drugs and synthetically viable [68].

CONCLUSION AND FUTURE OUTLOOK

The development of a new drug for the treatment of neglected tropical diseases is a complex process that requires knowledge in different areas of concentration, from the design of compounds, requiring knowledge in computational methods, to biological testing to validate the entire modeling protocol. It is an expensive process for both the pharmaceutical industry and research institutions; there is a clear need for new protocols based on computational techniques to predict the activity of the target compound, decreasing the number of inactive molecules and thus increasing the chances to obtain promising new compounds, saving million dollars in research and industrial production.

Through this work, it was shown that most of the computational studies are carried out in front of large libraries of organic molecules, mainly due to facilities of commercially obtainment compounds, reducing the costs of synthesis, and often using compounds already approved as drugs. It is very interesting from the economic point of view, once a drug has already well-elucidated toxicity mechanism and can be used in clinical practice more easily when compared with other molecules that need several studies that prove their safety and efficacy. The prediction of the activity of these compounds against other diseases is an interesting and possible strategy through computational protocols.

There are increasing studies supported by different computational approaches, such as LBDD and SBDD, as well as the combination of both techniques to avoid obtaining false positives and/or false negatives. Several research groups around the world make use of such approaches, developing increasingly sharp and effective protocols for discriminating active compounds from inactive compounds. These techniques are particularly useful in the development of new cysteine protease inhibitors, targets of great interest for the discovery of new compounds for the treatment of neglected diseases, cancer, viral diseases, osteoporosis inflammation, as shown and discussed in this chapter.

CONSENT FOR PUBLICATION

Not applicable.

CONFLICT OF INTEREST

The authors has no conflicts of interest, financial or otherwise, to declare.

ACKNOWLEDGEMENTS

The authors thank the Coordenação de Aperfeiçoamento de Pessoal de Nível Superior – Brazil (CAPES) and the National Council for Scientific and Technological Development (CNPq) – Brazil. Furthermore, the authors would also like to give thanks to the Biozentrum – The Center of Molecular Life Science from the University of Basel to make the Swiss-Model server possible.

REFERENCES

[1]   WHO. *World Health Organization,*

[2]   Ferreira, L.L.G.; Andricopulo, A.D. Drugs and vaccines in the 21st century for neglected diseases. *Lancet Infect. Dis.,* **2019,** *19*(2), 125-127.
       [http://dx.doi.org/10.1016/S1473-3099(19)30005-2] [PMID: 30712832]

[3]   Bailey, F.; Eaton, J.; Jidda, M.; van Brakel, W.H.; Addiss, D.G.; Molyneux, D.H. Neglected Tropical Diseases and Mental Health: Progress, Partnerships, and Integration. *Trends Parasitol.,* **2019,** *35*(1), 23-31.
       [http://dx.doi.org/10.1016/j.pt.2018.11.001] [PMID: 30578149]

[4]   Santos-Valle, A. B. C.; Souza, G. R. R.; Paes, C. Q.; Miyazaki, T.; Silva, A. H.; Altube, M. J.; Morilla, M. J.; Romero, E. L.; Creczynski-Pasa, T. B.; Cabral, H. Annu. Rev. Control . Nanomedicine Strategies for Addressing Major Needs in Neglected Tropical Diseases. **2019.**

[5]   Belllera, C.L.; Sbaraglini, M.L.; Alberca, L.N. *In Silico* Modeling of FDA-Approved Drugs for Discovery of Therapies Against Neglected Diseases: A Drug Repurposing Approach. *Silico Drug Des.,* **2019,** 625-648.
       [http://dx.doi.org/10.1016/b978-0-12-816125-8.00021-3]

[6]   Liu, X.; Thomas, C.E.; Felder, C.C. The impact of external innovation on new drug approvals: A retrospective analysis. *Int. J. Pharm.,* **2019,** *563*(563), 273-281.
       [http://dx.doi.org/10.1016/j.ijpharm.2018.12.093] [PMID: 30664998]

[7]   Klug, D.M.; Gelb, M.H.; Pollastri, M.P. Repurposing strategies for tropical disease drug discovery. *Bioorg. Med. Chem. Lett.,* **2016,** *26*(11), 2569-2576.
       [http://dx.doi.org/10.1016/j.bmcl.2016.03.103] [PMID: 27080183]

[8]   Zhang, L.; Tan, J.; Han, D.; Zhu, H. From machine learning to deep learning: progress in machine intelligence for rational drug discovery. *Drug Discov. Today,* **2017,** *22*(11), 1680-1685.
       [http://dx.doi.org/10.1016/j.drudis.2017.08.010] [PMID: 28881183]

[9]   Kinch, M.S.; Griesenauer, R.H. 2017 in review: FDA approvals of new molecular entities. *Drug Discov. Today,* **2018,** *23*(8), 1469-1473.
       [http://dx.doi.org/10.1016/j.drudis.2018.05.011] [PMID: 29751111]

[10]   Campbell, I.B.; Macdonald, S.J.F.; Procopiou, P.A. Medicinal chemistry in drug discovery in big pharma: past, present and future. *Drug Discov. Today,* **2018,** *23*(2), 219-234.
       [http://dx.doi.org/10.1016/j.drudis.2017.10.007] [PMID: 29031621]

[11]   Martin, L.; Hutchens, M.; Hawkins, C.; Radnov, A. How much do clinical trials cost? *Nat. Rev. Drug Discov.,* **2017,** *16*(6), 381-382.
[http://dx.doi.org/10.1038/nrd.2017.70] [PMID: 28529317]

[12]   Miwa, G. T. *The Drug Discovery Process,* **2010**.https://doi.org/10.1002/9780470538951.ch1.

[13]   Prieto-Martínez, F. D.; López-López, E.; Eurídice Juárez-Mercado, K.; Medina-Franco, J. L. L. Computational Drug Design Methods—Current and Future Perspectives. *Silico Drug Des,* **2019,** *3,* 19-14.

[14]   Katsila, T.; Spyroulias, G.A.; Patrinos, G.P.; Matsoukas, M.T. Computational approaches in target identification and drug discovery. *Comput. Struct. Biotechnol. J.,* **2016,** *14,* 177-184.
[http://dx.doi.org/10.1016/j.csbj.2016.04.004] [PMID: 27293534]

[15]   Khanna, V.; Ranganathan, S.; Petrovsky, N. *Rational Structure-Based Drug Design;* Elsevier Ltd., **2019**.
[http://dx.doi.org/10.1016/B978-0-12-809633-8.20275-6]

[16]   Ferreira, L.L.G.; Andricopulo, A.D. ADMET modeling approaches in drug discovery. *Drug Discov. Today,* **2019,** *24*(5), 1157-1165.
[http://dx.doi.org/10.1016/j.drudis.2019.03.015] [PMID: 30890362]

[17]   Hongmao, S. *Homology Modeling and Ligand-Based Molecule Design;* Elsevier Ltd., **2016**.
[http://dx.doi.org/10.1016/B978-0-08-100098-4.00004-1]

[18]   Wiltgen, M. *Algorithms for Structure Comparison and Analysis: Homology Modelling of Proteins;* Elsevier Ltd., **2019,** Vol. 1, .

[19]   Siklos, M.; BenAissa, M.; Thatcher, G.R.J. Cysteine proteases as therapeutic targets: does selectivity matter? A systematic review of calpain and cathepsin inhibitors. *Acta Pharm. Sin. B,* **2015,** *5*(6), 506-519.
[http://dx.doi.org/10.1016/j.apsb.2015.08.001] [PMID: 26713267]

[20]   Park, S.Y.; Jeong, M.S.; Park, S.A.; Ha, S.C.; Na, B.K.; Jang, S.B. Structural basis of the cystein protease inhibitor Clonorchis sinensis Stefin-1. *Biochem. Biophys. Res. Commun.,* **2018,** *498*(1), 9-17.
[http://dx.doi.org/10.1016/j.bbrc.2018.02.196] [PMID: 29499196]

[21]   Ward, D.J.; Van de Langemheen, H.; Koehne, E.; Kreidenweiss, A.; Liskamp, R.M.J. Highly tunable thiosulfonates as a novel class of cysteine protease inhibitors with anti-parasitic activity against Schistosoma mansoni. *Bioorg. Med. Chem.,* **2019,** *27*(13), 2857-2870.
[http://dx.doi.org/10.1016/j.bmc.2019.05.014] [PMID: 31126821]

[22]   Alam, B.; Biswas, S. Inhibition of Plasmodium falciparum cysteine protease falcipain-2 by a human cross-class inhibitor serpinB3: A mechanistic insight. *Biochim. Biophys. Acta. Proteins Proteomics,* **2019,** *1867*(9), 854-865.
[http://dx.doi.org/10.1016/j.bbapap.2019.06.012] [PMID: 31247344]

[23]   Ferreira, L.G.; Andricopulo, A.D. Targeting cysteine proteases in trypanosomatid disease drug discovery. *Pharmacol. Ther.,* **2017,** *180,* 49-61.
[http://dx.doi.org/10.1016/j.pharmthera.2017.06.004] [PMID: 28579388]

[24]   Danishuddin, M.; Khan, A.U. Structure based virtual screening to discover putative drug candidates: necessary considerations and successful case studies. *Methods,* **2015,** *71*(C), 135-145.
[http://dx.doi.org/10.1016/j.ymeth.2014.10.019] [PMID: 25448480]

[25]   McInnes, C. Virtual screening strategies in drug discovery. *Curr. Opin. Chem. Biol.,* **2007,** *11*(5), 494-502.
[http://dx.doi.org/10.1016/j.cbpa.2007.08.033] [PMID: 17936059]

[26]   Lengauer, T.; Lemmen, C.; Rarey, M.; Zimmermann, M. Novel technologies for virtual screening. *Drug Discov. Today,* **2004,** *9*(1), 27-34.
[http://dx.doi.org/10.1016/S1359-6446(04)02939-3] [PMID: 14761803]

[27]   Tanrikulu, Y.; Krüger, B.; Proschak, E. The holistic integration of virtual screening in drug discovery. *Drug Discov. Today,* **2013**, *18*(7-8), 358-364.
[http://dx.doi.org/10.1016/j.drudis.2013.01.007] [PMID: 23340112]

[28]   Kumar, A.; Zhang, K.Y.J. Hierarchical virtual screening approaches in small molecule drug discovery. *Methods,* **2015**, *71*(C), 26-37.
[http://dx.doi.org/10.1016/j.ymeth.2014.07.007] [PMID: 25072167]

[29]   Kontoyianni, M. *Docking and Virtual Screening in Drug Discovery,* **2017**.
[http://dx.doi.org/10.1007/978-1-4939-7201-2_18]

[30]   Ghosh, S.; Nie, A.; An, J.; Huang, Z. Structure-based virtual screening of chemical libraries for drug discovery. *Curr. Opin. Chem. Biol.,* **2006**, *10*(3), 194-202.
[http://dx.doi.org/10.1016/j.cbpa.2006.04.002] [PMID: 16675286]

[31]   Cerqueira, N.M.F.S.A.; Gesto, D.; Oliveira, E.F.; Santos-Martins, D.; Brás, N.F.; Sousa, S.F.; Fernandes, P.A.; Ramos, M.J. Receptor-based virtual screening protocol for drug discovery. *Arch. Biochem. Biophys.,* **2015**, *582*, 56-67.
[http://dx.doi.org/10.1016/j.abb.2015.05.011] [PMID: 26045247]

[32]   Vasiljeva, O.; Reinheckel, T.; Peters, C.; Turk, D.; Turk, V.; Turk, B. Emerging roles of cysteine cathepsins in disease and their potential as drug targets. *Curr. Pharm. Des.,* **2007**, *13*(4), 387-403.
[http://dx.doi.org/10.2174/138161207780162962] [PMID: 17311556]

[33]   Gurumallesh, P.; Alagu, K.; Ramakrishnan, B.; Muthusamy, S. A systematic reconsideration on proteases. *Int. J. Biol. Macromol.,* **2019**, *128*, 254-267.
[http://dx.doi.org/10.1016/j.ijbiomac.2019.01.081] [PMID: 30664968]

[34]   Coulombe, R.; Grochulski, P.; Sivaraman, J.; Ménard, R.; Mort, J.S.; Cygler, M. Structure of human procathepsin L reveals the molecular basis of inhibition by the prosegment. *EMBO J.,* **1996**, *15*(20), 5492-5503.
[http://dx.doi.org/10.1002/j.1460-2075.1996.tb00934.x] [PMID: 8896443]

[35]   Sijwali, P.S.; Rosenthal, P.J. Gene disruption confirms a critical role for the cysteine protease falcipain-2 in hemoglobin hydrolysis by Plasmodium falciparum. *Proc. Natl. Acad. Sci. USA,* **2004**, *101*(13), 4384-4389.
[http://dx.doi.org/10.1073/pnas.0307720101] [PMID: 15070727]

[36]   Mason, R.W.; Massey, S.D. Surface activation of pro-cathepsin L. *Biochem. Biophys. Res. Commun.,* **1992**, *189*(3), 1659-1666.
[http://dx.doi.org/10.1016/0006-291X(92)90268-P] [PMID: 1482371]

[37]   Verma, S.; Dixit, R.; Pandey, K.C. Cysteine Proteases: Modes of Activation and Future Prospects as Pharmacological Targets. *Front. Pharmacol.,* **2016**, *7*(107), 107.
[http://dx.doi.org/10.3389/fphar.2016.00107] [PMID: 27199750]

[38]   Bassetto, M.; De Burghgraeve, T.; Delang, L.; Massarotti, A.; Coluccia, A.; Zonta, N.; Gatti, V.; Colombano, G.; Sorba, G.; Silvestri, R.; Tron, G.C.; Neyts, J.; Leyssen, P.; Brancale, A. Computer-aided identification, design and synthesis of a novel series of compounds with selective antiviral activity against chikungunya virus. *Antiviral Res.,* **2013**, *98*(1), 12-18.
[http://dx.doi.org/10.1016/j.antiviral.2013.01.002] [PMID: 23380636]

[39]   Machado, P.A.; Carneiro, M.P.D.; Sousa-Batista, A.J.; Lopes, F.J.P.; Lima, A.P.C.A.; Chaves, S.P.; Sodero, A.C.R.; de Matos Guedes, H.L. Leishmanicidal therapy targeted to parasite proteases. *Life Sci.,* **2019**, *219*, 163-181.
[http://dx.doi.org/10.1016/j.lfs.2019.01.015] [PMID: 30641084]

[40]   Maurais, A.J.; Weerapana, E. Reactive-cysteine profiling for drug discovery. *Curr. Opin. Chem. Biol.,* **2019**, *50*, 29-36.
[http://dx.doi.org/10.1016/j.cbpa.2019.02.010] [PMID: 30897495]

[41]   Arafet, K.; Ferrer, S.; Martí, S.; Moliner, V. Quantum mechanics/molecular mechanics studies of the

mechanism of falcipain-2 inhibition by the epoxysuccinate E64. *Biochemistry*, **2014**, *53*(20), 3336-3346.
[http://dx.doi.org/10.1021/bi500060h] [PMID: 24811524]

[42]   Arafet, K.; Ferrer, S.; Moliner, V. First quantum mechanics/molecular mechanics studies of the inhibition mechanism of cruzain by peptidyl halomethyl ketones. *Biochemistry*, **2015**, *54*(21), 3381-3391.
[http://dx.doi.org/10.1021/bi501551g] [PMID: 25965914]

[43]   Arafet, K.; Ferrer, S.; Moliner, V. Computational Study of the Catalytic Mechanism of the Cruzain Cysteine Protease. *ACS Catal.*, **2017**, *7*(2), 1207-1215.
[http://dx.doi.org/10.1021/acscatal.6b03096]

[44]   Arafet, K.; Ferrer, S.; González, F.V.; Moliner, V. Quantum mechanics/molecular mechanics studies of the mechanism of cysteine protease inhibition by peptidyl-2,3-epoxyketones. *Phys. Chem. Chem. Phys.*, **2017**, *19*(20), 12740-12748.
[http://dx.doi.org/10.1039/C7CP01726J] [PMID: 28480929]

[45]   WHO | What is Chagas disease?

[46]   Kratz, J.M. Drug discovery for chagas disease: A viewpoint. *Acta Trop.*, **2019**, *198*(April)105107
[http://dx.doi.org/10.1016/j.actatropica.2019.105107] [PMID: 31351074]

[47]   Scarim, C.B.; Jornada, D.H.; Chelucci, R.C.; de Almeida, L.; Dos Santos, J.L.; Chung, M.C. Current advances in drug discovery for Chagas disease. *Eur. J. Med. Chem.*, **2018**, *155*, 824-838.
[http://dx.doi.org/10.1016/j.ejmech.2018.06.040] [PMID: 30033393]

[48]   da Silva-Junior, E.F.; Barcellos Franca, P.H.; Ribeiro, F.F.; Bezerra Mendonca-Junior, F.J.; Scotti, L.; Scotti, M.T.; de Aquino, T.M.; de Araujo-Junior, J.X. Molecular Docking Studies Applied to a Dataset of Cruzain Inhibitors. *Curr Comput Aided Drug Des*, **2018**, *14*(1), 68-78.
[http://dx.doi.org/10.2174/1573409913666170519112758] [PMID: 28523999]

[49]   Choe, Y.; Brinen, L.S.; Price, M.S.; Engel, J.C.; Lange, M.; Grisostomi, C.; Weston, S.G.; Pallai, P.V.; Cheng, H.; Hardy, L.W.; Hartsough, D.S.; McMakin, M.; Tilton, R.F.; Baldino, C.M.; Craik, C.S. Development of alpha-keto-based inhibitors of cruzain, a cysteine protease implicated in Chagas disease. *Bioorg. Med. Chem.*, **2005**, *13*(6), 2141-2156.
[http://dx.doi.org/10.1016/j.bmc.2004.12.053] [PMID: 15727867]

[50]   Freitas, R.F.; Oprea, T.I.; Montanari, C.A. 2D QSAR and similarity studies on cruzain inhibitors aimed at improving selectivity over cathepsin L. *Bioorg. Med. Chem.*, **2008**, *16*(2), 838-853.
[http://dx.doi.org/10.1016/j.bmc.2007.10.048] [PMID: 17996450]

[51]   Malvezzi, A.; de Rezende, L.; Izidoro, M.A.; Cezari, M.H.S.; Juliano, L.; do Amaral, A. Uncovering false positives on a virtual screening search for cruzain inhibitors. *Bioorg. Med. Chem. Lett.*, **2008**, *18*(1), 350-354.
[http://dx.doi.org/10.1016/j.bmcl.2007.10.068] [PMID: 17981033]

[52]   Wiggers, H.J.; Rocha, J.R.; Cheleski, J.; Montanari, C.A. Integration of Ligand- and Target-Based Virtual Screening for the Discovery of Cruzain Inhibitors. *Mol. Inform.*, **2011**, *30*(6-7), 565-578.
[http://dx.doi.org/10.1002/minf.201000146] [PMID: 27467157]

[53]   Wiggers, H.J.; Rocha, J.R.; Fernandes, W.B.; Sesti-Costa, R.; Carneiro, Z.A.; Cheleski, J.; da Silva, A.B.F.; Juliano, L.; Cezari, M.H.S.; Silva, J.S.; McKerrow, J.H.; Montanari, C.A. Non-peptidic cruzain inhibitors with trypanocidal activity discovered by virtual screening and in vitro assay. *PLoS Negl. Trop. Dis.*, **2013**, *7*(8)e2370
[http://dx.doi.org/10.1371/journal.pntd.0002370] [PMID: 23991231]

[54]   Bellera, C.L.; Balcazar, D.E.; Alberca, L.; Labriola, C.A.; Talevi, A.; Carrillo, C. Application of computer-aided drug repurposing in the search of new cruzipain inhibitors: discovery of amiodarone and bromocriptine inhibitory effects. *J. Chem. Inf. Model.*, **2013**, *53*(9), 2402-2408.
[http://dx.doi.org/10.1021/ci400284v] [PMID: 23906322]

[55] Bellera, C.L.; Balcazar, D.E.; Vanrell, M.C.; Casassa, A.F.; Palestro, P.H.; Gavernet, L.; Labriola, C.A.; Gálvez, J.; Bruno-Blanch, L.E.; Romano, P.S.; Carrillo, C.; Talevi, A. Computer-guided drug repurposing: identification of trypanocidal activity of clofazimine, benidipine and saquinavir. *Eur. J. Med. Chem.*, **2015**, *93*, 338-348.
[http://dx.doi.org/10.1016/j.ejmech.2015.01.065] [PMID: 25707014]

[56] Herrera-Mayorga, V.; Lara-Ramírez, E.E.; Chacón-Vargas, K.F.; Aguirre-Alvarado, C.; Rodríguez-Páez, L.; Alcántara-Farfán, V.; Cordero-Martínez, J.; Nogueda-Torres, B.; Reyes-Espinosa, F.; Bocanegra-García, V.; Rivera, G. Structure-Based Virtual Screening and In Vitro Evaluation of New *Trypanosoma cruzi* Cruzain Inhibitors. *Int. J. Mol. Sci.*, **2019**, *20*(7), 1-13.
[http://dx.doi.org/10.3390/ijms20071742] [PMID: 30970549]

[57] Deng, N.; Forli, S.; He, P.; Perryman, A.; Wickstrom, L.; Vijayan, R.S.K.; Tiefenbrunn, T.; Stout, D.; Gallicchio, E.; Olson, A.J.; Levy, R.M. Distinguishing binders from false positives by free energy calculations: fragment screening against the flap site of HIV protease. *J. Phys. Chem. B*, **2015**, *119*(3), 976-988.
[http://dx.doi.org/10.1021/jp506376z] [PMID: 25189630]

[58] Ferreira, R.S.; Simeonov, A.; Jadhav, A.; Eidam, O.; Mott, B.T.; Keiser, M.J.; McKerrow, J.H.; Maloney, D.J.; Irwin, J.J.; Shoichet, B.K. Complementarity between a docking and a high-throughput screen in discovering new cruzain inhibitors. *J. Med. Chem.*, **2010**, *53*(13), 4891-4905.
[http://dx.doi.org/10.1021/jm100488w] [PMID: 20540517]

[59] CDC - Parasites - Malaria. Available at: https://www.cdc.gov/parasites/malaria/index.html

[60] Qidwai, T.; Khan, F. *Antimalarial Drugs and Drug Targets Specific to Fatty Acid Metabolic Pathway of Plasmodium Falciparum.*, **2012**.
[http://dx.doi.org/10.1111/j.1747-0285.2012.01389.x]

[61] Kumar, S.; Bhardwaj, T.R.; Prasad, D.N.; Singh, R.K. Drug targets for resistant malaria: Historic to future perspectives. *Biomed. Pharmacother.*, **2018**, *104*(May), 8-27.
[http://dx.doi.org/10.1016/j.biopha.2018.05.009] [PMID: 29758416]

[62] Roy, K.K. Targeting the active sites of malarial proteases for antimalarial drug discovery: approaches, progress and challenges. *Int. J. Antimicrob. Agents*, **2017**, *50*(3), 287-302.
[http://dx.doi.org/10.1016/j.ijantimicag.2017.04.006] [PMID: 28668681]

[63] Desai, P.V.; Patny, A.; Sabnis, Y.; Tekwani, B.; Gut, J.; Rosenthal, P.; Srivastava, A.; Avery, M. Identification of novel parasitic cysteine protease inhibitors using virtual screening. 1. The ChemBridge database. *J. Med. Chem.*, **2004**, *47*(26), 6609-6615.
[http://dx.doi.org/10.1021/jm0493717] [PMID: 15588096]

[64] Desai, P.V.; Patny, A.; Gut, J.; Rosenthal, P.J.; Tekwani, B.; Srivastava, A.; Avery, M. Identification of novel parasitic cysteine protease inhibitors by use of virtual screening. 2. The available chemical directory. *J. Med. Chem.*, **2006**, *49*(5), 1576-1584.
[http://dx.doi.org/10.1021/jm0505765] [PMID: 16509575]

[65] Li, H.; Huang, J.; Chen, L.; Liu, X.; Chen, T.; Zhu, J.; Lu, W.; Shen, X.; Li, J.; Hilgenfeld, R.; Jiang, H. Identification of novel falcipain-2 inhibitors as potential antimalarial agents through structure-based virtual screening. *J. Med. Chem.*, **2009**, *52*(15), 4936-4940.
[http://dx.doi.org/10.1021/jm801622x] [PMID: 19586036]

[66] Shah, F.; Mukherjee, P.; Gut, J.; Legac, J.; Rosenthal, P.J.; Tekwani, B.L.; Avery, M.A. Identification of novel malarial cysteine protease inhibitors using structure-based virtual screening of a focused cysteine protease inhibitor library. *J. Chem. Inf. Model.*, **2011**, *51*(4), 852-864.
[http://dx.doi.org/10.1021/ci200029y] [PMID: 21428453]

[67] Shah, F.; Gut, J.; Legac, J.; Shivakumar, D.; Sherman, W.; Rosenthal, P.J.; Avery, M.A. Computer-aided drug design of falcipain inhibitors: virtual screening, structure-activity relationships, hydration site thermodynamics, and reactivity analysis. *J. Chem. Inf. Model.*, **2012**, *52*(3), 696-710.
[http://dx.doi.org/10.1021/ci2005516] [PMID: 22332946]

[68]   Rognan, D. The impact of in silico screening in the discovery of novel and safer drug candidates. *Pharmacol. Ther.,* **2017**, *175*, 47-66.
       [http://dx.doi.org/10.1016/j.pharmthera.2017.02.034] [PMID: 28223231]

[69]   Wingert, B.M.; Camacho, C.J. Improving small molecule virtual screening strategies for the next generation of therapeutics. *Curr. Opin. Chem. Biol.,* **2018**, *44*, 87-92.
       [http://dx.doi.org/10.1016/j.cbpa.2018.06.006] [PMID: 29920436]

[70]   Irwin, J.J.; Shoichet, B.K. Docking Screens for Novel Ligands Conferring New Biology. *J. Med. Chem.,* **2016**, *59*(9), 4103-4120.
       [http://dx.doi.org/10.1021/acs.jmedchem.5b02008] [PMID: 26913380]

[71]   Wang, L.; Zhang, S.; Zhu, J.; Zhu, L.; Liu, X.; Shan, L.; Huang, J.; Zhang, W.; Li, H. Identification of diverse natural products as falcipain-2 inhibitors through structure-based virtual screening. *Bioorg. Med. Chem. Lett.,* **2014**, *24*(5), 1261-1264.
       [http://dx.doi.org/10.1016/j.bmcl.2014.01.074] [PMID: 24530004]

[72]   Chakka, S.K.; Kalamuddin, M.; Sundararaman, S.; Wei, L.; Mundra, S.; Mahesh, R.; Malhotra, P.; Mohmmed, A.; Kotra, L.P. Identification of novel class of falcipain-2 inhibitors as potential antimalarial agents. *Bioorg. Med. Chem.,* **2015**, *23*(9), 2221-2240.
       [http://dx.doi.org/10.1016/j.bmc.2015.02.062] [PMID: 25840796]

[73]   Hillisch, A.; Pineda, L.F.; Hilgenfeld, R. Utility of homology models in the drug discovery process. *Drug Discov. Today,* **2004**, *9*(15), 659-669.
       [http://dx.doi.org/10.1016/S1359-6446(04)03196-4] [PMID: 15279849]

[74]   Cavasotto, C.N.; Phatak, S.S. Homology modeling in drug discovery: current trends and applications. *Drug Discov. Today,* **2009**, *14*(13-14), 676-683.
       [http://dx.doi.org/10.1016/j.drudis.2009.04.006] [PMID: 19422931]

[75]   Saddala, M.S.; Adi, P.J. Discovery of small molecules through pharmacophore modeling, docking and molecular dynamics simulation against *Plasmodium vivax* Vivapain-3 (VP-3). *Heliyon,* **2018**, *4*(5)e00612
       [http://dx.doi.org/10.1016/j.heliyon.2018.e00612] [PMID: 29756074]

[76]   Hernández-González, J.E.; Salas-Sarduy, E.; Hernández Ramírez, L.F.; Pascual, M.J.; Álvarez, D.E.; Pabón, A.; Leite, V.B.P.; Pascutti, P.G.; Valiente, P.A. Identification of (4-(9H-fluoren-9-yl) piperazin-1-yl) methanone derivatives as falcipain 2 inhibitors active against Plasmodium falciparum cultures. *Biochim. Biophys. Acta, Gen. Subj.,* **2018**, *1862*(12), 2911-2923.
       [http://dx.doi.org/10.1016/j.bbagen.2018.09.015] [PMID: 30253205]

[77]   Chikungunya virus | CDC. Available at: https://www.cdc.gov/chikungunya/index.html

[78]   Chikungunya. Available at: https://www.who.int/health-topics/chikungunya/#tab=tab_1

[79]   da Silva-Júnior, E.F.; Leoncini, G.O.; Rodrigues, É.E.S.; Aquino, T.M.; Araújo-Júnior, J.X. The medicinal chemistry of Chikungunya virus. *Bioorg. Med. Chem.,* **2017**, *25*(16), 4219-4244.
       [http://dx.doi.org/10.1016/j.bmc.2017.06.049] [PMID: 28689975]

[80]   Saisawang, C.; Kuadkitkan, A.; Smith, D.R.; Ubol, S.; Ketterman, A.J. Glutathionylation of chikungunya nsP2 protein affects protease activity. *Biochim. Biophys. Acta, Gen. Subj.,* **2017**, *1861*(2), 106-111.
       [http://dx.doi.org/10.1016/j.bbagen.2016.10.024] [PMID: 27984114]

[81]   Nguyen, P.T.V.; Yu, H.; Keller, P.A. Identification of chikungunya virus nsP2 protease inhibitors using structure-base approaches. *J. Mol. Graph. Model.,* **2015**, *57*, 1-8.
       [http://dx.doi.org/10.1016/j.jmgm.2015.01.001] [PMID: 25622129]

[82]   Meshram, C.D.; Lukash, T.; Phillips, A.T.; Akhrymuk, I.; Frolova, E.I.; Frolov, I. Lack of nsP2-specific nuclear functions attenuates chikungunya virus replication both in vitro and in vivo. *Virology,* **2019**, *534*(May), 14-24.
       [http://dx.doi.org/10.1016/j.virol.2019.05.016] [PMID: 31163352]

[83]   Tardugno, R.; Giancotti, G.; De Burghgraeve, T.; Delang, L.; Neyts, J.; Leyssen, P.; Brancale, A.; Bassetto, M. Design, synthesis and evaluation against Chikungunya virus of novel small-molecule antiviral agents. *Bioorg. Med. Chem.,* **2018**, *26*(4), 869-874.
[http://dx.doi.org/10.1016/j.bmc.2018.01.002] [PMID: 29336951]

[84]   Giancotti, G.; Cancellieri, M.; Balboni, A.; Giustiniano, M.; Novellino, E.; Delang, L.; Neyts, J.; Leyssen, P.; Brancale, A.; Bassetto, M. Rational modifications on a benzylidene-acrylohydrazide antiviral scaffold, synthesis and evaluation of bioactivity against Chikungunya virus. *Eur. J. Med. Chem.,* **2018**, *149*, 56-68.
[http://dx.doi.org/10.1016/j.ejmech.2018.02.054] [PMID: 29499487]

[85]   Schistosomiasis. Available at: https://www.who.int/news-room/fact-sheets/detail/schistosomiasis

[86]   CDC - Schistosomiasis. Available at: https://www.cdc.gov/parasites/schistosomiasis/index.html

[87]   Mafud, A.C.; Ferreira, L.G.; Mascarenhas, Y.P.; Andricopulo, A.D.; de Moraes, J. Discovery of Novel Antischistosomal Agents by Molecular Modeling Approaches. *Trends Parasitol.,* **2016**, *32*(11), 874-886.
[http://dx.doi.org/10.1016/j.pt.2016.08.002] [PMID: 27593339]

[88]   da Silva, V.B.R.; Campos, B.R.K.L.; de Oliveira, J.F.; Decout, J.L.; do Carmo Alves de Lima, M. Medicinal chemistry of antischistosomal drugs: Praziquantel and oxamniquine. *Bioorg. Med. Chem.,* **2017**, *25*(13), 3259-3277.
[http://dx.doi.org/10.1016/j.bmc.2017.04.031] [PMID: 28495384]

[89]   Chakrabarti, A.; Oehme, I.; Witt, O.; Oliveira, G.; Sippl, W.; Romier, C.; Pierce, R.J.; Jung, M. HDAC8: a multifaceted target for therapeutic interventions. *Trends Pharmacol. Sci.,* **2015**, *36*(7), 481-492.
[http://dx.doi.org/10.1016/j.tips.2015.04.013] [PMID: 26013035]

[90]   Kannan, S.; Melesina, J.; Hauser, A.T.; Chakrabarti, A.; Heimburg, T.; Schmidtkunz, K.; Walter, A.; Marek, M.; Pierce, R.J.; Romier, C.; Jung, M.; Sippl, W. Discovery of inhibitors of Schistosoma mansoni HDAC8 by combining homology modeling, virtual screening, and in vitro validation. *J. Chem. Inf. Model.,* **2014**, *54*(10), 3005-3019.
[http://dx.doi.org/10.1021/ci5004653] [PMID: 25243797]

[91]   Heimburg, T.; Chakrabarti, A.; Lancelot, J.; Marek, M.; Melesina, J.; Hauser, A.T.; Shaik, T.B.; Duclaud, S.; Robaa, D.; Erdmann, F.; Schmidt, M.; Romier, C.; Pierce, R.J.; Jung, M.; Sippl, W. Structure-Based Design and Synthesis of Novel Inhibitors Targeting HDAC8 from Schistosoma mansoni for the Treatment of Schistosomiasis. *J. Med. Chem.,* **2016**, *59*(6), 2423-2435.
[http://dx.doi.org/10.1021/acs.jmedchem.5b01478] [PMID: 26937828]

[92]   Khaket, T.P.; Kwon, T.K.; Kang, S.C. Cathepsins: Potent regulators in carcinogenesis. *Pharmacol. Ther.,* **2019**, *198*, 1-19.
[http://dx.doi.org/10.1016/j.pharmthera.2019.02.003] [PMID: 30763594]

[93]   Pogorzelska, A.; Żołnowska, B.; Bartoszewski, R. Cysteine cathepsins as a prospective target for anticancer therapies-current progress and prospects. *Biochimie,* **2018**, *151*, 85-106.
[http://dx.doi.org/10.1016/j.biochi.2018.05.023] [PMID: 29870804]

[94]   Deaton, D.N.; Kumar, S.; Cathepsin, K. 6. Cathepsin K inhibitors: their potential as anti-osteoporosis agents. *Prog. Med. Chem.,* **2004**, *42*, 245-375.
[http://dx.doi.org/10.1016/S0079-6468(04)42006-2] [PMID: 15003723]

[95]   Schmitz, J.; Gilberg, E.; Löser, R.; Bajorath, J.; Bartz, U.; Gütschow, M.; Cathepsin, B. Cathepsin B: Active site mapping with peptidic substrates and inhibitors. *Bioorg. Med. Chem.,* **2019**, *27*(1), 1-15.
[http://dx.doi.org/10.1016/j.bmc.2018.10.017] [PMID: 30473362]

[96]   Barrett, D.G.; Catalano, J.G.; Deaton, D.N.; Long, S.T.; McFadyen, R.B.; Miller, A.B.; Miller, L.R.; Wells-Knecht, K.J.; Wright, L.L. A structural screening approach to ketoamide-based inhibitors of cathepsin K. *Bioorg. Med. Chem. Lett.,* **2005**, *15*(9), 2209-2213.

[http://dx.doi.org/10.1016/j.bmcl.2005.03.023] [PMID: 15837295]

[97]   Schröder, J.; Klinger, A.; Oellien, F.; Marhöfer, R.J.; Duszenko, M.; Selzer, P.M. Docking-based virtual screening of covalently binding ligands: an orthogonal lead discovery approach. *J. Med. Chem.,* **2013**, *56*(4), 1478-1490.
     [http://dx.doi.org/10.1021/jm3013932] [PMID: 23350811]

[98]   Markt, P.; McGoohan, C.; Walker, B.; Kirchmair, J.; Feldmann, C.; De Martino, G.; Spitzer, G.; Distinto, S.; Schuster, D.; Wolber, G.; Laggner, C.; Langer, T. Discovery of novel cathepsin S inhibitors by pharmacophore-based virtual high-throughput screening. *J. Chem. Inf. Model.,* **2008**, *48*(8), 1693-1705.
     [http://dx.doi.org/10.1021/ci800101j] [PMID: 18637674]

[99]   Pinitglang, S.; Saiprajong, R.; Dussadee, T.; Ratanakhanokchai, K. *Structural Bioinformatics and Molecular Dynamics Simulations Studies of Cathepsins as a Potential Target for Drug Discovery.,* **2012**.
     [http://dx.doi.org/10.1016/j.procs.2012.09.008]

[100] Chowdhury, S.R.; Kennedy, S.; Zhu, K.; Mishra, R.; Chuong, P.; Nguyen, A.U.; Kathman, S.G.; Statsyuk, A.V. Discovery of covalent enzyme inhibitors using virtual docking of covalent fragments. *Bioorg. Med. Chem. Lett.,* **2019**, *29*(1), 36-39.
     [http://dx.doi.org/10.1016/j.bmcl.2018.11.019] [PMID: 30455147]

[101] Pourianfar, H.R.; Grollo, L. Development of antiviral agents toward enterovirus 71 infection. *J. Microbiol. Immunol. Infect.,* **2015**, *48*(1), 1-8.
     [http://dx.doi.org/10.1016/j.jmii.2013.11.011] [PMID: 24560700]

[102] Wu, K.X.; Ng, M.M.L.; Chu, J.J.H. Developments towards antiviral therapies against enterovirus 71. *Drug Discov. Today,* **2010**, *15*(23-24), 1041-1051.
     [http://dx.doi.org/10.1016/j.drudis.2010.10.008] [PMID: 20974282]

[103] Shang, L.; Xu, M.; Yin, Z. Antiviral drug discovery for the treatment of enterovirus 71 infections. *Antiviral Res.,* **2013**, *97*(2), 183-194.
     [http://dx.doi.org/10.1016/j.antiviral.2012.12.005] [PMID: 23261847]

[104] Ma, G.H.; Ye, Y.; Zhang, D.; Xu, X.; Si, P.; Peng, J.L.; Xiao, Y.L.; Cao, R.Y.; Yin, Y.L.; Chen, J.; Zhao, L.X.; Zhou, Y.; Zhong, W.; Liu, H.; Luo, X.M.; Chen, L.L.; Shen, X. Identification and biochemical characterization of DC07090 as a novel potent small molecule inhibitor against human enterovirus 71 3C protease by structure-based virtual screening. *Eur. J. Med. Chem.,* **2016**, *124*, 981-991.
     [http://dx.doi.org/10.1016/j.ejmech.2016.10.019] [PMID: 27776325]

[105] Zhang, Q.; Cao, R.; Liu, A.; Lei, S.; Li, Y.; Yang, J.; Li, S.; Xiao, J. Design, synthesis and evaluation of 2,2-dimethyl-1,3-dioxolane derivatives as human rhinovirus 3C protease inhibitors. *Bioorg. Med. Chem. Lett.,* **2017**, *27*(17), 4061-4065.
     [http://dx.doi.org/10.1016/j.bmcl.2017.07.049] [PMID: 28778471]

[106] Ghosh, A.K.; Xi, K.; Johnson, M.E.; Baker, S.C.; Mesecar, A.D. Progress in Anti-SARS Coronavirus Chemistry, Biology and Chemotherapy. *Annu. Rep. Med. Chem.,* **2007**, *41*(06), 183-196.
     [http://dx.doi.org/10.1016/S0065-7743(06)41011-3] [PMID: 19649165]

[107] Li, Y.H.; Hu, C.Y.; Wu, N.P.; Yao, H.P.; Li, L.J. Molecular Characteristics, Functions, and Related Pathogenicity of MERS-CoV Proteins. *Engineering (Beijing),* **2019**, *5*(5), 940-947.
     [http://dx.doi.org/10.1016/j.eng.2018.11.035] [PMID: 32288963]

[108] Cinatl, J., Jr; Michaelis, M.; Hoever, G.; Preiser, W.; Doerr, H.W. Development of antiviral therapy for severe acute respiratory syndrome. *Antiviral Res.,* **2005**, *66*(2-3), 81-97.
     [http://dx.doi.org/10.1016/j.antiviral.2005.03.002] [PMID: 15878786]

[109] Lee, H.; Mittal, A.; Patel, K.; Gatuz, J.L.; Truong, L.; Torres, J.; Mulhearn, D.C.; Johnson, M.E. Identification of novel drug scaffolds for inhibition of SARS-CoV 3-Chymotrypsin-like protease using virtual and high-throughput screenings. *Bioorg. Med. Chem.,* **2014**, *22*(1), 167-177.

[http://dx.doi.org/10.1016/j.bmc.2013.11.041] [PMID: 24332657]

[110]   Van Opdenbosch, N.; Lamkanfi, M. Caspases in Cell Death, Inflammation, and Disease. *Immunity,* **2019**, *50*(6), 1352-1364.
[http://dx.doi.org/10.1016/j.immuni.2019.05.020] [PMID: 31216460]

[111]   Songane, M.; Khair, M.; Saleh, M. An updated view on the functions of caspases in inflammation and immunity. *Semin. Cell Dev. Biol.,* **2018**, *82*, 137-149.
[http://dx.doi.org/10.1016/j.semcdb.2018.01.001] [PMID: 29366812]

[112]   McArthur, K.; Kile, B.T. Apoptotic Caspases: Multiple or Mistaken Identities? *Trends Cell Biol.,* **2018**, *28*(6), 475-493.
[http://dx.doi.org/10.1016/j.tcb.2018.02.003] [PMID: 29551258]

[113]   Khalil, H.; Bertrand, M.J.M.; Vandenabeele, P.; Widmann, C. Caspase-3 and RasGAP: a stress-sensing survival/demise switch. *Trends Cell Biol.,* **2014**, *24*(2), 83-89.
[http://dx.doi.org/10.1016/j.tcb.2013.08.002] [PMID: 24007977]

[114]   Thawkar, B.S.; Kaur, G. Inhibitors of NF-κB and P2X7/NLRP3/Caspase 1 pathway in microglia: Novel therapeutic opportunities in neuroinflammation induced early-stage Alzheimer's disease. *J. Neuroimmunol.,* **2019**, *326*(326), 62-74.
[http://dx.doi.org/10.1016/j.jneuroim.2018.11.010] [PMID: 30502599]

[115]   Patel, S.; Modi, P.; Chhabria, M. Rational approach to identify newer caspase-1 inhibitors using pharmacophore based virtual screening, docking and molecular dynamic simulation studies. *J. Mol. Graph. Model.,* **2018**, *81*, 106-115.
[http://dx.doi.org/10.1016/j.jmgm.2018.02.017] [PMID: 29549805]

[116]   Lakshmi, P.J.; Kumar, B.V.; Nayana, R.S.; Mohan, M.S.; Bolligarla, R.; Das, S.K.; Bhanu, M.U.; Kondapi, A.K.; Ravikumar, M. Design, synthesis, and discovery of novel non-peptide inhibitor of Caspase-3 using ligand based and structure based virtual screening approach. *Bioorg. Med. Chem.,* **2009**, *17*(16), 6040-6047.
[http://dx.doi.org/10.1016/j.bmc.2009.06.069] [PMID: 19631549]

[117]   Klebe, G. Virtual ligand screening: strategies, perspectives and limitations. *Drug Discov. Today,* **2006**, *11*(13-14), 580-594.
[http://dx.doi.org/10.1016/j.drudis.2006.05.012] [PMID: 16793526]

CHAPTER 3

Application of Systems Biology Methods in Understanding the Molecular Mechanism of Signalling Pathways in the Eukaryotic System

Aditya Rao S.J.[1] and **M. Paramesha**[1,2,*]

[1] *Dept. of Plant Cell Biotechnology, CSIR-Central Food Technological Research Institute, Mysuru, India*

[2] *Dept. of Food Technology, Shivagangotri, Davangere University, Davangere, Karnataka, India*

Abstract: A signalling pathway is a cascade of reactions carried out together by a group of molecules in a system to bring out the metabolic functions starting with cell division to cell death. When signalling pathways interact with one another, they form networks, which allow cellular responses to be coordinated, often by combinatorial signalling events. Any abnormal change affecting the activation or deactivation of such signalling events leads to the abnormal physiological cellular functions. The understanding of the molecular mechanism of signalling pathways is beneficial to understand the pathological scenery and treatment. Recent advancements in computational methods have given a new insight to understand the molecular mechanism involved in signalling pathways. The use of systemic and computational tools is crucial in systems biology as the complexity of the biological system is more, a vast amount of data is being generated and the scattered pieces of information has to be integrated into a meaningful order. The present chapter deals with the utilization of systems biology tools, and data mining techniques to understand the molecular mechanism of 'Wnt signalling pathway; an intercellular pathway which regulates critical aspects of cell fate determination, cell migration, cell polarity, neural patterning and organogenesis during embryonic development', with the integrated software platforms, which could help to address the future research problems in biology and medicine.

Keywords: Axin, Cancer, β-catenin, GSK-3β, Network analysis, Pathway prediction, Systems biology, Tankyrase, Therapeutics, Wnt signalling pathway, Wnt ligands, Wnt protein.

* **Corresponding author Dr.Paramesha M:** Associate Professor, Department of Food Technology, Shivagangotri, Davangere University, Davangere, Karnataka, India; Tel: +919481535168; E-mails: parameshbt@gmail.com; parameshaam8@davangereuniversity.ac.in

Zaheer Ul-Haq and Angela K. Wilson (Eds.)

INTRODUCTION

Intercellular Signalling Pathways

Cells communicate through cellular pathways, which play a significant role in their development. These pathways can be considered as 'control knobs' as they not only induce or block cellular differentiation during growth [1] but also control cellular behaviour [2, 3]. This can be exploited for biomedical applications in regenerative medicine [4, 5] as well as in providing valid drug targets [6, 7]. The intercellular signalling pathways have attracted many researchers due to their prevalent and diverse roles. The conservation of these intercellular signalling pathways across different species has made them the best-studied systems in biology in recent times. For many intercellular pathways, their association with ligands, receptors, intracellular effectors, transcription factors, and modulators have been identified, and their interactions have also been characterized. However, with the existing knowledge on the behaviour of intercellular pathways, some of the most basic operational questions still remained unanswered. These answers help in understanding specific molecular interactions, including how cell receives, processes and treats extracellular signals within the cell.

WHAT IS WNT?

The term Wnt stands for 'wingless- related integration site' which is derived from two words, namely 'wingless' and 'int'. The sequencing studies in Drosophila revealed the homogenicity of its wingless gene with oncogene int-1. Hence int/wingless family was considered as Wnt family. Wnts are involved in many crucial cellular processes like regulating the cell growth, motility, differentiation during embryonic development, controlling cell proliferation and cell fate apart from activating diverse signalling cascades. The degradation rate of β-catenin, a secondary messenger that acts as a transcriptional co-regulator in cell proliferation, is controlled by Wnt signalling. β-catenin will be degraded by multiple protein complexes when the Wnt pathway is inactive. This protein complex can be inactivated by the activity of receptor-targeted inhibitors, which results in accumulation of β-catenin [8]. This suggests that the β-catenin level can be externally controlled by the involvement of ligands, thus controlling Wnt signalling. The concentration of β-catenin is found to be different from cell to cell due to the variations in biochemical pathways and their fold-change resulted by the stimulation from external ligand but found to be uniform across cells [9]. This adaptive response of cells to external stimuli to produce β-catenin is controlled by Wnt genes [9]. Thus, it is possible to control the response of Wnt signals in turn controlling the cell fate decision by controlling the level of stimulation [10].

Types of Wnt Signalling Pathway

Based on the involvement of β-catenin, the Wnt pathway can be classified as follows:

1. Canonical Signalling Pathway
   ○ β-catenin dependent
   ○ β-catenin independent
2. Non-Canonical Signalling Pathway

CANONICAL PATHWAY

β-Catenin Dependent Canonical Pathway

The canonical pathway is also known as Wnt/β-catenin dependent signalling pathway [11] and also considered as a first identified pathway, which involved in the vital process such as cell proliferation, ribosome biogenesis, and inhibition of apoptosis [12]. Maintenance of intracellular β–catenin is associated with the canonical Wnt pathways. The activation of the canonical pathway starts with the binding of Wnt ligand to its receptor (Fig. **1**). The 'β-catenin destruction complex' is composed of scaffold protein Axin, glycogen synthase kinase-3β (GSK-3β), Casein Kinase-1α (CK-1α), and the tumour suppressor adenomatous polyposis coli (APC) [13, 14]. This destruction complex exhibits the sequential phospho-rylation by the CK-1α and GSK-3β enzymes present in the destruction complex and is targeted by ubiquitination and proteasome degradation [12, 15 - 17]. Thus, it controls the concentration of β-catenin accumulation in the cytoplasm and, in result, it will impede the signalling activity of Wnt/ β-catenin in the nucleus. Meanwhile, in the absence of β-catenin, transcriptional repressor Groucho binds to TCF (T-cell factor)/LEF (lymphoid enhancer factor), which inhibits the transcription of genes. Stabilization of β-catenin is initiated by Wnt protein, when it binds to the transmembrane receptor Fz (Frizzled) and the co-receptor LRP5/6 (low-density lipoprotein receptor-related protein-5/6) [18, 19]. The Wnt protein, Fz and LRP5/6, forms receptor complex, which facilitates the Dishevelled (Dvl) protein to the membrane, which leads to the formation of Signalosome [12]. The Signalosome provides the platform to the "β-catenin destruction complex," resulting in disassembly of "β-catenin destruction complex" initiated by CK-1α, which phosphorylates LRP5/6. As a result of this, cytoplasm abundant of β-catenin translocates to the nucleus, which ultimately displaces the transcriptional repressor Groucho and binds to TCF (T-cell factor)/ LEF (lymphoid enhancer factor) and triggers the expression of target genes [12, 17, 20 - 22].

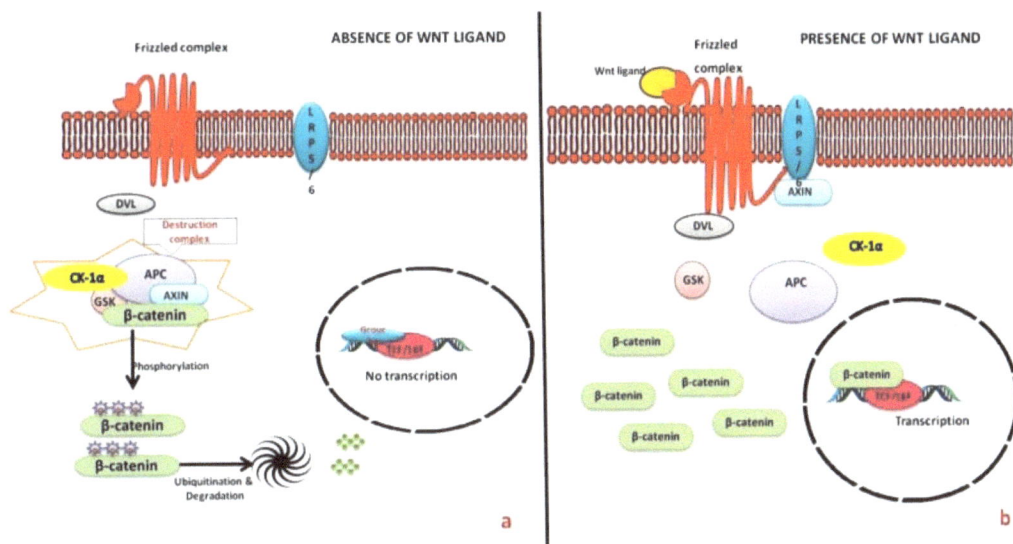

Fig. (1). A graphical representation of β-catenin dependent WNT Signalling pathway.

β-Catenin Independent Canonical Pathway

Recently, two different pathways have been identified, which are considered as parallel pathways to the canonical Wnt signalling pathway; Wnt/mTOR (mammalian target of rapamycin) and Wnt/STOP (Wnt-dependent stabilization of proteins) pathways. Initiation of these pathways is very similar to that of canonical β-catenin dependent pathway (as described above), *i.e.*, the Wnt protein ligands bind to the Fz–LRP5/6 complex, which allows the β-catenin destruction complex to the membrane along with GSK-3β (Fig. **2**). After this stage, the two distinctly different pathways start, which is deviated from the conservative canonical pathway and the steps are independent of the β-catenin role. Therefore, these two pathways are said to be a β-catenin independent canonical pathway. In the Wnt/mTOR signalling pathway, the GSK-3β is recruited by the Wnt-LRP5/6 complex. The phosphorylation mediated by GSK-3β of tuberous sclerosis protein 2 (TSC2; also known as tuberin) is avoided as a negative regulator of mTOR complex1(TORC1), upregulating mTOR signalling. This results in, the upregulation of protein translation and modulation of cytoskeleton dynamics. Therefore, mTOR pathway leads to an increase in the synthesis of proteins essential for cell cycle, such as cyclinD1, coupling cell growth and the progression into the cell cycle [23 - 26].

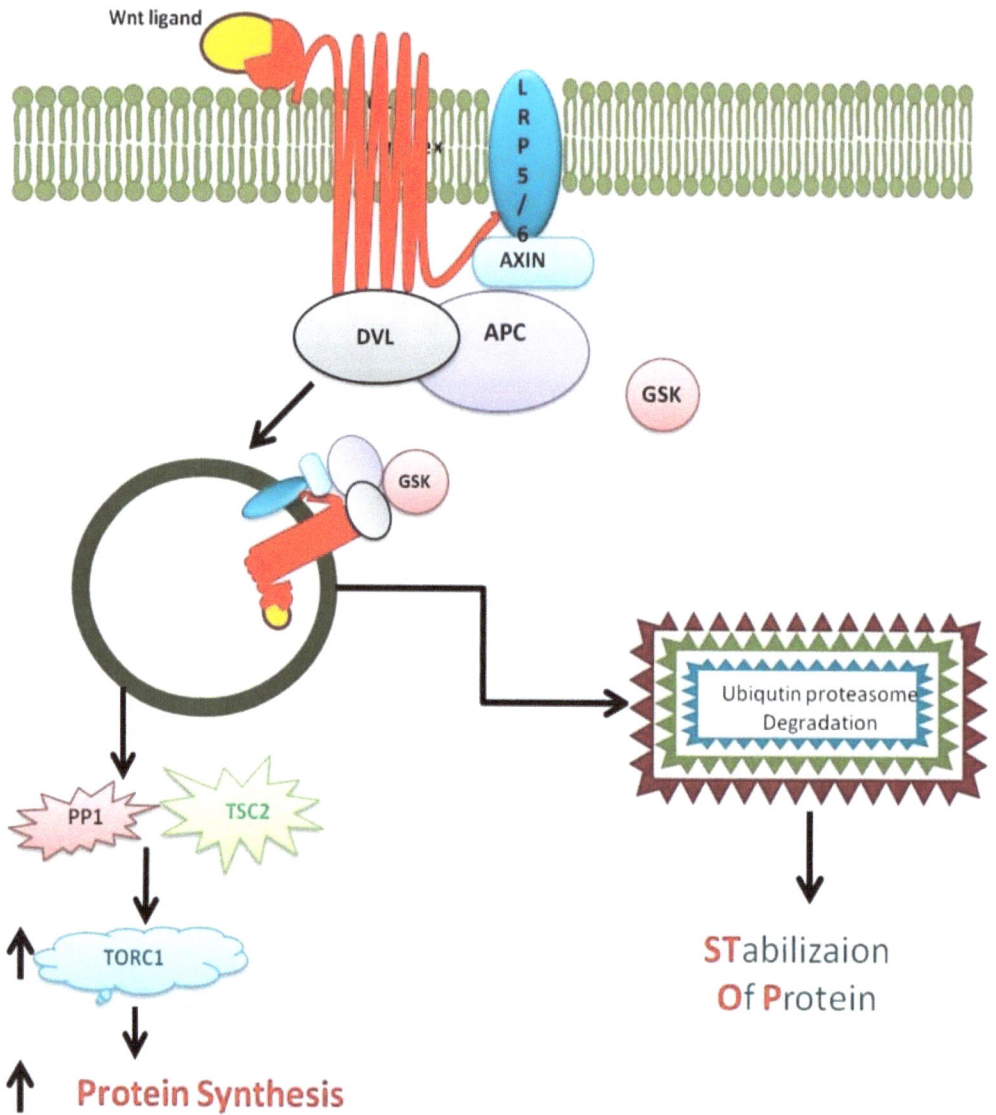

Fig. (2). A graphical representation of β-catenin independent Wnt/mTOR and STOP pathway.

The other alternate canonical β-catenin independent pathway is Wnt/STOP (Wnt-dependent stabilization of proteins) pathway. It works entirely in a unique way, as Wnt- LRP5/6 adopts GSK-3β, which phosphorylate several other proteins besides β-catenin [12]. Therefore, Wnt/STOP is independent of β-catenin and peaks during mitosis [27]. Wnt/STOP has a crucial role in the stabilization of proteins during mitosis. Wnt/STOP signalling protects the proteins like c-MYC through

GSK3-dependent polyubiquitination and degradation. It also helps to increase in cellular protein level and cell size [27 - 29]. GSK-3β phosphorylates many other proteins, in addition to the β-catenin [30]. This phosphorylation generates the degrons, which are detected by ubiquitin ligases such as β-TrCP, FBXW7, and NEDD4L, which helps to degrade these proteins [27, 31 - 34]. Due to the action of GSK-3β, the large protein molecules in addition to the β-catenin stabilizes with the acquaintance of half-life, and ultimately it slows down the degradation of protein during the peak of mitosis [27, 28, 35 - 37]. Wnt/STOP signalling pathway is also involved in many cellular processes, including cell-cycle progression, endolysosomal biogenesis, crosstalk with other pathways, DNA remodelling, and the cytoskeleton [27].

NON-CANONICAL SIGNALLING PATHWAY

The presence or absence of intermediary β-catenin or GSK-3β is considered as the main difference between canonical and non-canonical pathway. Non-canonical pathways play a vital role to unify various intracellular signalling cascades by triggering the binding of Wnt ligands to the Fz receptor. Wnt4, Wnt5a, Wnt11 *etc.*, are considered as non-canonical Wnt ligands which activate a cascade of reaction which does not involve the GSK-3β or β-catenin as intermediary molecules. The Wnt/Planar Cell Polarity (PCP) pathway, or Wnt/c-Jun-N-terminal Kinase (JNK), and the Wnt/Ca2+ pathway are considered to be two important divisions of non-canonical pathways [12, 18, 38].

Wnt/Planar Cell Polarity (PCP) Pathway

Planar Cell Polarity (PCP) pathway was initially identified in drosophila [39, 40], and it was considered to be the first non-canonical pathway identified [12]. Wnt/PCP pathway is initiated when Wnt ligand binds to Fz receptor on the membrane, which activates the small GTPases of the Rho and Rac subfamily which eventually stimulates the Rho-associated kinase (ROK) and JNK respectively as shown in Fig. (3) [12, 41, 42].

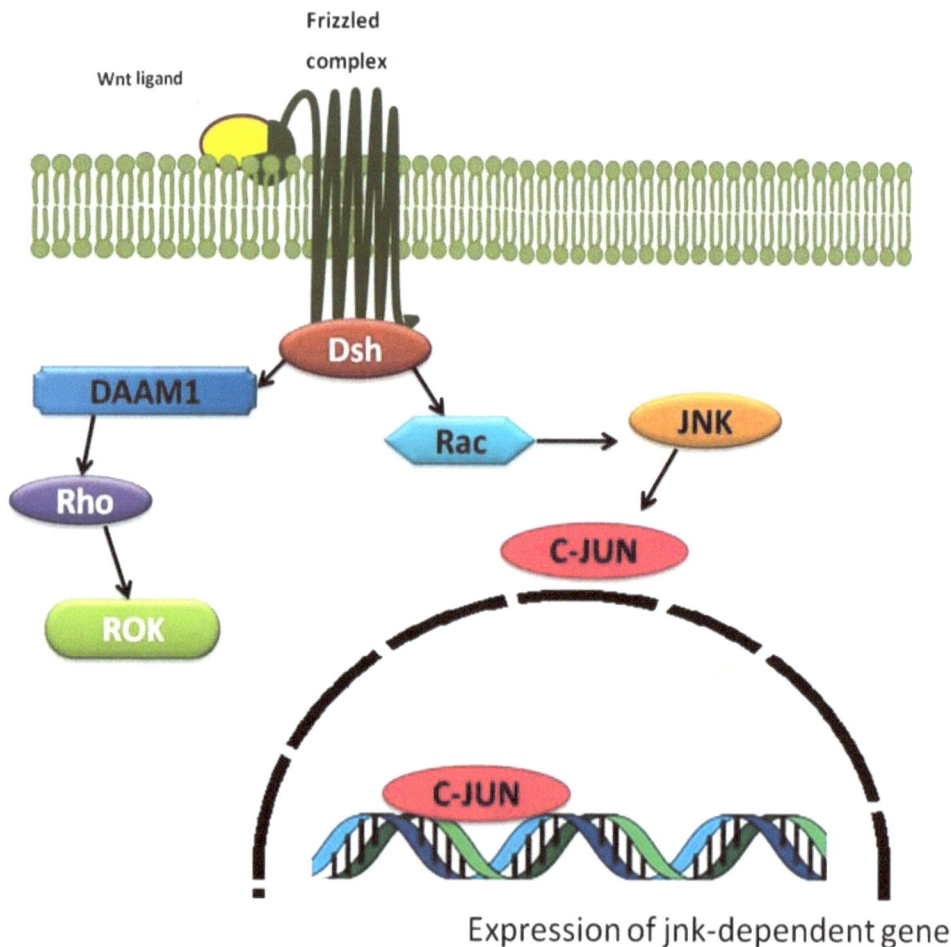

Fig. (3). A graphical representation of Non-Canonical Signalling Pathway, Wnt PCP Pathway.

The consequences of PCP pathway, is the regulation of the Cytoskeletal organization, cell motility [43, 44], and expression of JNK-dependent gene transcription factors such as ATF2 (activating transcription factor 2) along with the activation of associated and its target genes [45 - 49]. In mammals, Wnt/PCP regulates the development of left-right asymmetry in a coordinate manner [12, 50, 51]. Loss of function or deregulation of PCP genes can lead to a failed convergent extension of severe neural tube defects in animal models and human [48, 52].

The Wnt/Ca2+ Pathway

The Wnt/Ca2+ signalling pathway is a G-protein dependent signalling pathway which plays a crucial mediator role in the decoding the normal development [53,

54]. The activation of Phospholipase C (PLC) enzyme in the plasma membrane takes place due to the receptor-ligand interaction. The concentration of intracellular inositol-1,4,5- triphosphate [Ins [1, 4, 5]P3] [IP3] [55] and diacylglycerol (DAG) increases due to the cleavage of PIP2 by the activated of PLC enzyme. These increased IP3 acts as a second messenger and binds to the IP3 receptors (IP3R) on the RER (Rough Endoplasmic Reticulum) which leads to the diffusion of Ca2+ from the RER lumen to the cytoplasm of the cell, thus increase in the concentration of intracellular Ca2+ and activating the CaMKII (calcium calmodulin-dependent protein kinase II) [53, 56, 57], the main downstream effector. Meanwhile, DAG also favours the diffusion of Ca2+ from the RER and PKC (protein kinase C) activation [58]. These activated CaMKII and PKC activates different transcription factor, like CREB (cAMP response element-binding protein-1) and NFκB (nuclear factor kappa-light chain-enhancer of activating B cells), which helps to express the several genes which are considered as Wnt-target genes which have significant relevance in the Central nerve system [59]. The activated CaMKII persist long and typically increases the strength of nerve impulses (Fig. **4**).

SIGNIFICANCE OF THE WNT SIGNALLING PATHWAY

The discovery of the first member of the WNT family by Nusse & Varmus (1982) [60], opened a new arena to understand the life process, growth control pathway to pathological condition and, range from the physiological defects to cancer [61]. Wnt signal transduction pathway is not a simple pathway, and it triggers multiple discrete pathways elicited by different Wnt ligands and receptor interaction. Hence, the WNT signal transduction pathway is considered as a chief regulatory pathway in the animal kingdom. Any anomalous change occurring in Wnt/β-catenin signalling pathway or to its components leads to the abnormal growth and development in the animals. It may also lead to the irregular growth of cells which ultimately result in cancer [18, 62].

Wnt proteins are significant signal transduction molecule which triggers the cascade of metabolic reactions which helps in maintaining the normal health in multi-cellular animals [18, 22, 63, 64]. Till now, 19 Wnt ligands (*i.e.*, Wnt1, Wnt3a, Wnt7a/b, *etc.*,) have been identified [38], and these are extracellular secreted glycoprotein made up of approximately 350-400 highly conserved amino acid sequence, which can stimulate many intracellular pathways to regulate the proliferation and differentiation of embryonic and stem cells during growth and development [22, 63 - 65]. Therefore, a large group of researchers consider Wnt proteins as a therapeutic target. Consequently, many research groups are working on Wnt inhibition while others are targeting Wnt pathway enhancer to find solutions for many complicated disorders associated with Wnt signalling.

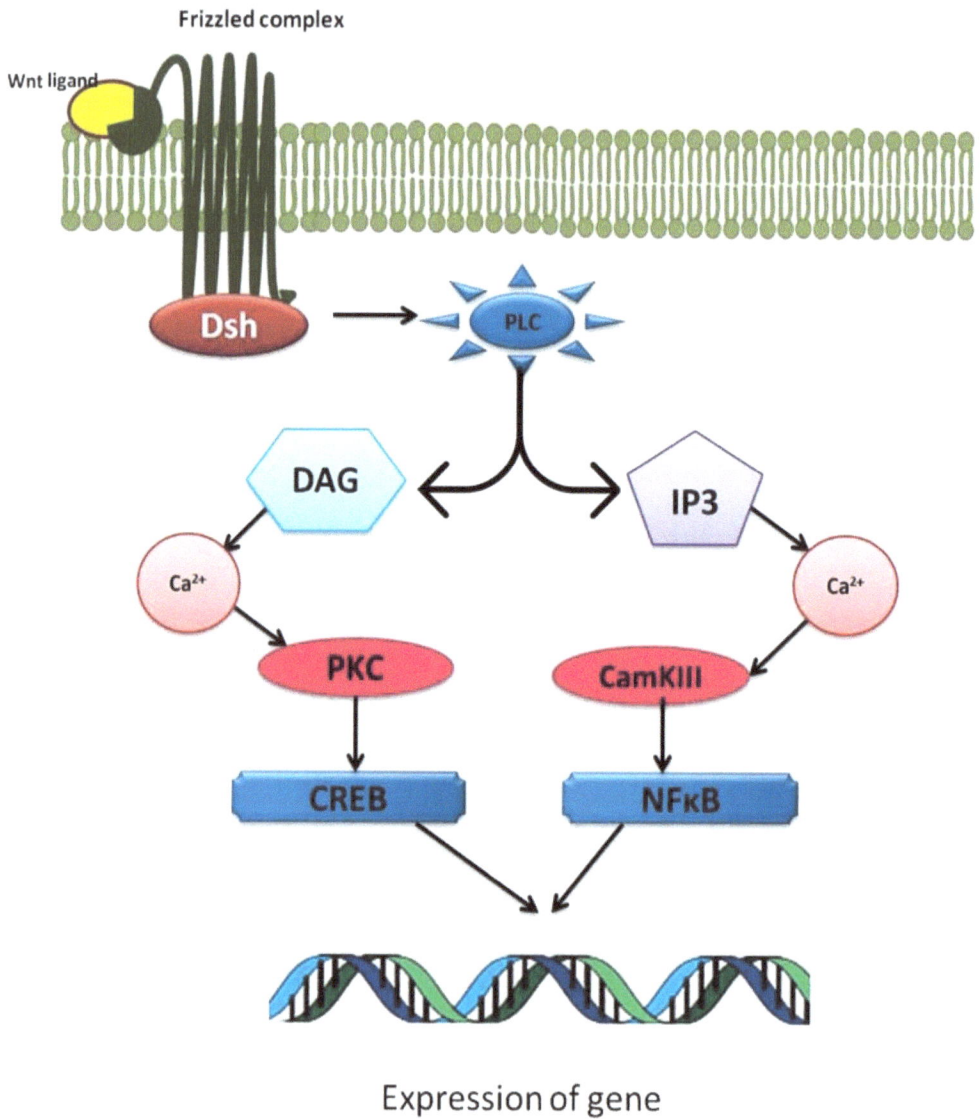

Fig. (4). A graphical representation of Non-Canonical Signalling Pathway, Wnt/Ca2+ Pathway.

IMPORTANCE OF WNT PATHWAYS IN DEFENCE

Wnt pathway plays a crucial role in the growth and development of organisms along with cell fate determination during embryonic development, cellular polarity, cell proliferation, cell cycle arrest and differentiation as well as in maintaining the tissue homeostasis and apoptosis [66]. Therefore, any aberrance in canonical or non-canonical Wnt signalling pathway will be a component for various genetic and non-genetic diseases. Wnt signalling pathway is run by a highly evolutionary conserved family of genes/proteins. A number of factors influence the Wnt pathway, any modulation in the proteins that either interact with the Wnt ligand or LRP 5/6 or Frizzled proteins function will bring a huge change leading to a wide spectrum of diseases.

Wnt Pathways and Cancers

Probably, cancer is one of the most rigorously associated diseases with the Wnt pathway since the first member of the WNT family was discovered by Nusse & Varmus (1982). Wnt/β-catenin signalling pathway has received considerable importance as a therapeutic target for cancer after the discovery of inactivated Familial adenomatous polyposis (FAP), a gene associated with the adenomatous polyposis coli (APC) [67 - 69]. Approximately above 85% of colorectal carcinomas come across due to the inactivation of FAP gene, which affects the translocation of β-catenin from the cytoplasm to nucleus [67 - 72]. Further, the mutation in Axin (important scaffold proteins present in β-catenin destruction complex) causes various types of cancer, for example, human adrenocortical adenomas tumour and adrenocortical carcinomas tumour samples [73, 74]. The incomplete closure of the neural tube or malfunctioning of the head folds may be the result of mutated Axin [69, 75]. Any changes like hyperactivity, hypoactivity and a missense mutation in LRP6; a co-receptor of the canonical pathway can cause neural tube defects [76 - 79]. Wnt5A act as a ligand to the non-canonical pathway, by binding to the receptors like ROR2, ROR1, *etc.* and any mutations in this ligand leads to cancer development [80]. Enomoto *et al.*, (2009) [81] showed that the increase in the human osteosarcoma, Wnt 5A binds to its receptor ROR2. WNT5A has also been shown to be involved with cancer cell metabolism and inflammation [82, 83]. Though, in the absence of mutations in Wnt/β-catenin, researchers noticed many tumours and hematologic malignancies, which may be due to the deregulation in the components of the Wnt/β-catenin signalling pathway. In many cancer conditions, it was noticed that the altered levels of expression in regulators in Wnt/β-catenin pathway without any mutations in the respective genes [69]. For example, secreted frizzled-related proteins (SFRPs) add their contribution to cancer, like colon, breast, prostate, lung and other cancers, due to their deregulation [84 - 88].

Wnt Pathways and Other Diseases

Wnt signalling pathway involved in the development of a central nervous system and is intensively involved in the Alzheimer's disease, which is an age-associated neurodegenerative disease (ND) characterized by progressive loss of cognitive function, memory and other associated neurobehavioral issues [89 - 91]. Recent studies described the role of receptor molecules such as LRP6 and the alteration in the concentration of messenger molecule like decreased β-catenin and increased GSK3β in the cortical lobes of Alzheimer's disease brains compared to age-matched controls [92]. Wnt signalling pathway potentially involved in the heart development, and any aberration within the pathway brings cardiovascular diseases [93, 94]. Autosomal dominant early coronary disease, hypertension, hyperlipidemia and osteoporosis may be the result of a mutation in the LRP6 gene [95]. DKK1 (Dickkopf-related protein 1), and LRP5/6 inhibitor found elevated in plasma and lesion of patients suffering from coronary artery disease and carotid plaques [96]. Endothelial dysfunction was also observed due to the increase in DKK1 in the endothelial cells to inhibit the platelets [96]. Inflammatory disorders are associated with the Wnt5A, which expressed intensively in macrophage rich regions of human and murine atherosclerotic lesions [83, 97]. In normal adult heart condition, Wnt signalling is very negligible; however, it was noticed in a mouse model that there is a deviation in the expression of FZD and Wnt in myocardial infarction [93, 98 - 101]. Several researchers established a link between Wnt signalling pathway and neuronal diseases like autism due to the mutation in Dvl1, and Dvl3 [102], Parkinson's disease- due to the dysregulation of Wnt components [103] and Schizophrenia due to the altered GSK3β activity [104].

The liver is the organ which has the ability to regenerate after injury and it was driven by the Wnt/β-catenin signalling pathway. Any dysregulation or loss of the signalling pathway leads to the delayed regeneration followed by a partial hepatectomy [105 - 111]. Due to the genetic polymorphism, the key regulators of Wnt signalling pathway may lead to the inflammation and fibrosis in patients with hepatitis C [112], a mutation in LRP6 develops fatty liver disease [113]. Dysregulation and mutations in Wnt signalling genes associated with oral diseases such as salivary gland tumour are due to the mutation in WIF1, which is a Wnt inhibitor [114], Gardner syndrome is due to the mutation in APC gene [115] and the agenesis is because of Axin2 gene mutation [68, 116]. Mutation in Wnt10A is responsible for a rare ectodermal dysplasia [117], and mutation in PORCN (porcupine) gene results in focal dermal hypoplasia [118]. An elevated level of matrix metalloproteinases (MMPs) and pro-inflammatory cytokines are the hallmarks of lung inflammation [119] and idiopathic pulmonary fibrosis [120] and expression of MMPs are regulated by the signalling pathway. Mutation in Wnt signalling proteins leads to the osteoporosis-pseudoglioma syndrome (OPPG) [121], osteogenesis imperfecta and early-onset osteoporosis [122, 123].

APPLICATION OF COMPUTATIONAL METHODS TO UNDERSTAND SIGNALLING PATHWAYS

With the advancements in system biology, it is now possible for the system-level understanding of living organisms, also extending the knowledge to various fields of medicine and biotechnology [124 - 128]. In recent decades, significant progress has also been made in deciphering intercellular signalling mechanisms, for example, in understanding the dynamics of AMPK (5' adenosine monophosphate-activated protein kinase) signalling and mitogen-activated protein kinase (MAPK) signalling [129]. Modern tools and resources of systems biology can tailor in achieving the objectives of novel biological discoveries, drug designing and many unanswered problems of biological research [130]. The main challenges in systems biology are the complexity of the system, the vastness of the generated data, scattered pieces of knowledge which all has to be integrated to generate meaningful information and hence the development of new computational strategies and implementation of computational tools are crucially important in systems biology [131].

The knowledge of DNA microarray, protein expressions and protein interactions can be integrated to identify metabolic networks and build cellular pathway models. This enables a systems biology researcher to perform integrative functional data analysis using biological networks. Further, it is also possible to understand the regulatory mechanism underlying a specific biological process extending the focus from an individual protein to proteome. A proteomic analysis including the prior knowledge of a group of proteins working in association with each other or with other genes or metabolites has enabled the understanding of complex cellular functions [132]. Newer techniques in network biology and systems biology integrated with proteomic studies resulted not only in better understanding of complex disease mechanism but also in rapid accumulation of data [132 - 135].

General Concept and Principle of Pathway-based Analysis

A pathway can be most generally defined as a set of related functional genes or proteins. A pathway-based analysis can provide a comprehensive outlook on the molecular interactions underneath a complex disease mechanism [136]. The pathway analysis is more specific and more detailed in terms of providing connecting links between the functionality of genes or proteins, and a disease phenotype (Fig. **5**). With time, pathway-based analysis has demonstrated improvements in power and robustness in deciphering complex biological interactions among different genes or proteins [137, 138].

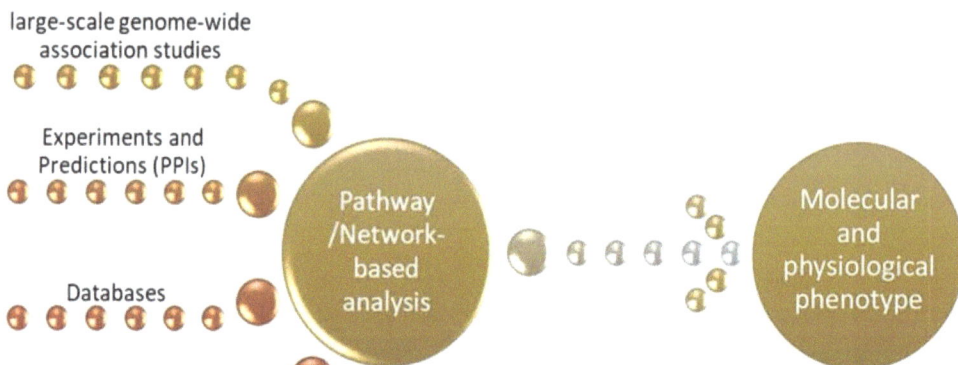

Fig. (5). An overview of the pathway and network analysis.

The pathway-based analysis originated from the GeneSet Enrichment Analysis (GSEA) in microarray data analysis of large-scale genome-wide association studies (GWAS) [136, 139] and with no time it has evolved in different directions [140, 141]. This includes linear combination test (LCT) [142] and supervised principal component analysis (SPCA) [143] which relies on the original data of principal components or joint distribution of multi-loci instead of statistical results [144]. Some modern methods like signalling pathway impact analysis (SPIA) [145] and CliPPER [146] have also emerged in recent times which parses the topological information of the pathway. These methods are increasingly applied in analysing complex disease mechanism [147]. With advancements in methodologies, the applications of pathway-based analysis in unravelling the complex human diseases has entered a new era [148, 149]. Several studies have demonstrated the in-depth details when applied to large-scale genetic datasets for the diseases like rheumatoid arthritis [142, 149], type 2 diabetes mellites [150], schizophrenia [140], Parkinson's disease [151], and others. Apart from these, the pathway analysis can also track the shared pathway among different pathological conditions, for example, the connection between T2D and cancer [152].

It is recognized that for most of the phenotypic changes, variations occurring at multiple loci are responsible, which often perturb signal transduction and regulation of metabolic pathways [142]. Hence, pathway-based methods analyse the gene sets or SNPs associated with specific functional units supported with biological knowledge. To analyse genotypic or single-point SNP data methods like over-representation analysis (ORA), gene set analysis [153], principal component or regression analysis can be applied for individual data [143, 154] whereas for multiple genetic loci topology-based analysis [145] can be performed to assess the overall association with the phenotype within a pathway [144]. For large-scale genome-wide association studies (GWAS) it is preferred to use pathway-based methods than single gene-based analysis method for better biological and statistical relevance [148]. The pathway-based methods focus on the expression of multiple genes instead of a single gene as it is unlikely to suffer from multiple test correlations resulting from large SNPs. Though the contribution of a single gene is minute and cannot be neglected, the disease often arises from the joint action of multiple genes/SNPs rather than a single gene. In comparison with single-gene analysis methods, pathway-based methods also consider locus heterogenicity in which alleles at different loci cause disease in the different population [155, 156].

In spite of having a huge amount of data on genes associated with complex diseases, it is still not possible to explain the connection between DNA variations and complex phenotypes which is essential for understanding the pathogenesis. In this view, the pathway-based analysis may play a complementary role in providing essential information on the molecular mechanism behind complex human diseases.

Pathway and Network Analysis for Proteomics

The success of proteomic analysis using systems biology approaches has not reached their optimum due to various reasons. Some of the most possible reasons are the data variations generated due to the heterogenicity in the biological sample, sample preparation errors, protein separation errors, the results are difficult to repeat, limits of detection methods and also quantification inaccuracy of proteomic data management tools. Though some level of data accuracy can be found in DNA microarray and NGS instruments, the statistical methods developed for their proteomic applications are mostly inaccurate. These factors increase the difficulties for a systems biologist to design intercellular pathways and interacting protein networks using proteomics data [157 - 159]. The ultimate challenge, however, is to annotate the functions of a list of proteins identified in order to provide biological insights into the underlying molecular mechanism [147]. Nevertheless, some statistical machine learning approaches like Support Vector Machine (SMV) [160], Markov clustering [161], ant colony optimization [162], and semi-supervised learning [163] have also been developed.

The interpretation of proteomics results can be addressed by pathway and network analysis techniques. Biological pathways can be correlated with the signalling as well as metabolic pathways. This helps in the grouping of a large list of proteins into a small list of closely associate proteins making it easier to interpret their altered expression [147]. Network analysis techniques rely on graph theory, information theory or Bayesian theory for their calculations. These techniques can be used to build, overlay, visualize and infer protein-protein interactions, their functional analysis and other systems biology data [164]. While pathway analysis helps to interpret the proteins at their expression level, network analysis uses the comprehensive networks connecting prior experimentation results with new *in silico* predictions concentrating biological significance at systems level [165]. To assess the biological significance, information of Gene Ontology (GO) can also be used [164] as the outputs from high-throughput proteomic instruments also suffers from false discovery rates mostly due to the inherent high-level variance in proteomic data [157].

Resources for Pathway-based Analysis

A rapid drift from gene-based data analysis methods to pathway-based methods resulted in increased accumulation of pathway resources as these methods analyse not just a single gene or SNP data but a complete pathway. The pathway composition can be retrieved from any public pathway databases, and to make this effortless these resources are categorized based on functional information (Table 1) [144].

Table 1. Categorization of pathway databases based on the type of information stored in them.

Category	Resources	Link
Metabolic pathway databases	KEGG BioCyc BIOPATH EMP	http://www.genome.ad.jp/kegg/ http://www.biocyc.org/ http://www.mol-net.de/databases/biopath.html http://emp.mcs.anl.gov/
Signal transduction pathway databases	CSNDB SPAD TransPath BBID	http://geo.nihs.go.jp/csndb/ http://www.grt.kyushu-u.ac.jp/spad/ http://transpath.gbf.de/ http://bbid.grc.nia.nih.gov/
Protein-protein interaction pathway databases	CYGD DIP GRID HPRD	http://mips.gsf.de/genre/proj/yeast/index.jsp http://dip.doe-mbi.ucla.edu/ http://biodata.mshri.on.ca/grid/servlet/Index http://www.hprd.org/
Transcriptional regulation pathway database	STKE BTITE TRANSFAC CST	http://www.stke.org/ http://www.genome.ad.jp/brite/ http://transfac.gbf.de/ http://www.cellsignal.com/

Apart from these resources many online pathway and network analysis knowledgebases and tools have also been developed in the recent times [147, 166] of which some have a direct association with proteomics [135, 167]. Some of the commonly used resources for pathway and network analysis studies are BioGRID

[168], STRING [169], KEGG [170], Reactome [171], BioCarta [172], PID [173], HAPPI [174], HPD [175], and PAGED [176] databases (Table **2**).

Table 2. List of some common Pathway/Network analysis resources.

Name	Description	Link	Reference
KEGG	Kyoto Encyclopedia of Genes and Genomes	https://www.genome.jp/kegg/	[170]
DAVID	The Database for Annotation, Visualization and Integrated Discovery	https://david.ncifcrf.gov/	[177]
GSEA	Gene Set Enrichment Analysis	http://software.broadinstitute.org/gsea/msigdb/index.jsp	[139]
IPA	Ingenuity Pathway Analysis	https://www.qiagenbioinformatics.com/products/ingenuity-pathway-analysis/	[178]
MetaCore	Integrated software suite for functional analysis of experimental data.	https://libguides.mit.edu/bioinfo/metacore	N/A
SPIA	Signalling Pathway Impact Analysis	http://bioconductor.org/packages/release/bioc/html/SPIA.html	[179]
PAGED	Pathway and Gene Enrichment Database	https://bio.informatics.iupui.edu/PAGED	[176]
HAPPI	Human Annotated and Predicted Protein Interactions	http://discovery.informatics.uab.edu/HAPPI/	[174]
STRING	Protein-Protein Interaction Networks Functional Enrichment Analysis	https://string-db.org/	[169]
Cytoscape	Open-source software platform for visualizing molecular interaction networks and biological pathways	https://cytoscape.org/	[180]
PIR	Protein Information Resource	https://proteininformationresource.org/	[181]
SMPDB	Small Molecule Pathway Database	http://smpdb.ca/	[182]
HumanCyc	Encyclopedic reference on Human metabolic pathways	https://humancyc.org/	[183]
MetaCyc	Metabolic Pathway Database	https://metacyc.org/	[184]
CellDesigner	A modelling tool of biochemical networks	http://www.celldesigner.org/	[185]
COPASI	Software application for simulation and analysis of biochemical networks and their dynamics	http://copasi.org/	[186]
Morpheus	A modelling and simulation environment for the study of multiscale and multicellular systems.	https://morpheus.gitlab.io/	[187]
Network Portal	Analysis and visualization tools for selected gene regulatory networks	http://networks.systemsbiology.net/	
Reactome	Provide intuitive bioinformatics tools for the visualization, interpretation and analysis of pathway knowledge to support basic research, genome analysis, modelling, systems biology and education	https://reactome.org/	[188]
NetPath	Manually curated resource of signal transduction pathways in humans	http://www.netpath.org/	[189]
HPRD	Human protein reference database	http://www.hprd.org/	[190]
IntAct	Molecular Interaction Database	https://www.ebi.ac.uk/intact/main.xhtml	[191]
HumanNet	A probabilistic functional gene network of validated protein-encoding genes of Homo sapiens	http://www.functionalnet.org/humannet/about.html	[192]
Pathguide	Contains information about biological pathway related resources and molecular interaction related resources	http://www.pathguide.org/	[193]
LitInspector	Large scale text mining on more than 20 million PubMed entries	http://www.litinspector.org/	[194]

Application of Systems Biology Tools to Decipher WNT Signalling: Targeting Modulators of Wnt Signalling for Therapy

Wnt signalling is one of the critical pathways associated with the homeostasis of the body. Thus, any abnormal Wnt signalling leads to a different type of cancer including gastrointestinal cancers, leukaemia, melanoma, and breast cancer [80]. As Wnt signalling is also associated with the regulation of stem cells, medicinal research can also be benefited in understanding the injury repair mechanism by targeting Wnt signalling [195]. Moreover, with the advancements in small molecule and natural product drug research, it is expected to generate promising results to effectively regulate Wnt signalling.

Wnt signalling can be modulated by targeting some of the key Wnt proteins. Wnt proteins are secreted in canonical Wnt signalling pathway which binds to seven-transmembrane receptor Frizzled (Fzd) and lipoprotein receptor-related protein (LPR) after their acylation by Porcupine (PORCN) which later activates Dishevelled protein (Dvl) [196]. Dvl binds to Fzd protein being also associated with Axin of the β-catenin destruction complex. During Wnt signalling, LPRs undergo phosphorylation by casein kinase 1 and GSK-3β which are activated by Dvl bound Axin. In the absence of these interactions, β-catenin will be phosphorylated and removed by the ubiquitin-proteasome system [197]. In addition, proteins Tankyrase (TNKS) also play a crucial role in regulating Wnt signalling proteins. They phosphorylate Axin which is also be degraded by the ubiquitin-proteasome system [197]. In the absence of Axin, GSK-3β cannot phosphorylate β-catenin effectively, resulting in the stabilization of β-catenin. The stabilized β-catenin accumulates in the nucleus and initiates the transcription of Wnt target genes. An overview of the Wnt signalling pathway is shown in Fig. (**6**) as produced by the KEGG pathway database [170].

There are several specific Protein-protein interacting molecules (inhibitors and activators) designed to target the Wnt signalling pathway. These molecules can specifically inhibit Fzd protein, Dvl protein, the β-catenin destruction complex, β-catenin molecule and the enzymes like Prcn and TNKS [198]. Similarly, well-known activators affect the β-catenin destruction complex of Wnt signalling pathway through the GSK-3β, resulting in allowing the β-catenin into the nucleus and influencing essential gene expression [198].

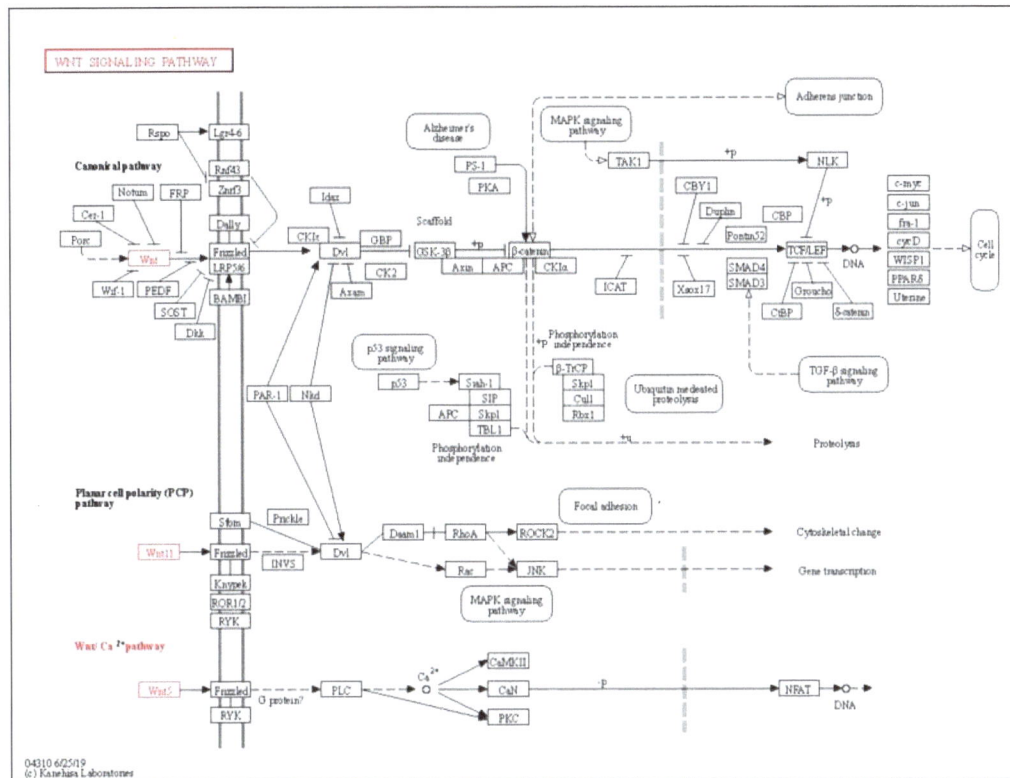

Fig. (6). Pathway representation of Wnt signalling (KEGG ID: map04310).

Inhibitors Blocking Fzd and Dvl Protein

Frizzled or Fzd antibody found to block the canonical Wnt pathway by binding to Fzd receptor which ultimately occupying the binding site of Wnt protein called Fzd cysteine-rich domains (CRD) essentially required to initiate the Wnt signalling pathway [199]. Due to the interaction between Fzd cysteine-rich domains (CRD) and Fzd antibody, the accumulation of β-catenin increases in the cytoplasm resulting in the inhibition of human tumour growth [199, 200].

Along with Fzd proteins, Dvl protein can also be considered as a potential target for the inhibition of Wnt signalling. Wong *et al.,* (2003) [201] described the interaction between C-terminal sequence of the seventh transmembrane helix of Fzd and PDZ domain of the Dvl protein which aids the transduction of downstream Wnt signals (Fig. 7).

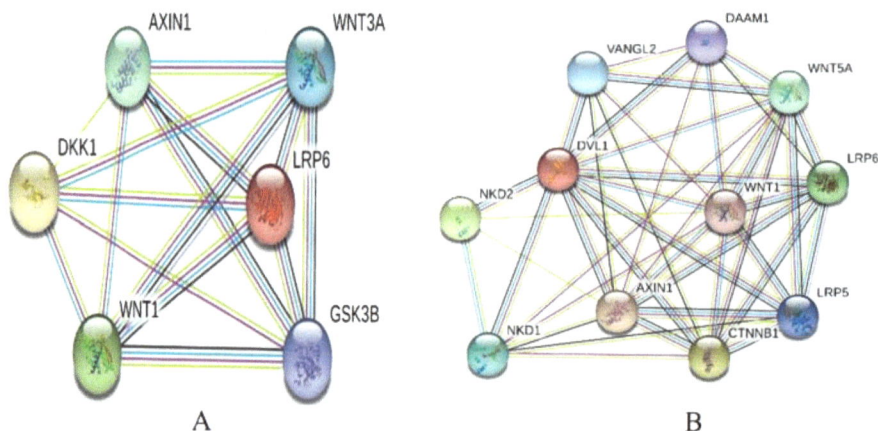

Fig. (7). String Network view of Wnt proteins association with Fzd (A) and Dlv (B) proteins.

Porcupine Enzyme Inhibitors

Wnt signalling plays a critical role in embryonic development and also in regulating homeostasis in adult tissues. Porcupine (PORCN) enzyme catalyses the palmitoylation of Wnt proteins leading to the activation of cellular response like transport, secretion and activity of Wnt proteins [202] (Fig. **8**). PORCN is the group of membrane-bound O-acyltransferases (MBOAT) family proteins which is targeted by a group of small molecules called Inhibitors of Wnt production (IWPs). It was shown that some of IPWs can reduce and also block the morphogenesis of tailfin contributing significantly to understand the role of Wnt signalling in morphogenesis [203].

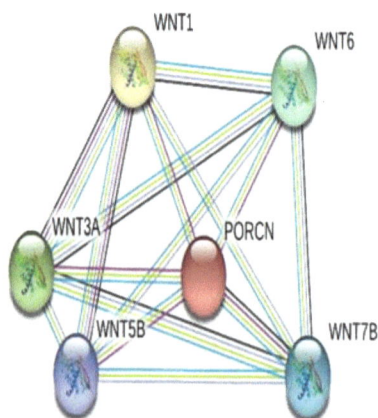

Fig. (8). String Network view of Wnt proteins association with porcupine.

TNKS Enzyme Inhibitors

TNKS of Tankyrase or TRF1-interacting ankyrin-related ADP-ribose polymerases are specific PARPs (Poly(ADP-ribose) polymerases) that enhance telomerase to access telomeres. They are involved in the destruction of β-catenin complex antagonistically by particularly targeting Axin [204] (Fig. **9**). By targeting TNKS it is possible to control the activity of β-catenin destruction complex, control the β-catenin accumulation and excess signalling of Wnt. Although it is because of the shielding effect provided by LEF1 and B9L inside the nucleus for the Axin of the β-catenin destruction complex, less success has been achieved in design inhibitors for TNKS [205].

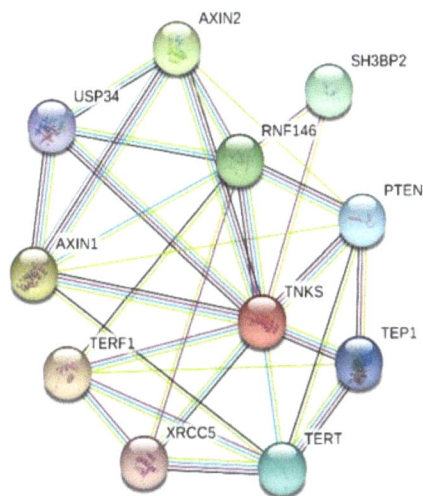

Fig. (9). String Network view of Wnt proteins association with Tankyrase.

GSK-3β targeted inhibitors for the activation of Wnt signalling

Wnt signalling can be controlled by using inhibitors particularly targeting GSK-3β (Fig. **10**). But excess inhibition will have effects similar to excess activation, which leads to several health complications. As in the case of Alzheimer's patients, β-amyloid dependent neurodegeneration in brain cells was due to the excess inhibition of Wnt signalling proteins [206]. Further, the down-regulation of β-catenin activity by the overexpression of GSK-3β found to disturb β-catenin complex which has a direct effect on diabetes pathways [207]. But the controlled activation or the inhibition can also be a good option in treating many health-associated problems.

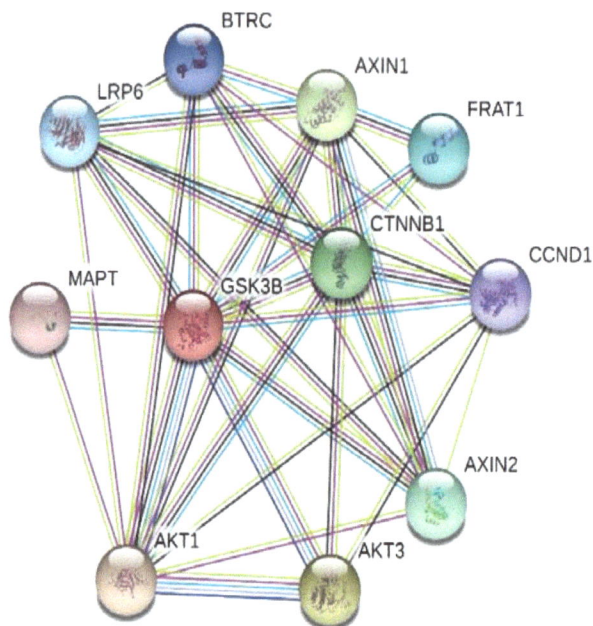

Fig. (10). String Network view of the association of GSK-3β with Wnt signalling proteins.

At this date, many of the Wnt activators are found to be effective in inhibiting the activity of GSK-3β; a part of β-catenin destruction complex. The inhibition of GSK-3β results in the dysfunctioning of the β-catenin degrading complex, allowing the accumulation of β-catenin in the nucleus and thus leading Wnt signals for further transcription. Thus, detailed research elucidating the mechanism of Wnt activators and their association with Wnt genes and their overexpression will be useful in designing effective strategies to control Wnt signalling.

Natural products as Wnt Inhibitors

Natural products are gaining the interest of the medicinal chemists due to their therapeutic benefits with no or fewer side effects. Chancones derivatives have recently discovered as effective inhibitors of the Wnt signalling pathway. Their mechanism ultimately involves lowering the amount of nuclear β-catenin. Though much is to explore on the action of natural products, the studies reported the low risk of mutagenesis as their interaction with DNA is minimum [208]. Further, the transcription of β-catenin was inhibited by the action of carnosic acid, a natural product derived from rosemary [209]. There are many natural products reported to be associated with Wnt signalling and producing encouraging results [210, 211], their effects are found to be prominent only in *in vitro* assays but for better understanding, their *in vivo* effects are yet to be examined.

CONCLUSION

Presently many research initiatives have been taken both by individual researches as well as research groups on establishing a correlation between Wnt signalling, Wnt inhibitors and Wnt associated diseases. On the other hand, many researches are being associated in understanding the molecular basis and mechanism of regulation of different Wnt pathways. The application of different computational techniques has influenced greatly in understanding molecular mechanisms of signalling pathways. Computational methods like structure-based screening, chemical genomics along with different 'omics' approaches have given a major boost in designing effective strategies in identifying small molecule ligands that can interact and modulate Wnt signalling proteins. As Wnt signalling is majorly associated with different types of cancer, Wnt inhibitors have gained tremendous importance in the recent days and the mechanism by which they regulate Wnt signalling has stimulated the insights with clinical significance. Further, the investigation of Wnt signalling activation can generate some critical information connected with injury repair, regulating the cell growth, motility, differentiation during embryonic development, controlling cell proliferation and other processes where Wnt signalling activation is required. Although several reports connecting small molecules, along with the natural products affecting Wnt signalling are available, still there exists some rift in the current knowledge on their regulation. A deeper understanding involving small molecules binding to Wnt proteins, their binding affinity and accuracy and determining their role in other pathways needs to be studied as their effects on healthy cells can be depicted.

With the application of different system biology tools, the in-depth understanding of protein-protein interactions in connection with a particular pathway, the hub protein identification in a network of the interacting protein associated with particular pathway can be achieved, resulting in simulating the cellular mechanisms *in silico* for further designing of effective strategy to control cellular activities and to develop effective medicine. Further, a structural insight on key molecules involved in Wnt signalling helps in elucidating the binding with different Wnt protein's binding sites allowing to improve binding and design more effective molecules that may produce effective results. More than these, it also helps in understanding and designing safe and effective medicines.

CONSENT FOR PUBLICATION

Not applicable.

CONFLICT OF INTEREST

The authors confirm that the contents of this chapter have no conflict of interest.

ACKNOWLEDGEMENTS

The authors express their gratitude to The Honourable Vice-Chancellor of Davangere University, Davangere, Karnataka, India for his kind support. The authors also wish to acknowledge, The Director, CSIR-CFTRI and Dr Nandini P Shetty, Principal Scientist, CSIR-CFTRI, Mysuru, Karnataka, India for their support. Thanks to Dr CK Ramesh, Professor, PG Department of studies and research in Biotechnology, Sahyadri Science College, Shimoga, Kuvempu University Karnataka, India for his valuable support and suggestion.

REFERENCES

[1]   Zúñiga-Pflücker, J.C. T-cell development made simple. *Nat. Rev. Immunol.,* **2004**, *4*(1), 67-72.
        [http://dx.doi.org/10.1038/nri1257] [PMID: 14704769]

[2]   Delaney, C.; Heimfeld, S.; Brashem-Stein, C.; Voorhies, H.; Manger, R.L.; Bernstein, I.D. Notch-mediated expansion of human cord blood progenitor cells capable of rapid myeloid reconstitution. *Nat. Med.,* **2010**, *16*(2), 232-236.
        [http://dx.doi.org/10.1038/nm.2080] [PMID: 20081862]

[3]   Kim, M.; Choe, S. BMPs and their clinical potentials. *BMB Rep.,* **2011**, *44*(10), 619-634.
        [http://dx.doi.org/10.5483/BMBRep.2011.44.10.619] [PMID: 22026995]

[4]   Date, S.; Sato, T. Mini-gut organoids: reconstitution of the stem cell niche. *Annu. Rev. Cell Dev. Biol.,* **2015**, *31*(1), 269-289.
        [http://dx.doi.org/10.1146/annurev-cellbio-100814-125218] [PMID: 26436704]

[5]   Fordham, R.P.; Yui, S.; Hannan, N.R.; Soendergaard, C.; Madgwick, A.; Schweiger, P.J.; Nielsen, O.H.; Vallier, L.; Pedersen, R.A.; Nakamura, T.; Watanabe, M.; Jensen, K.B. Transplantation of expanded fetal intestinal progenitors contributes to colon regeneration after injury. *Cell Stem Cell,* **2013**, *13*(6), 734-744.
        [http://dx.doi.org/10.1016/j.stem.2013.09.015] [PMID: 24139758]

[6]   Andersson, E.R.; Lendahl, U. Therapeutic modulation of Notch signalling--are we there yet? *Nat. Rev. Drug Discov.,* **2014**, *13*(5), 357-378.
        [http://dx.doi.org/10.1038/nrd4252] [PMID: 24781550]

[7]   Samatar, A.A.; Poulikakos, P.I. Targeting RAS-ERK signalling in cancer: promises and challenges. *Nat. Rev. Drug Discov.,* **2014**, *13*(12), 928-942.
        [http://dx.doi.org/10.1038/nrd4281] [PMID: 25435214]

[8]   MacDonald, B.T.; Tamai, K.; He, X. Wnt/β-catenin signaling: components, mechanisms, and diseases. *Dev. Cell,* **2009**, *17*(1), 9-26.
        [http://dx.doi.org/10.1016/j.devcel.2009.06.016] [PMID: 19619488]

[9]   Goentoro, L.; Kirschner, M.W. Evidence that fold-change, and not absolute level, of β-catenin dictates Wnt signaling. *Mol. Cell,* **2009**, *36*(5), 872-884.
        [http://dx.doi.org/10.1016/j.molcel.2009.11.017] [PMID: 20005849]

[10]   Antebi, Y.E.; Nandagopal, N.; Elowitz, M.B. An operational view of intercellular signaling pathways. *Curr. Opin. Syst. Biol.,* **2017**, *1*, 16-24.
        [http://dx.doi.org/10.1016/j.coisb.2016.12.003] [PMID: 29104946]

[11]   McMahon, A.P.; Moon, R.T. Ectopic expression of the proto-oncogene int-1 in Xenopus embryos leads to duplication of the embryonic axis. *Cell,* **1989**, *58*(6), 1075-1084.
[http://dx.doi.org/10.1016/0092-8674(89)90506-0] [PMID: 2673541]

[12]   Oliva, C.A.; Montecinos-Oliva, C.; Inestrosa, N.C. Wnt Signaling in the Central Nervous System: New Insights in Health and Disease. *Prog. Mol. Biol. Transl. Sci.,* **2018**, *153*, 81-130.
[http://dx.doi.org/10.1016/bs.pmbts.2017.11.018] [PMID: 29389523]

[13]   Cliffe, A.; Hamada, F.; Bienz, M. A role of Dishevelled in relocating Axin to the plasma membrane during wingless signaling. *Curr. Biol.,* **2003**, *13*(11), 960-966.
[http://dx.doi.org/10.1016/S0960-9822(03)00370-1] [PMID: 12781135]

[14]   Gao, C.; Chen, Y.G. Dishevelled: The hub of Wnt signaling. *Cell. Signal.,* **2010**, *22*(5), 717-727.
[http://dx.doi.org/10.1016/j.cellsig.2009.11.021] [PMID: 20006983]

[15]   Aberle, H.; Bauer, A.; Stappert, J.; Kispert, A.; Kemler, R. β-catenin is a target for the ubiquitin-proteasome pathway. *EMBO J.,* **1997**, *16*(13), 3797-3804.
[http://dx.doi.org/10.1093/emboj/16.13.3797] [PMID: 9233789]

[16]   Liu, C.; Li, Y.; Semenov, M.; Han, C.; Baeg, G.H.; Tan, Y.; Zhang, Z.; Lin, X.; He, X. Control of β-catenin phosphorylation/degradation by a dual-kinase mechanism. *Cell,* **2002**, *108*(6), 837-847.
[http://dx.doi.org/10.1016/S0092-8674(02)00685-2] [PMID: 11955436]

[17]   Clevers, H.; Nusse, R. Wnt/β-catenin signaling and disease. *Cell,* **2012**, *149*(6), 1192-1205.
[http://dx.doi.org/10.1016/j.cell.2012.05.012] [PMID: 22682243]

[18]   Morgan, R. Wnt Signaling as a Therapeutic Target in Cancer and Metastasis, **2017**.
[http://dx.doi.org/10.1016/B978-0-12-804003-4.00020-7]

[19]   Janda, C.Y.; Dang, L.T.; You, C.; Chang, J.; de Lau, W.; Zhong, Z.A.; Yan, K.S.; Marecic, O.; Siepe, D.; Li, X.; Moody, J.D.; Williams, B.O.; Clevers, H.; Piehler, J.; Baker, D.; Kuo, C.J.; Garcia, K.C. Surrogate Wnt agonists that phenocopy canonical Wnt and β-catenin signalling. *Nature,* **2017**, *545*(7653), 234-237.
[http://dx.doi.org/10.1038/nature22306] [PMID: 28467818]

[20]   Lustig, B. *Negative Feedback Loop of Wnt Signaling through Upregulation of Conductin/Axin2 in Colorectal and Liver Tumors,* **2002**.
[http://dx.doi.org/10.1128/MCB.22.4.1184-1193.2002]

[21]   Tan, S.H.; Barker, N. Wnt Signaling in Adult Epithelial Stem Cells and Cancer. *Prog. Mol. Biol. Transl. Sci.,* **2018**, *153*, 21-79.
[http://dx.doi.org/10.1016/bs.pmbts.2017.11.017] [PMID: 29389518]

[22]   Goldsberry, W.N.; Londoño, A.; Randall, T.D.; Norian, L.A.; Arend, R.C. A review of the role of wnt in cancer immunomodulation. *Cancers (Basel),* **2019**, *11*(6)E771
[http://dx.doi.org/10.3390/cancers11060771] [PMID: 31167446]

[23]   Inoki, K.; Ouyang, H.; Zhu, T.; Lindvall, C.; Wang, Y.; Zhang, X.; Yang, Q.; Bennett, C.; Harada, Y.; Stankunas, K.; Wang, C.Y.; He, X.; MacDougald, O.A.; You, M.; Williams, B.O.; Guan, K.L. TSC2 integrates Wnt and energy signals via a coordinated phosphorylation by AMPK and GSK3 to regulate cell growth. *Cell,* **2006**, *126*(5), 955-968.
[http://dx.doi.org/10.1016/j.cell.2006.06.055] [PMID: 16959574]

[24]   Saxton, R.A.; Sabatini, D.M. mTOR Signaling in Growth, Metabolism, and Disease. *Cell,* **2017**, *168*(6), 960-976.
[http://dx.doi.org/10.1016/j.cell.2017.02.004] [PMID: 28283069]

[25]   Meng, D.; Frank, A.R.; Jewell, J.L. mTOR signaling in stem and progenitor cells. *Development,* **2018**, *145*(1)dev152595
[http://dx.doi.org/10.1242/dev.152595] [PMID: 29311260]

[26]   Zeng, H.; Lu, B.; Zamponi, R.; Yang, Z.; Wetzel, K.; Loureiro, J.; Mohammadi, S.; Beibel, M.;

Bergling, S.; Reece-Hoyes, J.; Russ, C.; Roma, G.; Tchorz, J.S.; Capodieci, P.; Cong, F. mTORC1 signaling suppresses Wnt/β-catenin signaling through DVL-dependent regulation of Wnt receptor FZD level. *Proc. Natl. Acad. Sci. USA*, **2018**, *115*(44), E10362-E10369.
[http://dx.doi.org/10.1073/pnas.1808575115] [PMID: 30297426]

[27]   Acebron, S.P.; Niehrs, C. β-Catenin-Independent Roles of Wnt/LRP6 Signaling. *Trends Cell Biol.*, **2016**, *26*(12), 956-967.
[http://dx.doi.org/10.1016/j.tcb.2016.07.009] [PMID: 27568239]

[28]   Acebron, S.P.; Karaulanov, E.; Berger, B.S.; Huang, Y.L.; Niehrs, C. Mitotic wnt signaling promotes protein stabilization and regulates cell size. *Mol. Cell*, **2014**, *54*(4), 663-674.
[http://dx.doi.org/10.1016/j.molcel.2014.04.014] [PMID: 24837680]

[29]   Madan, B. Oncogenic Wnt/STOP signaling regulates ribosome biogenesis *in vivo*. *bioRxiv*, **2018**, *1922159*326819
[http://dx.doi.org/10.1101/326819]

[30]   Xu, C.; Kim, N.G.; Gumbiner, B.M. Regulation of protein stability by GSK3 mediated phosphorylation. *Cell Cycle*, **2009**, *8*(24), 4032-4039.
[http://dx.doi.org/10.4161/cc.8.24.10111] [PMID: 19923896]

[31]   Hart, M.; Concordet, J.P.; Lassot, I.; Albert, I.; del los Santos, R.; Durand, H.; Perret, C.; Rubinfeld, B.; Margottin, F.; Benarous, R.; Polakis, P. The F-box protein β-TrCP associates with phosphorylated β-catenin and regulates its activity in the cell. *Curr. Biol.*, **1999**, *9*(4), 207-210.
[http://dx.doi.org/10.1016/S0960-9822(99)80091-8] [PMID: 10074433]

[32]   Welcker, M.; Orian, A.; Jin, J.; Grim, J.E.; Harper, J.W.; Eisenman, R.N.; Clurman, B.E. The Fbw7 tumor suppressor regulates glycogen synthase kinase 3 phosphorylation-dependent c-Myc protein degradation. *Proc. Natl. Acad. Sci. USA*, **2004**, *101*(24), 9085-9090.
[http://dx.doi.org/10.1073/pnas.0402770101] [PMID: 15150404]

[33]   Fuentealba, L.C.; Eivers, E.; Ikeda, A.; Hurtado, C.; Kuroda, H.; Pera, E.M.; De Robertis, E.M. Integrating patterning signals: Wnt/GSK3 regulates the duration of the BMP/Smad1 signal. *Cell*, **2007**, *131*(5), 980-993.
[http://dx.doi.org/10.1016/j.cell.2007.09.027] [PMID: 18045539]

[34]   Aragón, E.; Goerner, N.; Zaromytidou, A.I.; Xi, Q.; Escobedo, A.; Massagué, J.; Macias, M.J. A Smad action turnover switch operated by WW domain readers of a phosphoserine code. *Genes Dev.*, **2011**, *25*(12), 1275-1288.
[http://dx.doi.org/10.1101/gad.2060811] [PMID: 21685363]

[35]   Taelman, V.F.; Dobrowolski, R.; Plouhinec, J.L.; Fuentealba, L.C.; Vorwald, P.P.; Gumper, I.; Sabatini, D.D.; De Robertis, E.M. Wnt signaling requires sequestration of glycogen synthase kinase 3 inside multivesicular endosomes. *Cell*, **2010**, *143*(7), 1136-1148.
[http://dx.doi.org/10.1016/j.cell.2010.11.034] [PMID: 21183076]

[36]   Vinyoles, M.; Del Valle-Pérez, B.; Curto, J.; Viñas-Castells, R.; Alba-Castellón, L.; García de Herreros, A.; Duñach, M. Multivesicular GSK3 sequestration upon Wnt signaling is controlled by p120-catenin/cadherin interaction with LRP5/6. *Mol. Cell*, **2014**, *53*(3), 444-457.
[http://dx.doi.org/10.1016/j.molcel.2013.12.010] [PMID: 24412065]

[37]   Kim, H.; Vick, P.; Hedtke, J.; Ploper, D.; De Robertis, E.M. Wnt Signaling Translocates Lys48-Linked Polyubiquitinated Proteins to the Lysosomal Pathway. *Cell Rep.*, **2015**, *11*(8), 1151-1159.
[http://dx.doi.org/10.1016/j.celrep.2015.04.048] [PMID: 26004177]

[38]   Gordon, M.D.; Nusse, R. Wnt signaling: multiple pathways, multiple receptors, and multiple transcription factors. *J. Biol. Chem.*, **2006**, *281*(32), 22429-22433.
[http://dx.doi.org/10.1074/jbc.R600015200] [PMID: 16793760]

[39]   Lawrence, P.A.; Shelton, P.M.J. The determination of polarity in the developing insect retina. *J. Embryol. Exp. Morphol.*, **1975**, *33*(2), 471-486.
[PMID: 1176856]

[40]   Vinson, C.R.; Adler, P.N. Directional non-cell autonomy and the transmission of polarity information by the frizzled gene of Drosophila. *Nature,* **1987**, *329*(6139), 549-551.
[http://dx.doi.org/10.1038/329549a0] [PMID: 3116434]

[41]   Mlodzik, M. Planar cell polarization: do the same mechanisms regulate Drosophila tissue polarity and vertebrate gastrulation? *Trends Genet.,* **2002**, *18*(11), 564-571.
[http://dx.doi.org/10.1016/S0168-9525(02)02770-1] [PMID: 12414186]

[42]   Veeman, M.T.; Axelrod, J.D.; Moon, R.T. A second canon. Functions and mechanisms of β-cateni--independent Wnt signaling. *Dev. Cell,* **2003**, *5*(3), 367-377.
[http://dx.doi.org/10.1016/S1534-5807(03)00266-1] [PMID: 12967557]

[43]   Marlow, F.; Topczewski, J.; Sepich, D.; Solnica-Krezel, L. Zebrafish Rho kinase 2 acts downstream of Wnt11 to mediate cell polarity and effective convergence and extension movements. *Curr. Biol.,* **2002**, *12*(11), 876-884.
[http://dx.doi.org/10.1016/S0960-9822(02)00864-3] [PMID: 12062050]

[44]   Winter, C.G.; Wang, B.; Ballew, A.; Royou, A.; Karess, R.; Axelrod, J.D.; Luo, L. Drosophila Rho-associated kinase (Drok) links Frizzled-mediated planar cell polarity signaling to the actin cytoskeleton. *Cell,* **2001**, *105*(1), 81-91.
[http://dx.doi.org/10.1016/S0092-8674(01)00298-7] [PMID: 11301004]

[45]   Boutros, M.; Paricio, N.; Strutt, D.I.; Mlodzik, M. Dishevelled activates JNK and discriminates between JNK pathways in planar polarity and wingless signaling. *Cell,* **1998**, *94*(1), 109-118.
[http://dx.doi.org/10.1016/S0092-8674(00)81226-X] [PMID: 9674432]

[46]   Paricio, N.; Feiguin, F.; Boutros, M.; Eaton, S.; Mlodzik, M. The Drosophila STE20-like kinase misshapen is required downstream of the Frizzled receptor in planar polarity signaling. *EMBO J.,* **1999**, *18*(17), 4669-4678.
[http://dx.doi.org/10.1093/emboj/18.17.4669] [PMID: 10469646]

[47]   Weber, U.; Paricio, N.; Mlodzik, M. Jun mediates Frizzled-induced R3/R4 cell fate distinction and planar polarity determination in the Drosophila eye. *Development,* **2000**, *127*(16), 3619-3629.
[PMID: 10903185]

[48]   Simons, M.; Mlodzik, M. Planar cell polarity signaling: from fly development to human disease. *Annu. Rev. Genet.,* **2008**, *42*(1), 517-540.
[http://dx.doi.org/10.1146/annurev.genet.42.110807.091432] [PMID: 18710302]

[49]   Weber, U.; Pataki, C.; Mihaly, J.; Mlodzik, M. Combinatorial signaling by the Frizzled/PCP and Egfr pathways during planar cell polarity establishment in the Drosophila eye. *Dev. Biol.,* **2008**, *316*(1), 110-123.
[http://dx.doi.org/10.1016/j.ydbio.2008.01.016] [PMID: 18291359]

[50]   Vladar, E.K.; Antic, D.; Axelrod, J.D. Planar cell polarity signaling: the developing cell's compass. *Cold Spring Harb. Perspect. Biol.,* **2009**, *1*(3)a002964
[http://dx.doi.org/10.1101/cshperspect.a002964] [PMID: 20066108]

[51]   Bayly, R.; Axelrod, J.D. Pointing in the right direction: new developments in the field of planar cell polarity. *Nat. Rev. Genet.,* **2011**, *12*(6), 385-391.
[http://dx.doi.org/10.1038/nrg2956] [PMID: 21502960]

[52]   Borg, J.-P. *Deregulation of the non-canonical pathway in triple-negative breast cancer;* , **2013**.

[53]   Slusarski, D.C.; Yang-Snyder, J.; Busa, W.B.; Moon, R.T. Modulation of embryonic intracellular Ca2+ signaling by Wnt-5A. *Dev. Biol.,* **1997**, *182*(1), 114-120.
[http://dx.doi.org/10.1006/dbio.1996.8463] [PMID: 9073455]

[54]   Kohn, A. D.; Moon, R. T. *Wnt and calcium signaling: β-Catenin-independent pathways,* **2005**.
[http://dx.doi.org/10.1016/j.ceca.2005.06.022]

[55]   Takei, K.; Shin, R. M.; Inoue, T.; Kato, K.; Mikoshiba, K. *Regulation of nerve growth mediated by*

inositol 1,4,5-trisphosphate receptors in growth cones, **1998**.
[http://dx.doi.org/10.1126/science.282.5394.1705]

[56]   Slusarski, D.C.; Corces, V.G.; Moon, R.T. Interaction of Wnt and a Frizzled homologue triggers G-protein-linked phosphatidylinositol signalling. *Nature,* **1997**, *390*(6658), 410-413.
[http://dx.doi.org/10.1038/37138] [PMID: 9389482]

[57]   Kühl, M.; Sheldahl, L.C.; Malbon, C.C.; Moon, R.T. Ca(2+)/calmodulin-dependent protein kinase II is stimulated by Wnt and Frizzled homologs and promotes ventral cell fates in Xenopus. *J. Biol. Chem.,* **2000**, *275*(17), 12701-12711.
[http://dx.doi.org/10.1074/jbc.275.17.12701] [PMID: 10777564]

[58]   Sheldahl, L.C.; Park, M.; Malbon, C.C.; Moon, R.T. Protein kinase C is differentially stimulated by Wnt and Frizzled homologs in a G-protein-dependent manner. *Curr. Biol.,* **1999**, *9*(13), 695-698.
[http://dx.doi.org/10.1016/S0960-9822(99)80310-8] [PMID: 10395542]

[59]   Inestrosa, N.C.; Montecinos-Oliva, C.; Fuenzalida, M. Wnt signaling: role in Alzheimer disease and schizophrenia. *J. Neuroimmune Pharmacol.,* **2012**, *7*(4), 788-807.
[http://dx.doi.org/10.1007/s11481-012-9417-5] [PMID: 23160851]

[60]   Nusse, R.; Varmus, H.E. Many tumors induced by the mouse mammary tumor virus contain a provirus integrated in the same region of the host genome. *Cell,* **1982**, *31*(1), 99-109.
[http://dx.doi.org/10.1016/0092-8674(82)90409-3] [PMID: 6297757]

[61]   Cadigan, K.M.; Peifer, M. Wnt signaling from development to disease: insights from model systems. *Cold Spring Harb. Perspect. Biol.,* **2009**, *1*(2)a002881
[http://dx.doi.org/10.1101/cshperspect.a002881] [PMID: 20066091]

[62]   Steinhart, Z.; Angers, S. Wnt signaling in development and tissue homeostasis. *Development,* **2018**, *145*(11)dev146589
[http://dx.doi.org/10.1242/dev.146589] [PMID: 29884654]

[63]   Logan, C.Y.; Nusse, R. The Wnt signaling pathway in development and disease. *Annu. Rev. Cell Dev. Biol.,* **2004**, *20*(1), 781-810.
[http://dx.doi.org/10.1146/annurev.cellbio.20.010403.113126] [PMID: 15473860]

[64]   Pfister, A.S.; Kühl, M. Of Wnts and Ribosomes. *Prog. Mol. Biol. Transl. Sci.,* **2018**, *153*, 131-155.
[http://dx.doi.org/10.1016/bs.pmbts.2017.11.006] [PMID: 29389514]

[65]   Kusserow, A.; Pang, K.; Sturm, C.; Hrouda, M.; Lentfer, J.; Schmidt, H.A.; Technau, U.; von Haeseler, A.; Hobmayer, B.; Martindale, M.Q.; Holstein, T.W. Unexpected complexity of the Wnt gene family in a sea anemone. *Nature,* **2005**, *433*(7022), 156-160.
[http://dx.doi.org/10.1038/nature03158] [PMID: 15650739]

[66]   Dickinson, M.E.; McMahon, A.P. The role of Wnt genes in vertebrate development. *Curr. Opin. Genet. Dev.,* **1992**, *2*(4), 562-566.
[http://dx.doi.org/10.1016/S0959-437X(05)80172-8] [PMID: 1388080]

[67]   Liu, W.; Dong, X.; Mai, M.; Seelan, R.S.; Taniguchi, K.; Krishnadath, K.K.; Halling, K.C.; Cunningham, J.M.; Boardman, L.A.; Qian, C.; Christensen, E.; Schmidt, S.S.; Roche, P.C.; Smith, D.I.; Thibodeau, S.N. Mutations in AXIN2 cause colorectal cancer with defective mismatch repair by activating β-catenin/TCF signalling. *Nat. Genet.,* **2000**, *26*(2), 146-147.
[http://dx.doi.org/10.1038/79859] [PMID: 11017067]

[68]   Lammi, L.; Arte, S.; Somer, M.; Jarvinen, H.; Lahermo, P.; Thesleff, I.; Pirinen, S.; Nieminen, P. Mutations in AXIN2 cause familial tooth agenesis and predispose to colorectal cancer. *Am. J. Hum. Genet.,* **2004**, *74*(5), 1043-1050.
[http://dx.doi.org/10.1086/386293] [PMID: 15042511]

[69]   Zimmerman, Z.F.; Moon, R.T.; Chien, A.J. Targeting Wnt pathways in disease. *Cold Spring Harb. Perspect. Biol.,* **2012**, *4*(11)a008086
[http://dx.doi.org/10.1101/cshperspect.a008086] [PMID: 23001988]

[70]   Kinzler, K. W. *Identification of FAP locus genes from chromosome 5q21,* **1991**.
        [http://dx.doi.org/10.1126/science.1651562]

[71]   Nishisho, I. *Mutations of chromosome 5q21 genes in FAP and colorectal cancer patients,* **1991**.
        [http://dx.doi.org/10.1126/science.1651563]

[72]   Su, L. K.; Vogelstein, B.; Kinzler, K. W. *Association of the APC tumor suppressor protein with catenins,* **1993**.
        [http://dx.doi.org/10.1126/science.8259519]

[73]   Chapman, A.; Durand, J.; Ouadi, L.; Bourdeau, I. Identification of genetic alterations of AXIN2 gene in adrenocortical tumors. *J. Clin. Endocrinol. Metab.,* **2011**, *96*(9), E1477-E1481.
        [http://dx.doi.org/10.1210/jc.2010-2987] [PMID: 21733995]

[74]   Thorvaldsen, T.E.; Pedersen, N.M.; Wenzel, E.M.; Stenmark, H. Differential roles of AXIN1 and AXIN2 in tankyrase inhibitor-induced formation of degradasomes and β-catenin degradation. *PLoS One,* **2017**, *12*(1)e0170508
        [http://dx.doi.org/10.1371/journal.pone.0170508] [PMID: 28107521]

[75]   Zeng, L.; Fagotto, F.; Zhang, T.; Hsu, W.; Vasicek, T.J.; Perry, W.L., III; Lee, J.J.; Tilghman, S.M.; Gumbiner, B.M.; Costantini, F. The mouse Fused locus encodes Axin, an inhibitor of the Wnt signaling pathway that regulates embryonic axis formation. *Cell,* **1997**, *90*(1), 181-192.
        [http://dx.doi.org/10.1016/S0092-8674(00)80324-4] [PMID: 9230313]

[76]   Carter, M.; Chen, X.; Slowinska, B.; Minnerath, S.; Glickstein, S.; Shi, L.; Campagne, F.; Weinstein, H.; Ross, M.E. Crooked tail (Cd) model of human folate-responsive neural tube defects is mutated in Wnt coreceptor lipoprotein receptor-related protein 6. *Proc. Natl. Acad. Sci. USA,* **2005**, *102*(36), 12843-12848.
        [http://dx.doi.org/10.1073/pnas.0501963102] [PMID: 16126904]

[77]   Kokubu, C.; Heinzmann, U.; Kokubu, T.; Sakai, N.; Kubota, T.; Kawai, M.; Wahl, M.B.; Galceran, J.; Grosschedl, R.; Ozono, K.; Imai, K. Skeletal defects in ringelschwanz mutant mice reveal that Lrp6 is required for proper somitogenesis and osteogenesis. *Development,* **2004**, *131*(21), 5469-5480.
        [http://dx.doi.org/10.1242/dev.01405] [PMID: 15469977]

[78]   Pinson, K.I.; Brennan, J.; Monkley, S.; Avery, B.J.; Skarnes, W.C. An LDL-receptor-related protein mediates Wnt signalling in mice. *Nature,* **2000**, *407*(6803), 535-538.
        [http://dx.doi.org/10.1038/35035124] [PMID: 11029008]

[79]   Freese, J.L.; Pino, D.; Pleasure, S.J. Wnt signaling in development and disease. *Neurobiol. Dis.,* **2010**, *38*(2), 148-153.
        [http://dx.doi.org/10.1016/j.nbd.2009.09.003] [PMID: 19765659]

[80]   Zhan, T.; Rindtorff, N.; Boutros, M. Wnt signaling in cancer. *Oncogene,* **2017**, *36*(11), 1461-1473.
        [http://dx.doi.org/10.1038/onc.2016.304] [PMID: 27617575]

[81]   Enomoto, M.; Hayakawa, S.; Itsukushima, S.; Ren, D.Y.; Matsuo, M.; Tamada, K.; Oneyama, C.; Okada, M.; Takumi, T.; Nishita, M.; Minami, Y. Autonomous regulation of osteosarcoma cell invasiveness by Wnt5a/Ror2 signaling. *Oncogene,* **2009**, *28*(36), 3197-3208.
        [http://dx.doi.org/10.1038/onc.2009.175] [PMID: 19561643]

[82]   Asem, M.S.; Buechler, S.; Wates, R.B.; Miller, D.L.; Stack, M.S. Wnt5a signaling in cancer. *Cancers (Basel),* **2016**, *8*(9)E79
        [http://dx.doi.org/10.3390/cancers8090079] [PMID: 27571105]

[83]   Ng, L.F.; Kaur, P.; Bunnag, N.; Suresh, J.; Sung, I.C.H.; Tan, Q.H.; Gruber, J.; Tolwinski, N.S. WNT Signaling in Disease. *Cells,* **2019**, *8*(8), 826.
        [http://dx.doi.org/10.3390/cells8080826] [PMID: 31382613]

[84]   Caldwell, G.M.; Jones, C.; Gensberg, K.; Jan, S.; Hardy, R.G.; Byrd, P.; Chughtai, S.; Wallis, Y.; Matthews, G.M.; Morton, D.G. The Wnt antagonist sFRP1 in colorectal tumorigenesis. *Cancer Res.,* **2004**, *64*(3), 883-888.

[http://dx.doi.org/10.1158/0008-5472.CAN-03-1346] [PMID: 14871816]

[85]   Lee, A.Y.; He, B.; You, L.; Dadfarmay, S.; Xu, Z.; Mazieres, J.; Mikami, I.; McCormick, F.; Jablons, D.M. Expression of the secreted frizzled-related protein gene family is downregulated in human mesothelioma. *Oncogene,* **2004**, *23*(39), 6672-6676.
[http://dx.doi.org/10.1038/sj.onc.1207881] [PMID: 15221014]

[86]   Suzuki, H.; Watkins, D.N.; Jair, K.W.; Schuebel, K.E.; Markowitz, S.D.; Chen, W.D.; Pretlow, T.P.; Yang, B.; Akiyama, Y.; Van Engeland, M.; Toyota, M.; Tokino, T.; Hinoda, Y.; Imai, K.; Herman, J.G.; Baylin, S.B. Epigenetic inactivation of SFRP genes allows constitutive WNT signaling in colorectal cancer. *Nat. Genet.,* **2004**, *36*(4), 417-422.
[http://dx.doi.org/10.1038/ng1330] [PMID: 15034581]

[87]   Fukui, T.; Kondo, M.; Ito, G.; Maeda, O.; Sato, N.; Yoshioka, H.; Yokoi, K.; Ueda, Y.; Shimokata, K.; Sekido, Y. Transcriptional silencing of secreted frizzled related protein 1 (SFRP 1) by promoter hypermethylation in non-small-cell lung cancer. *Oncogene,* **2005**, *24*(41), 6323-6327.
[http://dx.doi.org/10.1038/sj.onc.1208777] [PMID: 16007200]

[88]   Zou, H.; Molina, J.R.; Harrington, J.J.; Osborn, N.K.; Klatt, K.K.; Romero, Y.; Burgart, L.J.; Ahlquist, D.A. Aberrant methylation of secreted frizzled-related protein genes in esophageal adenocarcinoma and Barrett's esophagus. *Int. J. Cancer,* **2005**, *116*(4), 584-591.
[http://dx.doi.org/10.1002/ijc.21045] [PMID: 15825175]

[89]   Serrano-Pozo, A.; Frosch, M.P.; Masliah, E.; Hyman, B.T. Neuropathological alterations in Alzheimer disease. *Cold Spring Harb. Perspect. Med.,* **2011**, *1*(1)a006189
[http://dx.doi.org/10.1101/cshperspect.a006189] [PMID: 22229116]

[90]   Heneka, M.T.; Carson, M.J.; El Khoury, J.; Landreth, G.E.; Brosseron, F.; Feinstein, D.L.; Jacobs, A.H.; Wyss-Coray, T.; Vitorica, J.; Ransohoff, R.M.; Herrup, K.; Frautschy, S.A.; Finsen, B.; Brown, G.C.; Verkhratsky, A.; Yamanaka, K.; Koistinaho, J.; Latz, E.; Halle, A.; Petzold, G.C.; Town, T.; Morgan, D.; Shinohara, M.L.; Perry, V.H.; Holmes, C.; Bazan, N.G.; Brooks, D.J.; Hunot, S.; Joseph, B.; Deigendesch, N.; Garaschuk, O.; Boddeke, E.; Dinarello, C.A.; Breitner, J.C.; Cole, G.M.; Golenbock, D.T.; Kummer, M.P. Neuroinflammation in Alzheimer's disease. *Lancet Neurol.,* **2015**, *14*(4), 388-405.
[http://dx.doi.org/10.1016/S1474-4422(15)70016-5] [PMID: 25792098]

[91]   De Ferrari, G.V.; Avila, M.E.; Medina, M.A.; Perez-Palma, E.; Bustos, B.I.; Alarcon, M.A. Wnt/β-catenin signaling in Alzheimer's disease. *CNS Neurol. Disord. Drug Targets,* **2014**, *13*(5), 745-754.
[http://dx.doi.org/10.2174/1871527312666131223113900] [PMID: 24365184]

[92]   De Ferrari, G.V.; Papassotiropoulos, A.; Biechele, T.; Wavrant De-Vrieze, F.; Avila, M.E.; Major, M.B.; Myers, A.; Sáez, K.; Henríquez, J.P.; Zhao, A.; Wollmer, M.A.; Nitsch, R.M.; Hock, C.; Morris, C.M.; Hardy, J.; Moon, R.T. Common genetic variation within the low-density lipoprotein receptor-related protein 6 and late-onset Alzheimer's disease. *Proc. Natl. Acad. Sci. USA,* **2007**, *104*(22), 9434-9439.
[http://dx.doi.org/10.1073/pnas.0603523104] [PMID: 17517621]

[93]   van Gijn, M.E.; Daemen, M.J.A.P.; Smits, J.F.M.; Blankesteijn, W.M. The wnt-frizzled cascade in cardiovascular disease. *Cardiovasc. Res.,* **2002**, *55*(1), 16-24.
[http://dx.doi.org/10.1016/S0008-6363(02)00221-3] [PMID: 12062705]

[94]   Olson, E.N.; Schneider, M.D. Sizing up the heart: development redux in disease. *Genes Dev.,* **2003**, *17*(16), 1937-1956.
[http://dx.doi.org/10.1101/gad.1110103] [PMID: 12893779]

[95]   Mani, A. *LRP6 mutation in a family with early coronary disease and metabolic risk factors,* **2007**.
[http://dx.doi.org/10.1126/science.1136370]

[96]   Ueland, T.; Otterdal, K.; Lekva, T.; Halvorsen, B.; Gabrielsen, A.; Sandberg, W.J.; Paulsson-Berne, G.; Pedersen, T.M.; Folkersen, L.; Gullestad, L.; Oie, E.; Hansson, G.K.; Aukrust, P. Dickkopf-1 enhances inflammatory interaction between platelets and endothelial cells and shows increased

expression in atherosclerosis. *Arterioscler. Thromb. Vasc. Biol.,* **2009**, *29*(8), 1228-1234.
[http://dx.doi.org/10.1161/ATVBAHA.109.189761] [PMID: 19498175]

[97]   Christman, M.A., II; Goetz, D.J.; Dickerson, E.; McCall, K.D.; Lewis, C.J.; Benencia, F.; Silver, M.J.;
       Kohn, L.D.; Malgor, R. Wnt5a is expressed in murine and human atherosclerotic lesions. *Am. J.
       Physiol. Heart Circ. Physiol.,* **2008**, *294*(6), H2864-H2870.
       [http://dx.doi.org/10.1152/ajpheart.00982.2007] [PMID: 18456733]

[98]   Barandon, L.; Couffinhal, T.; Ezan, J.; Dufourcq, P.; Costet, P.; Alzieu, P.; Leroux, L.; Moreau, C.;
       Dare, D.; Duplàa, C. Reduction of infarct size and prevention of cardiac rupture in transgenic mice
       overexpressing FrzA. *Circulation,* **2003**, *108*(18), 2282-2289.
       [http://dx.doi.org/10.1161/01.CIR.0000093186.22847.4C] [PMID: 14581414]

[99]   Aisagbonhi, O.; Rai, M.; Ryzhov, S.; Atria, N.; Feoktistov, I.; Hatzopoulos, A.K. Experimental
       myocardial infarction triggers canonical Wnt signaling and endothelial-to-mesenchymal transition.
       Dis. Model. Mech., **2011**, *4*(4), 469-483.
       [http://dx.doi.org/10.1242/dmm.006510] [PMID: 21324930]

[100] Paik, D.T.; Rai, M.; Ryzhov, S.; Sanders, L.N.; Aisagbonhi, O.; Funke, M.J.; Feoktistov, I.;
       Hatzopoulos, A.K. Wnt10b gain-of-function improves cardiac repair by arteriole formation and
       attenuation of fibrosis. *Circ. Res.,* **2015**, *117*(9), 804-816.
       [http://dx.doi.org/10.1161/CIRCRESAHA.115.306886] [PMID: 26338900]

[101] Morishita, Y.; Kobayashi, K.; Klyachko, E.; Jujo, K.; Maeda, K.; Losordo, D.W.; Murohara, T. Wnt11
       Gene Therapy with Adeno-associated Virus 9 Improves Recovery from Myocardial Infarction by
       Modulating the Inflammatory Response. *Sci. Rep.,* **2016**, *6*, 21705.
       [http://dx.doi.org/10.1038/srep21705] [PMID: 26882996]

[102] Belinson, H.; Nakatani, J.; Babineau, B.A.; Birnbaum, R.Y.; Ellegood, J.; Bershteyn, M.; McEvilly,
       R.J.; Long, J.M.; Willert, K.; Klein, O.D.; Ahituv, N.; Lerch, J.P.; Rosenfeld, M.G.; Wynshaw-Boris,
       A. Prenatal β-catenin/Brn2/Tbr2 transcriptional cascade regulates adult social and stereotypic
       behaviors. *Mol. Psychiatry,* **2016**, *21*(10), 1417-1433.
       [http://dx.doi.org/10.1038/mp.2015.207] [PMID: 26830142]

[103] Zhang, L.; Deng, J.; Pan, Q.; Zhan, Y.; Fan, J.B.; Zhang, K.; Zhang, Z. Targeted methylation
       sequencing reveals dysregulated Wnt signaling in Parkinson disease. *J. Genet. Genomics,* **2016**,
       43(10), 587-592.
       [http://dx.doi.org/10.1016/j.jgg.2016.05.002] [PMID: 27692691]

[104] McGrath, J.J.; Féron, F.P.; Burne, T.H.J.; Mackay-Sim, A.; Eyles, D.W. The neurodevelopmental
       hypothesis of schizophrenia: a review of recent developments. *Ann. Med.,* **2003**, *35*(2), 86-93.
       [http://dx.doi.org/10.1080/07853890310010005] [PMID: 12795338]

[105] Thorgeirsson, S.S. Hepatic stem cells in liver regeneration. *FASEB J.,* **1996**, *10*(11), 1249-1256.
       [http://dx.doi.org/10.1096/fasebj.10.11.8836038] [PMID: 8836038]

[106] Michalopoulos, G.K.; DeFrances, M.C. Liver regeneration. *Science,* **1997**, *276*(5309), 60-66.
       [http://dx.doi.org/10.1126/science.276.5309.60] [PMID: 9082986]

[107] Monga, S.P.S.; Pediaditakis, P.; Mule, K.; Stolz, D.B.; Michalopoulos, G.K. Changes in WNT/β-
       catenin pathway during regulated growth in rat liver regeneration. *Hepatology,* **2001**, *33*(5), 1098-
       1109.
       [http://dx.doi.org/10.1053/jhep.2001.23786] [PMID: 11343237]

[108] Nelsen, C.J.; Rickheim, D.G.; Timchenko, N.A.; Stanley, M.W.; Albrecht, J.H. Transient expression
       of cyclin D1 is sufficient to promote hepatocyte replication and liver growth in vivo. *Cancer Res.,*
       2001, *61*(23), 8564-8568.
       [PMID: 11731443]

[109] Tan, X.; Behari, J.; Cieply, B.; Michalopoulos, G.K.; Monga, S.P.S. Conditional deletion of β-catenin
       reveals its role in liver growth and regeneration. *Gastroenterology,* **2006**, *131*(5), 1561-1572.
       [http://dx.doi.org/10.1053/j.gastro.2006.08.042] [PMID: 17101329]

[110] Yang, J.; Mowry, L.E.; Nejak-Bowen, K.N.; Okabe, H.; Diegel, C.R.; Lang, R.A.; Williams, B.O.; Monga, S.P. β-catenin signaling in murine liver zonation and regeneration: a Wnt-Wnt situation! *Hepatology,* **2014**, *60*(3), 964-976.
[http://dx.doi.org/10.1002/hep.27082] [PMID: 24700412]

[111] Planas-Paz, L.; Orsini, V.; Boulter, L.; Calabrese, D.; Pikiolek, M.; Nigsch, F.; Xie, Y.; Roma, G.; Donovan, A.; Marti, P.; Beckmann, N.; Dill, M.T.; Carbone, W.; Bergling, S.; Isken, A.; Mueller, M.; Kinzel, B.; Yang, Y.; Mao, X.; Nicholson, T.B.; Zamponi, R.; Capodieci, P.; Valdez, R.; Rivera, D.; Loew, A.; Ukomadu, C.; Terracciano, L.M.; Bouwmeester, T.; Cong, F.; Heim, M.H.; Forbes, S.J.; Ruffner, H.; Tchorz, J.S. The RSPO-LGR4/5-ZNRF3/RNF43 module controls liver zonation and size. *Nat. Cell Biol.,* **2016**, *18*(5), 467-479.
[http://dx.doi.org/10.1038/ncb3337] [PMID: 27088858]

[112] Liu, Y.; El-Serag, H.B.; Jiao, L.; Lee, J.; Moore, D.; Franco, L.M.; Tavakoli-Tabasi, S.; Tsavachidis, S.; Kuzniarek, J.; Ramsey, D.J.; White, D.L. WNT signaling pathway gene polymorphisms and risk of hepatic fibrosis and inflammation in HCV-infected patients. *PLoS One,* **2013**, *8*(12)e84407
[http://dx.doi.org/10.1371/journal.pone.0084407] [PMID: 24386373]

[113] Go, G.W.; Srivastava, R.; Hernandez-Ono, A.; Gang, G.; Smith, S.B.; Booth, C.J.; Ginsberg, H.N.; Mani, A. The combined hyperlipidemia caused by impaired Wnt-LRP6 signaling is reversed by Wnt3a rescue. *Cell Metab.,* **2014**, *19*(2), 209-220.
[http://dx.doi.org/10.1016/j.cmet.2013.11.023] [PMID: 24506864]

[114] Queimado, L.; Lopes, C.S.; Reis, A.M. WIF1, an inhibitor of the Wnt pathway, is rearranged in salivary gland tumors. *Genes Chromosomes Cancer,* **2007**, *46*(3), 215-225.
[http://dx.doi.org/10.1002/gcc.20402] [PMID: 17171686]

[115] Foulkes, W.D. A tale of four syndromes: familial adenomatous polyposis, Gardner syndrome, attenuated APC and Turcot syndrome. *QJM,* **1995**, *88*(12), 853-863.
[http://dx.doi.org/10.1093/oxfordjournals.qjmed.a069018] [PMID: 8593545]

[116] Mostowska, A.; Biedziak, B.; Jagodzinski, P.P. Axis inhibition protein 2 (AXIN2) polymorphisms may be a risk factor for selective tooth agenesis. *J. Hum. Genet.,* **2006**, *51*(3), 262-266.
[http://dx.doi.org/10.1007/s10038-005-0353-6] [PMID: 16432638]

[117] Adaimy, L.; Chouery, E.; Megarbane, H.; Mroueh, S.; Delague, V.; Nicolas, E.; Belguith, H.; de Mazancourt, P.; Megarbane, A. Mutation in WNT10A is associated with an autosomal recessive ectodermal dysplasia: the odonto-onycho-dermal dysplasia. *Am. J. Hum. Genet.,* **2007**, *81*(4), 821-828.
[http://dx.doi.org/10.1086/520064] [PMID: 17847007]

[118] Bornholdt, D.; Oeffner, F.; König, A.; Happle, R.; Alanay, Y.; Ascherman, J.; Benke, P.J.; Boente, Mdel.C.; van der Burgt, I.; Chassaing, N.; Ellis, I.; Francisco, C.R.; Della Giovanna, P.; Hamel, B.; Has, C.; Heinelt, K.; Janecke, A.; Kastrup, W.; Loeys, B.; Lohrisch, I.; Marcelis, C.; Mehraein, Y.; Nicolas, M.E.; Pagliarini, D.; Paradisi, M.; Patrizi, A.; Piccione, M.; Piza-Katzer, H.; Prager, B.; Prescott, K.; Strien, J.; Utine, G.E.; Zeller, M.S.; Grzeschik, K.H. PORCN mutations in focal dermal hypoplasia: coping with lethality. *Hum. Mutat.,* **2009**, *30*(5), E618-E628.
[http://dx.doi.org/10.1002/humu.20992] [PMID: 19309688]

[119] Tamamura, Y.; Otani, T.; Kanatani, N.; Koyama, E.; Kitagaki, J.; Komori, T.; Yamada, Y.; Costantini, F.; Wakisaka, S.; Pacifici, M.; Iwamoto, M.; Enomoto-Iwamoto, M. Developmental regulation of Wnt/β-catenin signals is required for growth plate assembly, cartilage integrity, and endochondral ossification. *J. Biol. Chem.,* **2005**, *280*(19), 19185-19195.
[http://dx.doi.org/10.1074/jbc.M414275200] [PMID: 15760903]

[120] Pardo, A.; Cabrera, S.; Maldonado, M.; Selman, M. Role of matrix metalloproteinases in the pathogenesis of idiopathic pulmonary fibrosis. *Respir. Res.,* **2016**, *17*(1), 23.
[http://dx.doi.org/10.1186/s12931-016-0343-6] [PMID: 26944412]

[121] Gong, Y.; Slee, R.B.; Fukai, N.; Rawadi, G.; Roman-Roman, S.; Reginato, A.M.; Wang, H.; Cundy, T.; Glorieux, F.H.; Lev, D.; Zacharin, M.; Oexle, K.; Marcelino, J.; Suwairi, W.; Heeger, S.;

Sabatakos, G.; Apte, S.; Adkins, W.N.; Allgrove, J.; Arslan-Kirchner, M.; Batch, J.A.; Beighton, P.; Black, G.C.; Boles, R.G.; Boon, L.M.; Borrone, C.; Brunner, H.G.; Carle, G.F.; Dallapiccola, B.; De Paepe, A.; Floege, B.; Halfhide, M.L.; Hall, B.; Hennekam, R.C.; Hirose, T.; Jans, A.; Jüppner, H.; Kim, C.A.; Keppler-Noreuil, K.; Kohlschuetter, A.; LaCombe, D.; Lambert, M.; Lemyre, E.; Letteboer, T.; Peltonen, L.; Ramesar, R.S.; Romanengo, M.; Somer, H.; Steichen-Gersdorf, E.; Steinmann, B.; Sullivan, B.; Superti-Furga, A.; Swoboda, W.; van den Boogaard, M.J.; Van Hul, W.; Vikkula, M.; Votruba, M.; Zabel, B.; Garcia, T.; Baron, R.; Olsen, B.R.; Warman, M.L. LDL receptor-related protein 5 (LRP5) affects bone accrual and eye development. *Cell,* **2001**, *107*(4), 513-523. [http://dx.doi.org/10.1016/S0092-8674(01)00571-2] [PMID: 11719191]

[122] Fahiminiya, S.; Majewski, J.; Mort, J.; Moffatt, P.; Glorieux, F.H.; Rauch, F. Mutations in WNT1 are a cause of osteogenesis imperfecta. *J. Med. Genet.,* **2013**, *50*(5), 345-348. [http://dx.doi.org/10.1136/jmedgenet-2013-101567] [PMID: 23434763]

[123] Laine, C.M.; Joeng, K.S.; Campeau, P.M.; Kiviranta, R.; Tarkkonen, K.; Grover, M.; Lu, J.T.; Pekkinen, M.; Wessman, M.; Heino, T.J.; Nieminen-Pihala, V.; Aronen, M.; Laine, T.; Kröger, H.; Cole, W.G.; Lehesjoki, A.E.; Nevarez, L.; Krakow, D.; Curry, C.J.; Cohn, D.H.; Gibbs, R.A.; Lee, B.H.; Mäkitie, O. WNT1 mutations in early-onset osteoporosis and osteogenesis imperfecta. *N. Engl. J. Med.,* **2013**, *368*(19), 1809-1816. [http://dx.doi.org/10.1056/NEJMoa1215458] [PMID: 23656646]

[124] Kitano, H. Systems biology: a brief overview. *Science,* **2002**, *295*(5560), 1662-1664. [http://dx.doi.org/10.1126/science.1069492] [PMID: 11872829]

[125] Kitano, H. Computational systems biology. *Nature,* **2002**, *420*(6912), 206-210. [http://dx.doi.org/10.1038/nature01254] [PMID: 12432404]

[126] Ideker, T.; Galitski, T.; Hood, L. A new approach to decoding life: systems biology. *Annu. Rev. Genomics Hum. Genet.,* **2001**, *2*, 343-372. [http://dx.doi.org/10.1146/annurev.genom.2.1.343] [PMID: 11701654]

[127] Kitano, H. Perspectives on systems biology. *New Gener. Comput.,* **2000**, *18*(3), 199-216. [http://dx.doi.org/10.1007/BF03037529]

[128] Aditya Rao, S.J.; Ramesh, C.K.; Raghavendra, S.; Paramesha, M. Dehydroabietylamine, a diterpene from Carthamus tinctorious L. showing antibacterial and anthelmintic effects with computational evidence. *Curr Comput Aided Drug Des,* **2019**, *15* [http://dx.doi.org/10.2174/1573409915666190301142811] [PMID: 30827256]

[129] Aoki, K.; Yamada, M.; Kunida, K.; Yasuda, S.; Matsuda, M. Processive phosphorylation of ERK MAP kinase in mammalian cells. *Proc. Natl. Acad. Sci. USA,* **2011**, *108*(31), 12675-12680. [http://dx.doi.org/10.1073/pnas.1104030108] [PMID: 21768338]

[130] Raghavendra, S.; Aditya Rao, S.J.; Kumar, V.; Ramesh, C.K. Multiple ligand simultaneous docking (MLSD): A novel approach to study the effect of inhibitors on substrate binding to PPO. *Comput. Biol. Chem.,* **2015**, *59*(Pt A), 81-86. [http://dx.doi.org/10.1016/j.compbiolchem.2015.09.008] [PMID: 26414950]

[131] Ghosh, S.; Matsuoka, Y.; Asai, Y.; Hsin, K.Y.; Kitano, H. Software for systems biology: from tools to integrated platforms. *Nat. Rev. Genet.,* **2011**, *12*(12), 821-832. [http://dx.doi.org/10.1038/nrg3096] [PMID: 22048662]

[132] MacBeath, G. Protein microarrays and proteomics. *Nat. Genet.,* **2002**, *32*(4S) Suppl., 526-532. [http://dx.doi.org/10.1038/ng1037] [PMID: 12454649]

[133] Bensimon, A.; Heck, A.J.R.; Aebersold, R. Mass spectrometry-based proteomics and network biology. *Annu. Rev. Biochem.,* **2012**, *81*(1), 379-405. [http://dx.doi.org/10.1146/annurev-biochem-072909-100424] [PMID: 22439968]

[134] Sabidó, E.; Selevsek, N.; Aebersold, R. Mass spectrometry-based proteomics for systems biology. *Curr. Opin. Biotechnol.,* **2012**, *23*(4), 591-597. [http://dx.doi.org/10.1016/j.copbio.2011.11.014] [PMID: 22169889]

[135] Goh, W.W.; Wong, L. Networks in proteomics analysis of cancer. *Curr. Opin. Biotechnol.,* **2013**, *24*(6), 1122-1128.
[http://dx.doi.org/10.1016/j.copbio.2013.02.011] [PMID: 23481377]

[136] Wang, K.; Li, M.; Bucan, M. Pathway-based approaches for analysis of genomewide association studies. *Am. J. Hum. Genet.,* **2007**, *81*(6), 1278-1283.
[http://dx.doi.org/10.1086/522374] [PMID: 17966091]

[137] Curtis, R.K.; Orešič, M.; Vidal-Puig, A. Pathways to the analysis of microarray data. *Trends Biotechnol.,* **2005**, *23*(8), 429-435.
[http://dx.doi.org/10.1016/j.tibtech.2005.05.011] [PMID: 15950303]

[138] Tilford, C.A.; Siemers, N.O. Gene set enrichment analysis. *Methods Mol. Biol.,* **2009**, *563*, 99-121.
[http://dx.doi.org/10.1007/978-1-60761-175-2_6] [PMID: 19597782]

[139] Subramanian, A.; Tamayo, P.; Mootha, V.K.; Mukherjee, S.; Ebert, B.L.; Gillette, M.A.; Paulovich, A.; Pomeroy, S.L.; Golub, T.R.; Lander, E.S.; Mesirov, J.P. Gene set enrichment analysis: a knowledge-based approach for interpreting genome-wide expression profiles. *Proc. Natl. Acad. Sci. USA,* **2005**, *102*(43), 15545-15550.
[http://dx.doi.org/10.1073/pnas.0506580102] [PMID: 16199517]

[140] Jia, P.; Wang, L.; Meltzer, H.Y.; Zhao, Z. Common variants conferring risk of schizophrenia: a pathway analysis of GWAS data. *Schizophr. Res.,* **2010**, *122*(1-3), 38-42.
[http://dx.doi.org/10.1016/j.schres.2010.07.001] [PMID: 20659789]

[141] Yang, W.; de las Fuentes, L.; Dávila-Román, V.G.; Charles Gu, C. Variable set enrichment analysis in genome-wide association studies. *Eur. J. Hum. Genet.,* **2011**, *19*(8), 893-900.
[http://dx.doi.org/10.1038/ejhg.2011.46] [PMID: 21427759]

[142] Luo, L.; Peng, G.; Zhu, Y.; Dong, H.; Amos, C.I.; Xiong, M. Genome-wide gene and pathway analysis. *Eur. J. Hum. Genet.,* **2010**, *18*(9), 1045-1053.
[http://dx.doi.org/10.1038/ejhg.2010.62] [PMID: 20442747]

[143] Chen, X.; Wang, L.; Hu, B.; Guo, M.; Barnard, J.; Zhu, X. Pathway-based analysis for genome-wide association studies using supervised principal components. *Genet. Epidemiol.,* **2010**, *34*(7), 716-724.
[http://dx.doi.org/10.1002/gepi.20532] [PMID: 20842628]

[144] Jin, L.; Zuo, X.Y.; Su, W.Y.; Zhao, X.L.; Yuan, M.Q.; Han, L.Z.; Zhao, X.; Chen, Y.D.; Rao, S.Q. Pathway-based analysis tools for complex diseases: a review. *Genomics Proteomics Bioinformatics,* **2014**, *12*(5), 210-220.
[http://dx.doi.org/10.1016/j.gpb.2014.10.002] [PMID: 25462153]

[145] Draghici, S.; Khatri, P.; Tarca, A.L.; Amin, K.; Done, A.; Voichita, C.; Georgescu, C.; Romero, R. A systems biology approach for pathway level analysis. *Genome Res.,* **2007**, *17*(10), 1537-1545.
[http://dx.doi.org/10.1101/gr.6202607] [PMID: 17785539]

[146] Martini, P.; Sales, G.; Massa, M.S.; Chiogna, M.; Romualdi, C. Along signal paths: an empirical gene set approach exploiting pathway topology. *Nucleic Acids Res.,* **2013**, *41*(1)e19
[http://dx.doi.org/10.1093/nar/gks866] [PMID: 23002139]

[147] Khatri, P.; Sirota, M.; Butte, A.J. Ten years of pathway analysis: current approaches and outstanding challenges. *PLOS Comput. Biol.,* **2012**, *8*(2)e1002375
[http://dx.doi.org/10.1371/journal.pcbi.1002375] [PMID: 22383865]

[148] Wilke, R.A.; Mareedu, R.K.; Moore, J.H. The Pathway Less Traveled: Moving from Candidate Genes to Candidate Pathways in the Analysis of Genome-Wide Data from Large Scale Pharmacogenetic Association Studies. *Curr. Pharmacogenomics Person. Med.,* **2008**, *6*(3), 150-159.
[http://dx.doi.org/10.2174/1875692110806030150] [PMID: 19421424]

[149] Giacomelli, L.; Covani, U. Bioinformatics and data mining studies in oral genomics and proteomics: new trends and challenges. *Open Dent. J.,* **2010**, *4*(1), 67-71.
[http://dx.doi.org/10.2174/1874210601004010067] [PMID: 20871759]

[150] Elbers, C.C.; van Eijk, K.R.; Franke, L.; Mulder, F.; van der Schouw, Y.T.; Wijmenga, C.; Onland-Moret, N.C. Using genome-wide pathway analysis to unravel the etiology of complex diseases. *Genet. Epidemiol.,* **2009**, *33*(5), 419-431.
[http://dx.doi.org/10.1002/gepi.20395] [PMID: 19235186]

[151] Lesnick, T. G. *A genomic pathway approach to a complex disease: Axon guidance and Parkinson disease,* **2007**.
[http://dx.doi.org/10.1371/journal.pgen.0030098]

[152] Wu, D.; Hu, D.; Chen, H.; Shi, G.; Fetahu, I.S.; Wu, F.; Rabidou, K.; Fang, R.; Tan, L.; Xu, S.; Liu, H.; Argueta, C.; Zhang, L.; Mao, F.; Yan, G.; Chen, J.; Dong, Z.; Lv, R.; Xu, Y.; Wang, M.; Ye, Y.; Zhang, S.; Duquette, D.; Geng, S.; Yin, C.; Lian, C.G.; Murphy, G.F.; Adler, G.K.; Garg, R.; Lynch, L.; Yang, P.; Li, Y.; Lan, F.; Fan, J.; Shi, Y.; Shi, Y.G. Glucose-regulated phosphorylation of TET2 by AMPK reveals a pathway linking diabetes to cancer. *Nature,* **2018**, *559*(7715), 637-641.
[http://dx.doi.org/10.1038/s41586-018-0350-5] [PMID: 30022161]

[153] Peng, G.; Luo, L.; Siu, H.; Zhu, Y.; Hu, P.; Hong, S.; Zhao, J.; Zhou, X.; Reveille, J.D.; Jin, L.; Amos, C.I.; Xiong, M. Gene and pathway-based second-wave analysis of genome-wide association studies. *Eur. J. Hum. Genet.,* **2010**, *18*(1), 111-117.
[http://dx.doi.org/10.1038/ejhg.2009.115] [PMID: 19584899]

[154] Ballard, D.; Abraham, C.; Cho, J.; Zhao, H. Pathway analysis comparison using Crohn's disease genome wide association studies. *BMC Med. Genomics,* **2010**, *3*, 25.
[http://dx.doi.org/10.1186/1755-8794-3-25] [PMID: 20584322]

[155] Thomas, D. Gene--environment-wide association studies: emerging approaches. *Nat. Rev. Genet.,* **2010**, *11*(4), 259-272.
[http://dx.doi.org/10.1038/nrg2764] [PMID: 20212493]

[156] Wu, M.C.; Lin, X. Prior biological knowledge-based approaches for the analysis of genome-wide expression profiles using gene sets and pathways. *Stat. Methods Med. Res.,* **2009**, *18*(6), 577-593.
[http://dx.doi.org/10.1177/0962280209351925] [PMID: 20048386]

[157] Vitek, O. Getting started in computational mass spectrometry-based proteomics. *PLOS Comput. Biol.,* **2009**, *5*(5)e1000366
[http://dx.doi.org/10.1371/journal.pcbi.1000366] [PMID: 19492072]

[158] Noble, W.S.; MacCoss, M.J. Computational and statistical analysis of protein mass spectrometry data. *PLOS Comput. Biol.,* **2012**, *8*(1)e1002296
[http://dx.doi.org/10.1371/journal.pcbi.1002296] [PMID: 22291580]

[159] Barla, A.; Jurman, G.; Riccadonna, S.; Merler, S.; Chierici, M.; Furlanello, C. Machine learning methods for predictive proteomics. *Brief. Bioinform.,* **2008**, *9*(2), 119-128.
[http://dx.doi.org/10.1093/bib/bbn008] [PMID: 18310105]

[160] Elias, J.E.; Gibbons, F.D.; King, O.D.; Roth, F.P.; Gygi, S.P. Intensity-based protein identification by machine learning from a library of tandem mass spectra. *Nat. Biotechnol.,* **2004**, *22*(2), 214-219.
[http://dx.doi.org/10.1038/nbt930] [PMID: 14730315]

[161] Krogan, N.J.; Cagney, G.; Yu, H. Global landscape of protein complexes in the yeast Saccharomyces cerevisiae. *Nature,* **2006**, *440*(7084), 637-643.
[http://dx.doi.org/10.1038/nature04670] [PMID: 16554755]

[162] Ressom, H.W.; Varghese, R.S.; Drake, S.K.; Hortin, G.L.; Abdel-Hamid, M.; Loffredo, C.A.; Goldman, R. Peak selection from MALDI-TOF mass spectra using ant colony optimization. *Bioinformatics,* **2007**, *23*(5), 619-626.
[http://dx.doi.org/10.1093/bioinformatics/btl678] [PMID: 17237065]

[163] Käll, L.; Canterbury, J.D.; Weston, J.; Noble, W.S.; MacCoss, M.J. Semi-supervised learning for peptide identification from shotgun proteomics datasets. *Nat. Methods,* **2007**, *4*(11), 923-925.
[http://dx.doi.org/10.1038/nmeth1113] [PMID: 17952086]

[164] Wu, X.; Hasan, M.A.; Chen, J.Y. Pathway and network analysis in proteomics. *J. Theor. Biol.,* **2014**, *362*, 44-52.
[http://dx.doi.org/10.1016/j.jtbi.2014.05.031] [PMID: 24911777]

[165] Wu, X.; Chen, J.Y. Molecular Interaction Networks: Topological and Functional Characterizations, **2009**.
[http://dx.doi.org/10.1002/9780470741191.ch6]

[166] Ramanan, V.K.; Shen, L.; Moore, J.H.; Saykin, A.J. Pathway analysis of genomic data: concepts, methods, and prospects for future development. *Trends Genet.,* **2012**, *28*(7), 323-332.
[http://dx.doi.org/10.1016/j.tig.2012.03.004] [PMID: 22480918]

[167] Goh, W.W.B.; Lee, Y.H.; Chung, M.; Wong, L. How advancement in biological network analysis methods empowers proteomics. *Proteomics,* **2012**, *12*(4-5), 550-563.
[http://dx.doi.org/10.1002/pmic.201100321] [PMID: 22247042]

[168] Chatr-Aryamontri, A.; Breitkreutz, B.J.; Heinicke, S.; Boucher, L.; Winter, A.; Stark, C.; Nixon, J.; Ramage, L.; Kolas, N.; O'Donnell, L.; Reguly, T.; Breitkreutz, A.; Sellam, A.; Chen, D.; Chang, C.; Rust, J.; Livstone, M.; Oughtred, R.; Dolinski, K.; Tyers, M. The BioGRID interaction database: 2013 update. *Nucleic Acids Res.,* **2013**, *41*(Database issue), D816-D823.
[http://dx.doi.org/10.1093/nar/gks1158] [PMID: 23203989]

[169] Franceschini, A.; Szklarczyk, D.; Frankild, S.; Kuhn, M.; Simonovic, M.; Roth, A.; Lin, J.; Minguez, P.; Bork, P.; von Mering, C.; Jensen, L.J. STRING v9.1: protein-protein interaction networks, with increased coverage and integration. *Nucleic Acids Res.,* **2013**, *41*(Database issue), D808-D815.
[http://dx.doi.org/10.1093/nar/gks1094] [PMID: 23203871]

[170] Kanehisa, M.; Goto, S. KEGG: kyoto encyclopedia of genes and genomes. *Nucleic Acids Res.,* **2000**, *28*(1), 27-30.
[http://dx.doi.org/10.1093/nar/28.1.27] [PMID: 10592173]

[171] Matthews, L. *Reactome knowledgebase of human biological pathways and processes,* **2009**.
[http://dx.doi.org/10.1093/nar/gkn863]

[172] Nishimura, D. BioCarta. *Biotech Softw. Internet Rep.,* **2001**, *2*(3), 117-120.
[http://dx.doi.org/10.1089/152791601750294344]

[173] Schaefer, C. F. *PID: The pathway interaction database,* **2009**.
[http://dx.doi.org/10.1093/nar/gkn653]

[174] Chen, J.Y.; Mamidipalli, S.R.; Huan, T. HAPPI: An online database of comprehensive human annotated and predicted protein interactions, **2009**.
[http://dx.doi.org/10.1186/1471-2164-10-S1-S16]

[175] Chowbina, S. R. *HPD: An online integrated human pathway database enabling systems biology studies,* **2009**.
[http://dx.doi.org/10.1186/1471-2105-10-S11-S5]

[176] Huang, H. PAGED: A pathway and gene-set enrichment database to enable molecular phenotype discoveries. *BMC Cancer,* **2016**, *16*(1)
[http://dx.doi.org/10.1186/1471-2105-13-S15-S2] [PMID: 23046413]

[177] Huang, W.; Sherman, B.T.; Lempicki, R.A. Systematic and integrative analysis of large gene lists using DAVID bioinformatics resources. *Nat. Protoc.,* **2009**, *4*(1), 44-57.
[http://dx.doi.org/10.1038/nprot.2008.211] [PMID: 19131956]

[178] Krämer, A.; Green, J.; Pollard, J., Jr; Tugendreich, S. Causal analysis approaches in ingenuity pathway analysis. *Bioinformatics,* **2014**, *30*(4), 523-530.
[http://dx.doi.org/10.1093/bioinformatics/btt703] [PMID: 24336805]

[179] Tarca, A.; Khatri, P.; Draghici, S. *Signaling Pathway Impact Analysis (SPIA) using combined evidence of pathway over-representation and unusual signaling perturbations. ,* **2011**.

[180] Shannon, P.; Markiel, A.; Ozier, O.; Baliga, N.S.; Wang, J.T.; Ramage, D.; Amin, N.; Schwikowski, B.; Ideker, T. Cytoscape: a software environment for integrated models of biomolecular interaction networks. *Genome Res.,* **2003**, *13*(11), 2498-2504.
[http://dx.doi.org/10.1101/gr.1239303] [PMID: 14597658]

[181] Wu, C.H.; Yeh, L.S.; Huang, H.; Arminski, L.; Castro-Alvear, J.; Chen, Y.; Hu, Z.; Kourtesis, P.; Ledley, R.S.; Suzek, B.E.; Vinayaka, C.R.; Zhang, J.; Barker, W.C. The protein information resource. *Nucleic Acids Res.,* **2003**, *31*(1), 345-347.
[http://dx.doi.org/10.1093/nar/gkg040] [PMID: 12520019]

[182] Frolkis, A. *SMPDB: The small molecule pathway database,* **2009**.

[183] Romero, P.; Wagg, J.; Green, M.L.; Kaiser, D.; Krummenacker, M.; Karp, P.D. Computational prediction of human metabolic pathways from the complete human genome. *Genome Biol.,* **2005**, *6*(1), R2.
[http://dx.doi.org/10.1186/gb-2004-6-1-r2] [PMID: 15642094]

[184] Caspi, R.; Billington, R.; Fulcher, C.A.; Keseler, I.M.; Kothari, A.; Krummenacker, M.; Latendresse, M.; Midford, P.E.; Ong, Q.; Ong, W.K.; Paley, S.; Subhraveti, P.; Karp, P.D. The MetaCyc database of metabolic pathways and enzymes. *Nucleic Acids Res.,* **2018**, *46*(D1), D633-D639.
[http://dx.doi.org/10.1093/nar/gkx935] [PMID: 29059334]

[185] Funahashi, A.; Morohashi, M.; Kitano, H.; Tanimura, N. CellDesigner: a process diagram editor for gene-regulatory and biochemical networks. *BIOSILICO,* **2003**, *1*(5), 159-162.
[http://dx.doi.org/10.1016/S1478-5382(03)02370-9]

[186] Hoops, S.; Sahle, S.; Gauges, R.; Lee, C.; Pahle, J.; Simus, N.; Singhal, M.; Xu, L.; Mendes, P.; Kummer, U. COPASI--a COmplex PAthway SImulator. *Bioinformatics,* **2006**, *22*(24), 3067-3074.
[http://dx.doi.org/10.1093/bioinformatics/btl485] [PMID: 17032683]

[187] Starruß, J.; de Back, W.; Brusch, L.; Deutsch, A. Morpheus: a user-friendly modeling environment for multiscale and multicellular systems biology. *Bioinformatics,* **2014**, *30*(9), 1331-1332.
[http://dx.doi.org/10.1093/bioinformatics/btt772] [PMID: 24443380]

[188] Fabregat, A.; Jupe, S.; Matthews, L. The reactome pathway knowledgebase. *Nucleic Acids Res.,* **2018**, *46*(D1), D649-D655.
[http://dx.doi.org/10.1093/nar/gkx1132] [PMID: 29145629]

[189] Kandasamy, K.; Mohan, S.S.; Raju, R. NetPath: a public resource of curated signal transduction pathways. *Genome Biol.,* **2010**, *11*(1), R3.
[http://dx.doi.org/10.1186/gb-2010-11-1-r3] [PMID: 20067622]

[190] Peri, S.; Navarro, J.D.; Amanchy, R. Development of human protein reference database as an initial platform for approaching systems biology in humans. *Genome Res.,* **2003**, *13*(10), 2363-2371.
[http://dx.doi.org/10.1101/gr.1680803] [PMID: 14525934]

[191] Orchard, S.; Ammari, M.; Aranda, B. The MIntAct project--IntAct as a common curation platform for 11 molecular interaction databases. *Nucleic Acids Res.,* **2014**, *42*(Database issue), D358-D363.
[http://dx.doi.org/10.1093/nar/gkt1115] [PMID: 24234451]

[192] Lee, I.; Blom, U.M.; Wang, P.I.; Shim, J.E.; Marcotte, E.M. Prioritizing candidate disease genes by network-based boosting of genome-wide association data. *Genome Res.,* **2011**, *21*(7), 1109-1121.
[http://dx.doi.org/10.1101/gr.118992.110] [PMID: 21536720]

[193] Bader, G.D.; Cary, M.P.; Sander, C. Pathguide: a pathway resource list. *Nucleic Acids Res.,* **2006**, *34*(Database issue), D504-D506.
[http://dx.doi.org/10.1093/nar/gkj126] [PMID: 16381921]

[194] Frisch, M.; Klocke, B.; Haltmeier, M.; Frech, K. *LitInspector: Literature and signal transduction pathway mining in PubMed abstracts,* **2009**.
[http://dx.doi.org/10.1093/nar/gkp303]

[195] Whyte, J.L.; Smith, A.A.; Helms, J.A. Wnt signaling and injury repair. *Cold Spring Harb. Perspect. Biol.*, **2012**, *4*(8)a008078 [http://dx.doi.org/10.1101/cshperspect.a008078] [PMID: 22723493]

[196] Huelsken, J.; Behrens, J. The Wnt signalling pathway. *J. Cell Sci.*, **2002**, *115*(Pt 21), 3977-3978. [http://dx.doi.org/10.1242/jcs.00089] [PMID: 12356903]

[197] Voronkov, A.; Krauss, S. Wnt/beta-catenin signaling and small molecule inhibitors. *Curr. Pharm. Des.*, **2013**, *19*(4), 634-664. [http://dx.doi.org/10.2174/138161213804581837] [PMID: 23016862]

[198] Tran, F.H.; Zheng, J.J. Modulating the wnt signaling pathway with small molecules. *Protein Sci.*, **2017**, *26*(4), 650-661. [http://dx.doi.org/10.1002/pro.3122] [PMID: 28120389]

[199] Gurney, A.; Axelrod, F.; Bond, C.J. Wnt pathway inhibition via the targeting of Frizzled receptors results in decreased growth and tumorigenicity of human tumors. *Proc. Natl. Acad. Sci. USA*, **2012**, *109*(29), 11717-11722. [http://dx.doi.org/10.1073/pnas.1120068109] [PMID: 22753465]

[200] Lee, H.J.; Bao, J.; Miller, A. Structure-based discovery of novel small molecule Wnt signaling inhibitors by targeting the cysteine-rich domain of Frizzled. *J. Biol. Chem.*, **2015**, *290*(51), 30596-30606. [http://dx.doi.org/10.1074/jbc.M115.673202] [PMID: 26504084]

[201] Wong, H.C.; Bourdelas, A.; Krauss, A. Direct binding of the PDZ domain of Dishevelled to a conserved internal sequence in the C-terminal region of Frizzled. *Mol. Cell*, **2003**, *12*(5), 1251-1260. [http://dx.doi.org/10.1016/S1097-2765(03)00427-1] [PMID: 14636582]

[202] Proffitt, K.D.; Virshup, D.M. Precise regulation of porcupine activity is required for physiological Wnt signaling. *J. Biol. Chem.*, **2012**, *287*(41), 34167-34178. [http://dx.doi.org/10.1074/jbc.M112.381970] [PMID: 22888000]

[203] Wang, X.; Moon, J.; Dodge, M.E. The development of highly potent inhibitors for porcupine. *J. Med. Chem.*, **2013**, *56*(6), 2700-2704. [http://dx.doi.org/10.1021/jm400159c] [PMID: 23477365]

[204] Madan, B.; Ke, Z.; Harmston, N. Wnt addiction of genetically defined cancers reversed by PORCN inhibition. *Oncogene*, **2016**, *35*(17), 2197-2207. [http://dx.doi.org/10.1038/onc.2015.280] [PMID: 26257057]

[205] Croy, H.E.; Fuller, C.N.; Giannotti, J. The poly(ADP-ribose) polymerase enzyme Tankyrase antagonizes activity of the β-catenin destruction complex through ADP-ribosylation of Axin and APC2. *J. Biol. Chem.*, **2016**, *291*(24), 12747-12760. [http://dx.doi.org/10.1074/jbc.M115.705442] [PMID: 27068743]

[206] McGonigle, S.; Chen, Z.; Wu, J. E7449: A dual inhibitor of PARP1/2 and tankyrase1/2 inhibits growth of DNA repair deficient tumors and antagonizes Wnt signaling. *Oncotarget*, **2015**, *6*(38), 41307-41323. [http://dx.doi.org/10.18632/oncotarget.5846] [PMID: 26513298]

[207] Inestrosa, N.C.; Toledo, E.M. The role of Wnt signaling in neuronal dysfunction in Alzheimer's Disease. *Mol. Neurodegener.*, **2008**, *3*(1), 9. [http://dx.doi.org/10.1186/1750-1326-3-9] [PMID: 18652670]

[208] Fonseca, B.F.; Predes, D.; Cerqueira, D.M. Derricin and derricidin inhibit Wnt/β-catenin signaling and suppress colon cancer cell growth in vitro. *PLoS One*, **2015**, *10*(3)e0120919 [http://dx.doi.org/10.1371/journal.pone.0120919] [PMID: 25775405]

[209] de la Roche, M.; Rutherford, T.J.; Gupta, D.; Veprintsev, D.B.; Saxty, B.; Freund, S.M.; Bienz, M. An intrinsically labile α-helix abutting the BCL9-binding site of β-catenin is required for its inhibition by carnosic acid. *Nat. Commun.*, **2012**, *3*, 680.

[http://dx.doi.org/10.1038/ncomms1680] [PMID: 22353711]

[210] Aditya Rao, S.J.; Jeevitha, B.; Smitha, R.; Ramesh, C.K.; Paramesha, M.; Jamuna, K.S. Wound healing activity from the leaf extracts of Morus laevigata and in silico binding studies from its isolates with gsk 3-β. *Int. J. Res. Dev. Pharm. Life Sci.,* **2015**, *4*(4), 1686-1696.

[211] Paramesha, M.; Ramesh, C.K.; Krishna, V.; Kumar Swamy, H.M.; Aditya Rao, S.J.; Hoskerri, J. Effect of dehydroabietylamine in angiogenesis and GSK3-β inhibition during wound healing activity in rats. *Med. Chem. Res.,* **2015**, *24*(1), 295-303.
[http://dx.doi.org/10.1007/s00044-014-1110-1]

Implementation of the Molecular Electrostatic Potential over GPUs: Large Systems as Main Target

J. César Cruz, Ponciano García-Gutierrez, Rafael A. Zubillaga, Rubicelia Vargas and **Jorge Garza**[*]

Departamento de Química, División de Ciencias Básicas e Ingeniería, Universidad Autónoma Metropolitana-Iztapalapa, Ciudad de México, México

Abstract: The molecular electrostatic potential (MEP) is a useful tool to design and develop drugs. However, the evaluation of this property using quantum chemistry methods presents a challenge for molecules of medium or large size since this is computationally expensive. In this chapter, we showed two implementations of this property over graphics processing units (GPU). In the first instance, we discussed some details that must be considered when GPUs are involved in high-performance computing. After this step, the algorithms considered to evaluate MEP over GPUs are exposed to observe the main differences between a method with minimal approximations and another one where usual approximations are implemented in many quantum chemistry codes. The benefits provided by these graphics cards are evidenced when our implementations are applied over molecules of considerable size like those found in protein-ligand complexes, where usually the electrostatic potential is modeled by a set of point charges.

Keywords: Cuda-C, GPUs, Molecular Electrostatic Potential, Multipolar Expansion, Proteins.

INTRODUCTION

Electrostatic interactions are crucial to form non-covalent interactions in biological systems where many biomacromolecules contain electrical charges in a great variety of fragments built with small charged molecules and monoatomic ions. As an example, we can mention globular proteins, which are polyelectrolytes with ionized amino acid residues like Asp and Glu with a negative charge, and Lys and Arg with a positive charge; these four types of

[*] **Corresponding author Jorge Garza:** Departamento de Química, División de Ciencias Básicas e Ingeniería, Universidad Autónoma Metropolitana-Iztapalapa, San Rafael Atlixco 186, Col. Vicentina. Iztapalapa. C. P. 09340, Ciudad de México, México; Tel: 525558044600; E-mail: jgo@xanum.uam.mx

Zaheer Ul-Haq and Angela K. Wilson (Eds.)

residues represent about a quarter of all residues in an average protein [1]. In general, the biological activity of a protein shows up by interacting selectively with another biomolecule, large or small, and the electrostatic contribution to the binding energy is crucial in many cases [2, 3]. To visualize better the tendency of favorable electrostatic interactions between two molecules, instead of using the representation of positive and negative surface charges it is more accurate to display the molecular electrostatic potential (MEP), since even in proteins with a small net charge, patterns of positive and negative electrostatic potential are over the molecular surface [1]. The complementarity of the MEP between different fragments dictates the formation of a complex, and sometimes such a complementary is more important than a different sign of their net charges [4]. Besides, the charged state of some polypeptides is modulated by post-translational modifications like phosphorylation, in response to extracellular signals, changing the MEP locally and creating new binding epitopes for specific targets. Many of the mutations that lead to cancer involve ionizable residues [4]. Nucleic acids are among the strongest natural polyelectrolytes, which contain a phosphate group (PO^-_4) for each base in their sequence so they can interact favorably with molecules displaying surfaces of positive MEP (see Fig. 3 of Honig and Nicholls [4]). Phospholipid membranes also have strong electrostatic interactions with proteins and other biomolecules. Therefore, the computation of the MEP is desirable for these cases.

Evaluation of the MEP is not an easy task if this is obtained from its definition [5, 6]

$$\Phi(\mathbf{r}_2) = \sum_A \frac{Z_A}{|\mathbf{r}_2 - \mathbf{R}_A|} - \int d\mathbf{r}_1 \frac{\rho(\mathbf{r}_1)}{|\mathbf{r}_1 - \mathbf{r}_2|}, \tag{1}$$

where precisely the charges in the system are involved in computing this property. In this case, Z_A represents the nuclear charge of each atom in the system, and the electron density is represented by $\rho(\mathbf{r})$. From definition (1), one can deduce that there are sites in a molecule where nuclei command and, therefore $\phi(\mathbf{r})$ is positive, and regions where the electron density rules and consequently $\phi(\mathbf{r})$ is negative; this characteristic has been widely used in quantum chemistry applications [7 - 11]. For molecular systems, we need the electron density from quantum chemistry methods, or from experimental information, to use equation (1). In this chapter, we will deal only with theoretical approaches related to the electronic structure of atoms and molecules. Thus, the evaluation of the MEP could be computationally expensive and some times prohibitive for systems of considerable size. This contribution aims to discuss two algorithms to obtain the MEP over graphics processing units (GPU). In our group we have developed some strategies to implement grid-based methods over GPUs to determine quantum chemistry scalar and vector fields like the electron density and derivatives of it. As we can see from equation (1) this property is directly involved with the MEP, and for that reason, we think that an analysis of its implementation over GPUs is appropriate currently, since this hardware has high availability.

GRAPHICS PROCESSING UNITS IN QUANTUM CHEMISTRY

Graphics processing units have gained a relevant role in the high-performance computing, particularly in quantum chemistry [12 - 19] or in computational modeling based on molecular dynamics [20 - 23]. The number of threads and the memory involved in a GPU of the last generation are desirable to developers of parallel computing applications. In our group, we have reported the evaluation of scalar and vector fields defined in quantum chemistry, over grids with the use of GPUs. In specific, we have developed the graphics processing units for atoms and molecules (GPUAM) [24, 25] code with the primary purpose of the development of an application where personal computers with an installed GPU can be used for demanding computational tasks. This code uses WFN or WFX files as input data, which are ASCII files, and they can be used in any computational system. These files are obtained from programs like GAMESS [12], Gaussian [13], or NWChem [14]. Besides, the Molden2AIM [26] code allows us to use other formats to convert the corresponding files in WFN or WFX formats, which contain nuclei coordinates, exponents of the basis set, and the coefficients $\{c_i\}$ that give the best representation of Hartree-Fock or Kohn-Sham orbitals in terms of a basis set, even correlated methods can be used if the coefficients of natural orbitals are in these formats. In this chapter, we do use Terachem [15, 16, 18], which has been designed from scratch to be executed on GPUs and the files delivered by this code are transformed to WFX format by using Molden2AIM. With wfn or wfx files, the GPUAM evaluates a scalar or vector field, assigning one thread of the GPU to one point of the grid. At the end of this process, GPUAM writes the final results in permanent memory through files with cube format, which is a standard extension introduced by Gaussian, Inc. This procedure is sketched in Fig. (**1**).

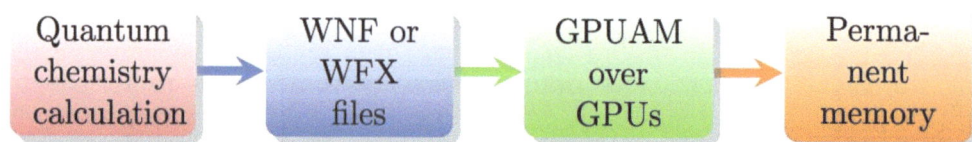

Fig. (1). Procedure to obtain a scalar or vector field defined in quantum chemistry using GPUAM.

In GPUAM, the grid is distributed over a GPU, or over the GPUs defined by a user. The kernel designed in GPUAM to allocate a grid over a GPU [24, 25] under CUDA-C programming techniques [27] is summarized in the following instructions:

global_thread_x = threadIdx.x + blockIdx.x * blockDim.x;

global_thread_y = threadIdx.y + blockIdx.y * blockDim.y;

sign_x = global_thread_x;

sign_y = global_thread_y/pts_z;

sign_z = global_thread_y%pts_z;

sign_f = global_thread_x*pts_y*pts_z + global_thread_y;

if (sign_x < pts_x && sign_y < pts_y && sign_z < pts_z) {

x = x0 + (float) sign_x*step;

y = y0 + (float) sign_y*step;

z = z0 + (float) sign_z*step;

buffer_MEP[sign_f] =

MEP(x,y,z,coordinates,exponents,coefficients);

}

In this case, **threadIdx.x, threadIdx.y, blockIdx.x, blockIdx.y, blockDim.x,** and **blockDim.y** are reserved words by CUDA-C. From these instructions, we stress that in our implementation we pass a 3D grid of points defined in (x,y,z) to a 2D grid of threads in the GPU. In our group, we have implemented several quantum chemistry fields in GPUAM [24, 25]. In this chapter, we present two implementations of the MEP over GPUs by using the kernel presented above. One of these implementations has been already reported [28] and a new approach is discussed in detail to take advantage of GPUs.

EVALUATING THE MEP USING QUANTUM CHEMISTRY METHODS

As we mentioned in the Introduction, the MEP is obtained from the electron density, $\rho(\mathbf{r})$, which can be computed by many quantum chemistry methods as [29, 30]

$$\rho(\mathbf{r}) = \sum_i \omega_i \psi_i^*(\mathbf{r})\psi_i(\mathbf{r}), \tag{2}$$

where $\psi_i(\mathbf{r})$ represents an orbital with occupancy ω_i. Besides, in many cases the orbital is expressed in terms of a basis set, $\{\varphi_\mu\}$, through the linear combination

$$\psi_i(\mathbf{r}) = \sum_{\mu=1}^{K} c_\mu^{(i)} \phi_\mu(\mathbf{r}). \tag{3}$$

In this equation, K represents the number of functions in the basis set, and $\{c_\mu^{(i)}\}$ are coefficients that minimize the total energy. In this chapter, Gaussian type functions (GTF)

$$\phi_\mu(\mathbf{r}) = e^{|\mathbf{r}-\mathbf{R}_A|^2}, \tag{4}$$

are the main target of the discussion. With the introduction of a basis set, equation (2) has the form

$$\rho(\mathbf{r}) = \sum_{\mu=1}^{K} \sum_{\nu=1}^{K} P_{\nu\mu} \phi_\mu(\mathbf{r})\phi_\nu(\mathbf{r}), \tag{5}$$

with

$$P_{\nu\mu} = \sum_i \omega_i c_\nu^{(i)} c_\mu^{(i)}. \tag{6}$$

Thus, the MEP has the expression

$$\Phi(\mathbf{r}_2) = \sum_A \frac{Z_A}{|\mathbf{r}_2-\mathbf{R}_A|} - \sum_{\mu,\nu=1}^{K} P_{\nu\mu} \int d\mathbf{r}_1 \frac{\phi_\mu(\mathbf{r}_1)\phi_\nu(\mathbf{r}_1)}{|\mathbf{r}_1-\mathbf{r}_2|}. \tag{7}$$

For the MEP analysis, commonly, this quantity is evaluated over a grid built with many internal points. The number of points in the grid depends on the system, and we can find examples where there are millions of points contained in the grid. From expression (7) we see that the main problem to evaluate $\phi(\mathbf{r}_2)$ is related to the integral of the last equation since it must be computed at each point $\mathbf{r}_2$. Naturally, there are proposals to evaluate this integral, in particular for GTFs [15, 17, 31 - 37], although this is an open issue in quantum chemistry to speedup the calculations. Many approaches use approximations to estimate this quantity [38 - 41]. We can find programs very fast to evaluate MEP. However, we must recognize that many times, such evaluation is obtained by using strong approximations to compute the corresponding integral, or in some cases, this integral is not evaluated, and consequently this information is used for qualitative interpretations. In summary, there is a compromise between the accuracy and the acceleration to evaluate the MEP since if we want high accuracy, then we will need a significant computational effort.

IMPLEMENTATIONS TO EVALUATE THE MOLECULAR ELECTROSTATIC POTENTIAL

Some methods evaluate very fast the MEP in the following way:

$$\Phi_{charge}(\mathbf{r}_2) = \sum_A \frac{Q_A}{|\mathbf{r}_2 - \mathbf{R}_A|}. \tag{8}$$

In this case there are only point charges where Q_A represents the charge of an atom A with localization $\mathbf{R}_A$. The main problem with this approach is the assignation of the charges Q_A for the atoms in a molecule [41]. Another way to obtain the MEP is from Poisson's equation [42].

$$\nabla^2 \Phi(\mathbf{r}) = 4\pi\rho(\mathbf{r}). \tag{9}$$

This approach has been tackled and discussed by several methods, and this is not part of this chapter since we want to evaluate $\phi(\mathbf{r})$ using the integral (7). In this sense, Gadre and collaborators have many reports working with this integral [43 - 45]. However, they split the electron density in pseudo-atomic contributions to accelerate the MEP evaluation.

In this chapter we do use expression (2) to compute the electron density with GTFs as basis set to evaluate each orbital. A GTF is written as:

$$\phi_\mu(r; A) = x_A^{i_\mu} y_A^{j_\mu} z_A^{k_\mu} e^{-\zeta_\mu r_A^2}, \tag{10}$$

with $i_\mu, j_\mu, k_\mu = 0,1,2...$

$$r_A^2 = |\mathbf{r} - \mathbf{R}_A|^2 = x_A^2 + y_A^2 + z_A^2, \tag{11}$$

$$x_A = x - X_A, \tag{12}$$

and so on for y and z.

The standard way to transform the integral starts writing the exponential as

$$e^{-\zeta_\mu r_A^2 - \zeta_\nu r_B^2} = M e^{-p|\mathbf{r}-\mathbf{P}|^2} \tag{13}$$

with

$$p = \zeta_\mu + \zeta_\nu, \tag{14}$$

$$\mathbf{P} = \frac{\zeta_\mu \mathbf{R_A} + \zeta_\nu \mathbf{R_B}}{p}, \tag{15}$$

and

$$M = e^{-\zeta_\mu \zeta_\nu |\mathbf{R_A} - \mathbf{R_B}|^2 / p}. \tag{16}$$

FULL EVALUATION OF THE MEP

Integral in equation (7) can be transformed by coupling the Boys [46] approach with that reported by Macmurchie and Davidson [47] to deal with powers of x_A, y_A and z_A. The integral becomes as

$$\int dr_1 \frac{\phi_\mu(r_1)\phi_\nu(r_1)}{|r_1 - r_2|} = \sum_{m=0}^{L} C_m F_m(X), \tag{17}$$

where $F_m(X)$ is

$$F_m(X) = \int_0^1 t^{2m} e^{-Xt^2} dt, \tag{18}$$

and C_m are expansion coefficients where m reaches its maximum value when $L = i_\mu + i_\nu + j_\mu + j_\nu + k_\mu + k_\nu$. The exponent X is defined as

$$X = pr_{2p}^2, \tag{19}$$

with

$$r_{2p} = |P - r_2|. \tag{20}$$

Another way to express equation (17) is in terms of a polynomial, $PL(t^2)$, of degree L as

$$\int dr_1 \frac{\phi_\mu(r_1)\phi_\nu(r_1)}{|r_1-r_2|} = \int_0^1 dt P_L(t^2)e^{-Xt^2}. \tag{21}$$

We want to remark that in our approach the integral involved in equation (21) is evaluated over a GPU. The first case of this integral is obtained for $L = 0$,

$$\int_0^1 dt e^{-Xt^2} = \frac{\sqrt{\pi}}{2\sqrt{pr_{2p}}} \operatorname{erf}(\sqrt{pr_{2p}}), \tag{22}$$

where erf represents the error function. For $L > 0$ the corresponding integral is evaluated numerically. In our case we use a N-point Gaussian quadrature rule

$$\int_0^1 dt P_L(t^2)e^{-Xt^2} = \sum_{\kappa=1}^N w_\kappa P_L(t_\kappa^2), \tag{23}$$

with $N = L/2 + 2$. Weights, w_k, and roots, t_k, of Rys polynomials are needed to evaluate exactly the integral (23). It is convenient that $PL(t^2)$ can be written in terms of one dimensional cartesian integrals $I_{i_\mu i_\nu}(t^2)$, $I_{j_\mu j_\nu}(t^2)$ and $I_{k_\mu k_\nu}(t^2)$ for x, y and z coordinates, respectively [48]. Thus, we have

$$P_L(t^2) = \frac{2\pi}{p}\sum_{\kappa=1}^N w_\kappa I_{i_\mu i_\nu}(t_\kappa^2)I_{j_\mu j_\nu}(t_\kappa^2)I_{k_\mu k_\nu}(t_\kappa^2), \tag{24}$$

from here it is clear the dependence of P_L on μ and ν.

To evaluate these integrals we use the vertical recurrence relations for each cartesian coordinate, *e.g.*, for the X coordinate

$$I_{i_\mu+1,i_\nu}(t^2) = (X_{PA} - t^2 X_{Pr_2})I_{i_\mu i_\nu} + \frac{1}{2p}(1 - t^2)(iI_{i_\mu-1,i_\nu} + jI_{i_\mu,i_\nu-1}), \tag{25}$$

$$I_{i_\mu,i_\nu+1}(t^2) = (X_{PB} - t^2 X_{Pr_2})I_{i_\mu i_\nu} + \frac{1}{2p}(1 - t^2)(iI_{i_\mu-1,i_\nu} + jI_{i_\mu,i_\nu-1}), \tag{26}$$

where $X_{PA} = P_x\text{-}X_A$ and $X_{PB} = P_x\text{-}X_B$. The process starts from $I_{00} = 1$.

The Rys polynomials form an orthogonal set of functions with the weight function $w(X,t) = e^{-Xt^2}$; the half-range Rys polynomials, $Rm(t)$, in the interval $t \in [0,1]$ and the full-range polynomials, $Jm(t)$, in the range $t \in [-1,1]$. From here, it is clear that roots and weights of Rys polynomials must be evaluated over each point of the grid where the MEP is computed. This task is computationally demanding, and for that reason, some approximations have been reported to evaluate roots and weights [49].

In general, a three-term recurrence relation

$$J_{n+1}(t,X) = (t - \alpha_n)J_n(t,X) - \beta_n J_{n-1}(t,X), \tag{27}$$

is involved for monic orthogonal polynomials, with α_n and β_n obtained from

$$\alpha_n = \frac{\int_a^b w(t,X)tJ_n^2(t,X)dt}{\int_a^b w(t,X)J_n^2(t,X)dt}, \tag{28}$$

and

$$\beta_n = \frac{\int_a^b w(t,X)J_n^2(t,X)dt}{\int_a^b w(t,X)J_{n-1}^2(t,X)dt}. \tag{29}$$

For the interval $t \in [-1,1]$ the recursion coefficient α_n is zero because this quantity has involved an odd function. For that reason, we do use full-range Rys polynomials to avoid computational work. In this way, the recurrence relation is written as [50]:

$$tJ_n(t,X) = \sqrt{\beta_{n+1}}J_{n+1} + \sqrt{\beta_n}J_{n-1}, n = 0,1,..N. \tag{30}$$

This equation is obtained from equation (27) by using the normalization condition of the monic orthogonal polynomials [51].

The starting point of the recurrence relation (30) is $J_{-1} = 0$, $J_0 = 1$ and

$$\beta_0 = \int_{-1}^1 w(t,X)dt = \int_{-1}^1 e^{-Xt^2}dt. \tag{31}$$

Moments of the weight function (μn) appear in the evaluation of β_n, which are defined as

$$\mu_n = \int_{-1}^1 dt e^{-Xt^2}t^n. \tag{32}$$

A recurrence relation is obtained for these moments in the integration by parts

$$\mu_n = \frac{n-1}{2X}\mu_{n-2} - \frac{1}{X}e^{-X}, n = 2,4,6..$$ (33)

with

$$\mu_0 = \sqrt{\frac{\pi}{X}}\,\mathrm{erf}\left(\sqrt{X}\right).$$ (34)

For large values of n, this method is numerically unstable [52]. However, in the evaluation of the MEP high values of n are not necessary since, in the worst case, angular moment values are around 12 and this method starts to be unstable for $n > 20$ [53]. To control possible numerical issues in the implementation over the GPU, we used two different methods to obtain β_n coefficients. For $X \leq 100$, integrals of equation (29) were solved numerically using a Lagrange Gaussian quadrature of one hundred points according to the discretized Gautschi-Stieltjes method [54]. For $X > 100$ we use the Gram-Schmidt procedure to construct the Rys polynomials; in this way, numerical errors are not present when $X \rightarrow \infty$. The use of these two methods is a new approach proposed in our implementation [28].

To obtain roots and weights of the Rys polynomials, a $N \times N$ Jacobi matrix, **J**, must be diagonalized, its eigenvalues are the roots (t_k), and the weights (W_k) are obtained from the Christoffel-Darboux relation [53]

$$w_\kappa = \frac{1}{\sum_{i=0}^{N-1} J_i(t_\kappa)^2}.$$ (35)

In our implementation, the QL algorithm was used to diagonalize **J** [55].

Finally, to evaluate the MEP only $K(K + 1)/2$ values are required since we have symmetric matrices to obtain

$$\int dr_1 \frac{\rho(r_1)}{|r_1-r_2|} = \sum_\mu^K \tilde{G}_{\mu\mu} \int_{-1}^1 dt P_L(t^2)e^{-Xt^2}$$

$$+2\sum_\mu^K \sum_{v>\mu}^K \tilde{G}_{\mu v} \int_{-1}^1 dt P_L(t^2)e^{-Xt^2},$$ (36)

where

$$\tilde{G}_{\mu\nu} = \sum_{\mu\nu}^{K} P_{\nu\mu} e^{\frac{\zeta_\mu \zeta_\nu}{p}|A-B|^2}. \tag{37}$$

Thus, this is the final expression coded in the GPUAM. Details of our implementation can be found in reference [28]. In summary, our implementation consists of a mix of two methods: Gram-Schmidt plus the discretized Gautschi-Stieltjes method, which can be implemented in GPUs or CPUs. The accuracy of the MEP depends on the number of points used to evaluate numerically the integrals involved in the recurrence relationships (25) and (26). In our implementation, the number of points is determined by L through the relation $N = L/2 + 2$, which we found by making many tests.

LINEAR SCALING TECHNIQUE

Our main interest in this chapter is the evaluation of the integral

$$\int d\mathbf{r}_1 \frac{\phi_\mu(\mathbf{r}_1)\phi_\nu(\mathbf{r}_1)}{|\mathbf{r}_1-\mathbf{r}_2|}, \tag{38}$$

by using linear scaling techniques since the full evaluation has been published recently [28]. An essential property of the Gaussian functions is that two of them give a new charge distribution different to zero if these two functions are close between themselves. In this way, the previous integral scales linearly with the size of the system. Besides, the electronic contribution to the MEP can be separated into two contributions, which depend on the distance between the charge distribution and the point where the MEP is evaluated. The discussion about the two contributions of the MEP can be developed if we use as an example a spherical Gaussian distribution centered at P. In this way we obtain

$$U_p(r_2) = \int d r_1 \frac{\exp(-p r_{1p}^2)}{|\mathbf{r}_1-\mathbf{r}_2|} = \frac{S_{\mu\nu}}{r_{2p}} \operatorname{erf}(\sqrt{p}r_{2p}), \tag{39}$$

where we have used the Laplace transformation of the bielectronic operator [56] and the Gaussian product rule. In this expression $S_{\mu\nu}$ represents an element of the overlap matrix. This expression can be written in another way if we use the definition of the complementary error function

$$1 = \operatorname{erf}(x) + \operatorname{erfc}(x), \tag{40}$$

to obtain

$$U_p(r_2) = U_p^{cls}(r_2) + U_p^{non}(r_2) = \frac{S_{\mu\nu}}{r_{2p}} - \frac{S_{\mu\nu}}{r_{2p}}\,\mathrm{erfc}(\sqrt{p}r_{2p}), \tag{41}$$

where the classical contribution is defined as

$$U_p^{cls} = \frac{S_{\mu\nu}}{r_{2p}}, \tag{42}$$

and the nonclassical part as

$$U_p^{non} = -\frac{S_{\mu\nu}}{r_{2p}}\,\mathrm{erfc}(\sqrt{p}r_{2p}). \tag{43}$$

We observe that both contributions go to zero when $r_{2p} \to \infty$. However, in this limit, the non-classical contribution goes to zero faster than the classical part due to the definition of the complementary error function.

The non-classical contribution, $U_p^{non}(r_2)$, is known as short-range interaction and the classical one, $U_p^{cls}(r_2)$, long-range interaction. The importance of separating in this way the integral (38) is because $U_p^{cls}(r_2)$ scales linearly with the size of the system.

The screening involved in this method is crucial since it opens a gate to evaluate the MEP through long-range interactions instead of a full integral with similar accuracy. The core of this method is based on the expansion

$$r_{12}^{-1} = \sum_{n=0}^{\infty} \frac{r_{1p}^n}{r_{2p}^{n+1}} P_n(\cos\theta), \tag{44}$$

where $P_n(\cos\theta)$ represent Legendre polynomials.

SCREENING OF THE MEP THROUGH SHORT- AND LONG-RANGE INTERACTIONS

The distance between a point $\mathbf{r}_2$ where the MEP is evaluated and a Gaussian distribution, $\Omega_{\mu\nu}(\mathbf{r}_{1p})$ centered at $\mathbf{P}$, is crucial to apply the screening in the MEP's evaluation. If the charge distribution does not overlap the evaluation point, then the interaction between them can be considered as a classical contribution, which satisfies the relation $r_{1p} - r_{2p}$. However, in general, a non-overlap definition between Gaussian functions is not simple. For a Gaussian function of exponent p, Head-Gordon and White proposed a definition of the extent [57] of this distribution as

$$r_p = p^{-1/2}\text{erfc}^{-1}(10^{-k}),\qquad(45)$$

where 10^{-k} represents a target accuracy, and erfc^{-1} is the inverse of the complementary error function. From this definition, two Gaussian distributions separated by more than the sum of their extents interact classically. For the MEP there is only one Gaussian distribution and consequently $r_{2p} > r_p$. Lambrecht *et* al. introduced the concept of absolute spherical multipoles [58] of order n, $M_A^{(n)}$, defined as

$$M_A^{(n)} = \int dr |\Omega_A(r)r^n|,\qquad(46)$$

which has been used to evaluate two-electron integrals. Following this procedure to evaluate the MEP [59], which depends on one Gaussian distribution, we obtained the final expression

$$\frac{M_p^{(0)}}{r_{2p}-\text{ext}_p} = \sum_{n=0}^{\infty} \frac{1}{r_{2p}^{n+1}} \sum_{i=0}^{n} \binom{n}{i} \text{ext}_p^{n-i} M_p^{(0)},\qquad(47)$$

where ext_p represents the extent of a Gaussian distribution with exponent p according to the notation employed by Maurer *et al.* [60]. Equation (47) is relevant for our implementation since it represents an upper bound of the integrals involved in the MEP's evaluation. Therefore, Gaussian distributions that satisfy

$$\frac{M_p^{(0)}}{r_{2p}-\text{ext}_p} \leqslant 10^{-k},\qquad(48)$$

can be treated as long-range interactions.

Using the addition theorem for the spherical harmonics $Y_n^m(\theta,\phi)$, the Legendre polynomials are expressed as

$$P_n(\cos\theta) = \sum_{m=-n}^{n} \frac{4\pi}{2n+1} Y_n^m(\theta_{1p},\phi_{1p})Y_n^m(\theta_{2p},\phi_{2p})\qquad(49)$$

When $r_{1p} - r_{2p}$ the multipole series has the expression

$$r_{12}^{-1} = \sum_{n=0}^{\infty} \sum_{m=-n}^{n} \left(\frac{4\pi}{2n+1}\right) \left(\frac{r_{1p}^n}{r_{2p}^{n+1}}\right) Y_n^m(\theta_{1p}, \phi_{1p}) Y_n^m(\theta_{2p}, \phi_{2p}). \tag{50}$$

For convenience this expansion is written as

$$r_{12}^{-1} = \sum_{n=0}^{\infty} \sum_{m=-n}^{n} I_{nm}^*(r_{2p}) R_{nm}(r_{1p}), \tag{51}$$

with

$$R_{nm}(r) = \frac{1}{\sqrt{(n-m)!(n+m)!}} r^n C_{nm}(\theta, \phi), \tag{51}$$

$$I_{nm}(r) = \sqrt{(n-m)!\,(n+m)!}\, r^{-n-1} C_{nm}(\theta, \phi). \tag{53}$$

and

$$C_{nm}(\theta\phi) = \sqrt{\frac{4\pi}{2n+1}} Y_n^m(\theta, \phi). \tag{54}$$

With these definitions the MEP acquires the form

$$U_p(r_2) = \sum_{n=0}^{\infty} \sum_{m=-n}^{n} I_{nm}^*(r_{2p}) q_{nm}^P(P), \tag{55}$$

with

$$q_{nm}^A(A) = \int dr \Omega_{\mu\nu}(r) R_{nm}(r_A). \tag{56}$$

From expression (55) it is clear that there are complexes quantities in the implementation, although the result must be real. Thus, in practical terms it is convenient to use

$$R_{nm}(r) = R_{nm}^c(r) + i R_{nm}^s(r), \tag{57}$$

$$I_{nm}(r) = I_{nm}^c(r) + iI_{nm}^s(r),$$ (58)

where c and s indicate real (cosine) and imaginary (sine) parts of each expression, respectively. Besides, we have the relations

$$R_{n,-m}^c(r) = (-1)^m R_{nm}^c(r),$$ (59)

$$R_{n,-m}^s(r) = -(-1)^m R_{nm}^s(r).$$ (60)

With these definitions

$$q_{nm}(P) = q_{nm}^c(P) + iq_{nm}^s(P).$$ (61)

Finally, if equations (58) and (57) are used in (44) then we obtain the expression to be implemented over GPUs

$$U_p(r_{2p}) = \sum_{n=0}^{\infty} I_{n0}^c(r_{2p})q_{n0}^c(P) + 2\left(\sum_{n=1}^{\infty}\sum_{m=1}^{n} I_{nm}^c(r_{2p})q_{nm}^c(P) + I_{nm}^s(r_{2p})q_{nm}^s(P)\right)$$ (62)

To obtain this result we need expressions (60) and (61).

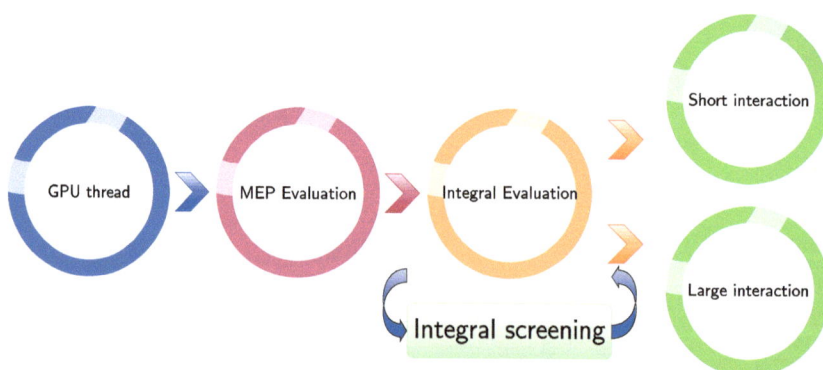

Fig. (2). Implementation of the screening method to evaluate the MEP over one thread of a GPU.

The screening method to evaluate the MEP was implemented in the GPUAM code using CUDA-C. In this implementation, each point of the grid is assigned to each thread of a GPU. Such an implementation is sketched in Fig. (**2**).

The results obtained for each implementation of the MEP on GPUAM are presented and discussed in the next section.

EVALUATION OF THE MEP OVER GPUS

In this section, we will illustrate our implementations with a few examples using exclusively personal computers: I) A laptop with an Intel Core i7-8750H CPU at 2.20 GHz and a GeForce GTX 1060 GPU with Max-Q design of NVIDIA. II) A desktop with an Intel Core i9-9900K CPU at 3.60 GHz and a Titan V GPU. With this hardware we studied five systems: a) Water (H_2O). b) Phenol (C_6H_5OH). c) A crown-ether (24-crown-8). d) The buckminsterfullerene (C_{60}). e) An inclusion adduct between the β-cyclodextrin (βCD) and the sertraline (SRT). The structures of all these systems are presented in Fig. (**3**). The PBE0/6-31+G* Kohn-Sham method was applied using TeraChem over the laptop described above without symmetry restrictions. Details to obtain the energy by TeraChem for each system are reported in Table **1**. The time consumed by TeraChem in the self-consistent process is also reported in the same table.

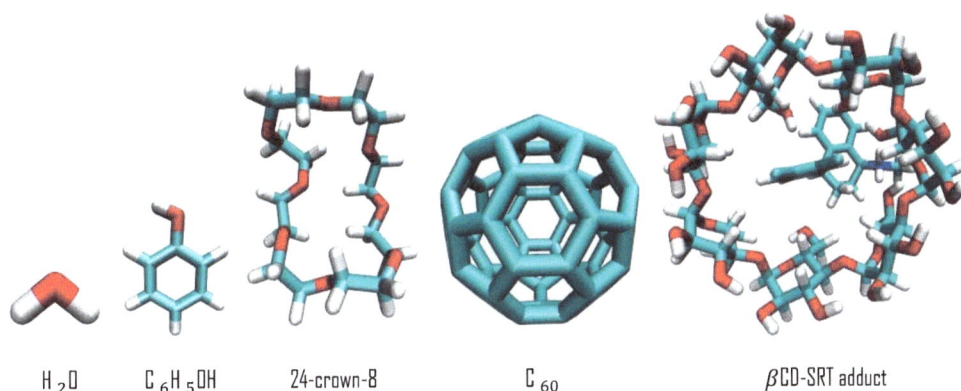

H_2O C_6H_5OH 24-crown-8 C_{60} βCD-SRT adduct

Fig. (3). Structures of five systems considered in this chapter.

Table 1. Details to perform a single point calculation for each system consider in this chapter with the PBE0/6-31+G* method.

System	\multicolumn{4}{c}{Number of}	Time (seconds) SCF
	nuclei	electrons	primitives functions	Contracted functions	
H_2O	3	10	40	23	0.3
Phenol	13	50	248	145	2.2
24-crown-8	56	192	896	520	26.8

(Table 1) cont.....

C_{60}	60	360	1920	1140	552.0
βCD-SRT adduct	185	762	3504	2027	728.8

It is worth to note that we could not obtain convergence in the self-consistent process for C_{60} and βCD-SRT adduct with the PBE0/6-31+G* method. For that reason we used firstly a small basis set to reach the convergence rapidly and its eigenvectors were projected over the 6-31+G* basis set to obtain the final energy. For C_{60} the small basis set was the 6-31G*, and the STO-3G for the βCD-SRT adduct. For these basis sets TeraChem spent 134.3 and 30.3 seconds, respectively. From these results we can say that a laptop of last generation can be used to study current chemical systems in reasonable times.

The electron density was computed for the five systems discussed above, this property is presented in Fig. (**4**). In this figure, the electron density is presented with an isosurface for $\rho(\mathbf{r}) = 0.01$ atomic units (au). The computational effort to obtain this property depends on the number of primitive functions and the number of points in the grid involved in each system. The number of points used in each grid and the time consumed by GPUAM to evaluate the electron density is reported in Table **1**. From these results, we can say that with the GPUAM code the evaluation of the electron density requires short times even for systems of considerable size like the βCD-SRT adduct over a GPU installed in a laptop of the last generation.

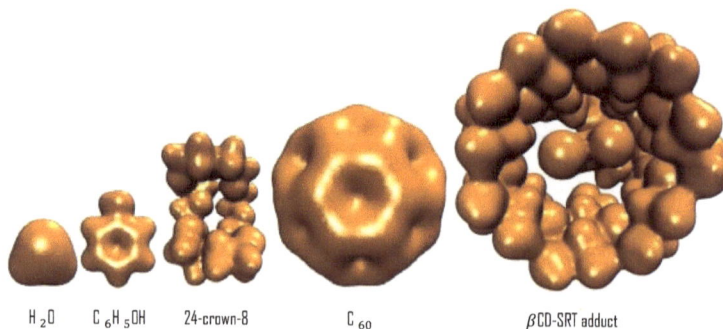

H$_2$O   C$_6$H$_5$OH   24-crown-8   C$_{60}$ βCD-SRT adduct

Fig. (4). Electron density of five systems considered in this chapter with isosurfaces obtained for $\rho(\mathbf{r}) = 0.01$ atomic units.

The MEP for these systems was evaluated over the same laptop used to solve the Kohn-Sham equations and to obtain the electron density. Thus, the MEP mapped over the electron density is presented in Fig. (**5**). For this property, the computational effort is enormous compared to that used for the electronic density, which is corroborated in Table **2**. We observe that the time consumed by the laptop GPU is appreciable for 24-crown-8, C_{60} and βCD-SRT adduct. If the desktop GPU is used for these three systems, then the time used by GPUAM is shorter than that exhibited by the laptop, which is corroborated in the same table (see results in parenthesis). From here it is evident the performance presented by the two GPUs tested in this chapter, where the Titan V GPU is faster than that used in a laptop (GeForce GTX 1060 GPU with Max-Q design).

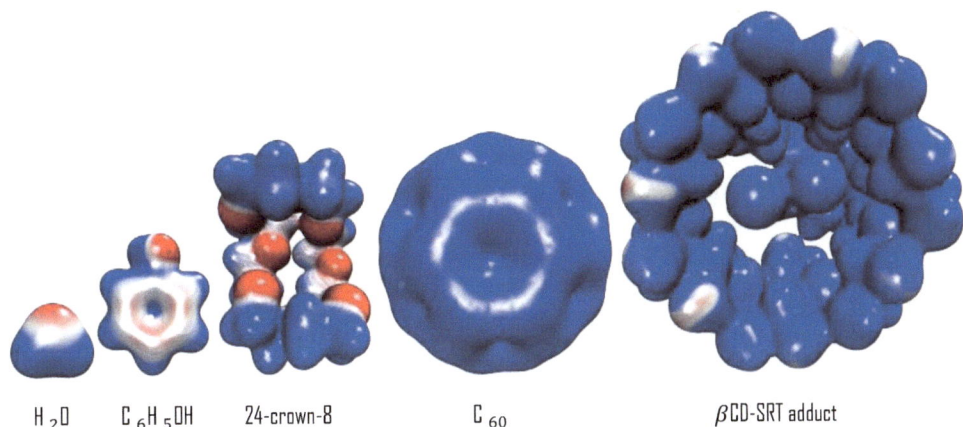

H $_2$O   C $_6$H $_5$OH   24-crown-8   C $_{60}$ βCD-SRT adduct

Fig. (5). Molecular electrostatic potential for the 5 systems presented in Fig. (**3**). Blue regions indicate positive values and negative values are in red, for this case for $\rho(\mathbf{r})$ atomic units.

Table 2. Time consumed to evaluate electron density and the MEP for the five systems presented in Figure 3. The time consumed by the Titan V GPU is reported in parenthesis.

		Time (seconds)	
System	Number of points	$\rho(\mathbf{r})$	MEP
H$_2$O	53664	0.1	0.3
Phenol	210012	0.4	22.9
24-crown-8	393162	2.4	175.7 (47.0).
C $_{60}$	245952	5.0	552.0 (152.0)
βCD-SRT	980869	88.0	1701.0 (410.0)

In our implementation over GPUs, we did not find appreciable differences in the elapsed time between the multipolar expansion and the full evaluation, for that reason, only one result has been reported.

Multiwfn is a widely used code to evaluate several scalar and vector fields defined within quantum chemistry [61]. This code is accesible because it is free, and it gives technical support for platforms based on CPUs. Furthermore, this code has been programmed over OpenMP to use all threads contained in a server. The MEP is a scalar field that multiwfn can compute, and its implementation is similar to the full evaluation presented in this chapter. We will compare some results obtained by multiwfn and GPUAM. For this comparison, multiwfn was executed over 32 Intel(R) Xeon E5-2863 CPU v4 @2.1 GHz, and GPUAM was used over the GPU installed in the laptop described above. The results delivered by these codes are presented in Table **3**. In this comparison, we used the four biggest systems of Fig. (**1**) with a small number of points in the grid of each system.

From the results of Table **3**, it is clear that our implementation over GPUs delivers short times for the MEP evaluation about the times exhibited by a code based on CPUs. It is worth to note the large difference between GPUAM and multiwfn for the βCD-SRT system. The main reason for such a difference is the way our code decides which integrals are evaluated.

Table 3. Evaluation of the MEP for the four biggest systems of Fig. (1). Time consumed by GPUAM over a GeForce GTX 1060 GPU with Max-Q design of NVIDIA installed in a laptop, and multiwfn using 32 Intel(R) Xeon E5-2863 CPU v4 @2.1 GHz. The number of points involved in the grid is the same for both applications.

-	-	Time (seconds)	
System	Number of points	GPUAM	Multiwfn
Phenol	131950	5	13
24-crown-8	131924	58	240
C_{60}	132651	412	1193
βCD-SRT	131924	415	5975

Molecules of large size are the main target of this chapter. For that reason we will apply our MEP implementations on a protein-peptide complex (Protein Data Bank ID: 2M3O) formed by a domain of the human Nedd4-1, a ubiquitin ligase which has 692 nuclei and 2648 electrons (chain W), and an undecapeptide from the epithelial human Na(+) channel with 150 nuclei and 566 electrons (chain P). For these systems there are 8416 primitive functions for the chain W, and 1800 primitive functions for the chain P, the grid for the former consisted of 24,883,000 points and 8,277,750 points for the second one. Naturally, such systems were analyzed using the desktop computer. Thus, the number of primitive functions and the number of points in the grid define an excellent problem to test our implementations. For the biggest system (chain W) the GPUAM consumed 1295 seconds to obtain the electron density and 13061 seconds to evaluate the MEP. These times are reasonable since the evaluation of the MEP contains small approximations with or without the multipolar expansion, which gives the same results with similar times. From these results, we can say that our implementation under the multipolar expansion does not show improvements over GPUs. The MEP for chain W, chain P and complex was projected over the electron density;

all of them are presented in Fig. (**6**). The contacting surfaces of the partners in the complex are also depicted in Fig. (**6**). From here it is clear how the central negative MEP on the chain P fits and complements the positive MEP on the chain W, evidencing an electrostatic contribution to the binding affinity in this complex whose association enthalpy is 58 kJ mol^{-1} [62]. The relevance of the MEP in this system is evident and justify the effort to implement this property over GPUs, which can be evaluated in a desktop with installed GPUs of the last generation.

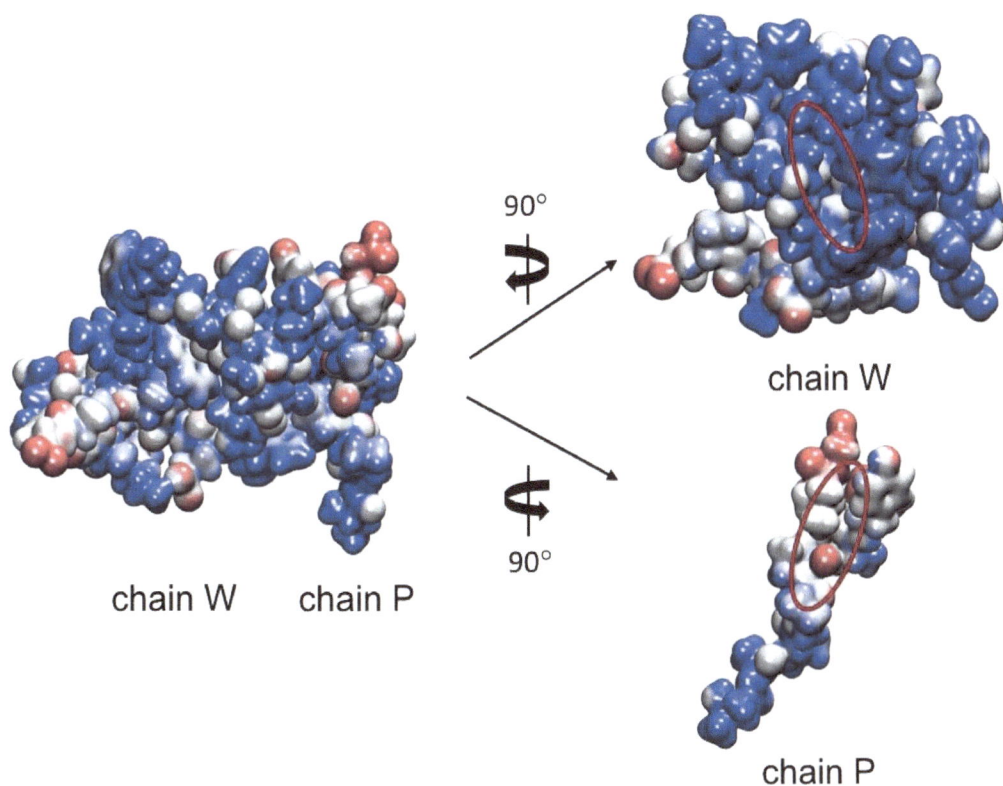

Fig. (6). Molecular electrostatic potential (positive regions in blue and negative regions in red) mapped over the electron density ($\rho(\mathbf{r}) = 0.01$ a.u.). Rotations of a protein (chain W) and a peptide (chain P) to form a complex with 842 nuclei.

CONCLUSION

In this chapter two implementations of the (MEP) over graphics processing units (GPU) were given in detail. Results from multipolar expansion were contrasted with those obtained by the full evaluation of the integral involved in this property. We found that there is no appreciable difference in the time consumed by both implementations in our GPU implementation. However, as we presented previously, these implementations show an impressive performance concerning other applications over CPUs. The convenience to use GPUs has been demonstrated for typical chemical systems with current interest by using a laptop involving CPU and GPU of the last generation. We think that if new GPUs offer more threads and bigger memory, then the elapsed times will be reduced, and the size of the chemical size could be increased. Without a doubt, desktops with GPUs like the Titan V100 are good alternatives to study the MEP of large systems.

The authors thank the facilities provided by the Laboratorio de Supercómputo y Visualización en Paralelo at the Universidad Autónoma Metropolitana-Iztapalapa. Partial funding was provided by CONACYT, México, through the project FC-2016/2412. J.-C. C. thanks CONACYT for the scholarship 815945.

CONSENT FOR PUBLICATION

Not applicable.

CONFLICT OF INTEREST

The authors confirm that the contents of this chapter have no conflict of interest.

ACKNOWLEDGEMENTS

Declare none.

REFERENCES

[1]   Kim, J.; Mao, J.; Gunner, M.R. Are acidic and basic groups in buried proteins predicted to be ionized? *J. Mol. Biol.,* **2005**, *348*(5), 1283-1298.
[http://dx.doi.org/10.1016/j.jmb.2005.03.051] [PMID: 15854661]

[2]   Semenyuk, P.; Muronetz, V. Protein interaction with charged macromolecules: From model polymers to unfolded proteins and post-translational modifications. *Int. J. Mol. Sci.,* **2019**, *20*(5), 1252.
[http://dx.doi.org/10.3390/ijms20051252] [PMID: 30871103]

[3]   Serratos, I.N.; Pérez-Hernández, G.; Garza-Ramos, G.; Hernández-Arana, A.; González-Mondragón, E.; Zubillaga, R.A. Binding thermodynamics of phosphorylated inhibitors to triosephosphate isomerase and the contribution of electrostatic interactions. *J. Mol. Biol.,* **2011**, *405*(1), 158-172.
[http://dx.doi.org/10.1016/j.jmb.2010.10.018] [PMID: 20970429]

[4]   Honig, B.; Nicholls, A. Classical electrostatics in biology and chemistry. *Science,* **1995**, *268*(5214), 1144-1149.
[http://dx.doi.org/10.1126/science.7761829] [PMID: 7761829]

[5]   Scrocco, E.; Tomasi, J. *The electrostatic molecular potential as a tool for the interpretation of molecular properties*; New Concepts II: Berlin, Heidelberg, **1973**, pp. 95-170.
[http://dx.doi.org/10.1007/3-540-06399-4_6]

[6]   Scrocco, E.; Tomasi, J. In: *Electronic Molecular Structure, Reactivity and Intermolecular Forces: An Euristic Interpretation by Means of Electrostatic Molecular Potentials*; Löwdi, Per-Olov., Ed.; Advances in Quantum Chemistry; Academic Press,, **1978**; 11, pp. 115-193.

[7]   Bauzá, A.; Seth, S.K.; Frontera, A. Molecular electrostatic potential and "atoms-in-molecules"

analyses of the interplay between π-hole and lone pair···π/X-H···π/metal···π interactions. *J. Comput. Chem.,* **2018**, *39*(9), 458-463.
[http://dx.doi.org/10.1002/jcc.24869] [PMID: 28667684]

[8]   Murray, J.S.; Politzer, P. Molecular electrostatic potentials and noncovalent interactions. *Wiley Interdiscip. Rev. Comput. Mol. Sci.,* **2017**, *7*e13260
[http://dx.doi.org/10.1002/wcms.1326]

[9]   Bijina, P.V.; Suresh, C.H.; Gadre, S.R. Electrostatics for probing lone pairs and their interactions. *J. Comput. Chem.,* **2018**, *39*(9), 488-499.
[http://dx.doi.org/10.1002/jcc.25082] [PMID: 29094379]

[10]   Liu, L.; Miao, L.; Li, L.; Li, F.; Lu, Y.; Shang, Z.; Chen, J. Molecular Electrostatic Potential: A New Tool to Predict the Lithiation Process of Organic Battery Materials. *J. Phys. Chem. Lett.,* **2018**, *9*(13), 3573-3579.
[http://dx.doi.org/10.1021/acs.jpclett.8b01123] [PMID: 29897763]

[11]   Mehmood, A.; Jones, S.I.; Tao, P.; Janesko, B.G. An Orbital-Overlap Complement to Ligand and Binding Site Electrostatic Potential Maps. *J. Chem. Inf. Model.,* **2018**, *58*(9), 1836-1846.
[http://dx.doi.org/10.1021/acs.jcim.8b00370] [PMID: 30160959]

[12]   Schmidt, M.W.; Baldridge, K.K.; Boatz, J.A.; Elbert, S.T.; Gordon, M.S.; Jensen, J.H.; Koseki, S.; Matsunaga, N.; Nguyen, K.A.; Su, S.J.; Windus, T.L.; Dupuis, M.; Montgomery, J.A. General Atomic and Molecular Electronic-Structure System. *J. Comput. Chem.,* **1993**, *14*, 1347-1363.
[http://dx.doi.org/10.1002/jcc.540141112]

[13]   Frisch, M.J. *Gaussian 16 Revision A.03*; Gaussian Inc.: Wallingford, CT, **2016**.

[14]   Valiev, M.; Bylaska, E.; Govind, N. NWChem: A comprehensive and scalable open-source solution for large scale molecular simulations. *Comput. Phys. Commun.,* **2010**, *181*, 1477-1489.
[http://dx.doi.org/10.1016/j.cpc.2010.04.018]

[15]   Ufimtsev, I.S.; Martínez, T.J. Quantum Chemistry on Graphical Processing Units. 1. Strategies for Two-Electron Integral Evaluation. *J. Chem. Theory Comput.,* **2008**, *4*(2), 222-231.
[http://dx.doi.org/10.1021/ct700268q] [PMID: 26620654]

[16]   Liu, F.; Luehr, N.; Kulik, H.J.; Martínez, T.J. Quantum Chemistry for Solvated Molecules on Graphical Processing Units Using Polarizable Continuum Models. *J. Chem. Theory Comput.,* **2015**, *11*(7), 3131-3144.
[http://dx.doi.org/10.1021/acs.jctc.5b00370] [PMID: 26575750]

[17]   Kussmann, J.; Ochsenfeld, C. Preselective screening for linear-scaling exact exchange-gradient calculations for graphics processing units and general strong-scaling massively parallel calculations. *J. Chem. Theory Comput.,* **2015**, *11*(3), 918-922.
[http://dx.doi.org/10.1021/ct501189u] [PMID: 26579745]

[18]   Liu, F.; Sanchez, D.M.; Kulik, H.J.; Martínez, T.J. Exploiting graphical processing units to enable quantum chemistry calculation of large solvated molecules with conductor-like polarizable continuum models. *Int. J. Quantum Chem.,* **2019**, *119*e25760
[http://dx.doi.org/10.1002/qua.25760]

[19]   Kussmann, J.; Ochsenfeld, C. Employing OpenCL to Accelerate Ab Initio Calculations on Graphics Processing Units. *J. Chem. Theory Comput.,* **2017**, *13*(6), 2712-2716.
[http://dx.doi.org/10.1021/acs.jctc.7b00515] [PMID: 28561575]

[20]   Lee, T-S.; Cerutti, D.S.; Mermelstein, D. GPU-Accelerated Molecular Dynamics and Free Energy Methods in Amber18: Performance Enhancements and New Features. *J. Chem. Inf. Model.,* **2018**, *58*(10), 2043-2050.
[http://dx.doi.org/10.1021/acs.jcim.8b00462] [PMID: 30199633]

[21]   Abraham, M.J.; Murtola, T.; Schulz, R.; Páll, S.; Smith, J.C.; Hess, B.; Lindahl, E. GROMACS: High performance molecular simulations through multi-level parallelism from laptops to supercomputers.

SoftwareX, **2015**, *1-2*, 19-25.
[http://dx.doi.org/10.1016/j.softx.2015.06.001]

[22]   Stone, J.E.; Hardy, D.J.; Ufimtsev, I.S.; Schulten, K. GPU-accelerated molecular modeling coming of age. *J. Mol. Graph. Model.*, **2010**, *29*(2), 116-125.
[http://dx.doi.org/10.1016/j.jmgm.2010.06.010] [PMID: 20675161]

[23]   Anderson, J.A.; Lorenz, C.D.; Travesset, A. General purpose molecular dynamics simulations fully implemented on graphics processing units. *J. Comput. Phys.*, **2008**, *227*, 5342.
[http://dx.doi.org/10.1016/j.jcp.2008.01.047]

[24]   Hernández-Esparza, R.; Mejía-Chica, S.M.; Zapata-Escobar, A.D.; Guevara-García, A.; Martínez-Melchor, A.; Hernández-Pérez, J.M.; Vargas, R.; Garza, J. Grid-based algorithm to search critical points, in the electron density, accelerated by graphics processing units. *J. Comput. Chem.*, **2014**, *35*(31), 2272-2278.
[http://dx.doi.org/10.1002/jcc.23752] [PMID: 25345784]

[25]   Hernández-Esparza, R.; Vázquez-Mayagoitia, A.; Soriano-Agueda, L-A.; Vargas, R.; Garza, J. GPUs as boosters to analyze scalar and vector fields in quantum chemistry. *Int. J. Quantum Chem.*, **2018**, *119*: e25671.
[http://dx.doi.org/10.1002/qua.25671]

[26]   Zou, W.; Nori-Shargh, D.; Boggs, J.E. On the covalent character of rare gas bonding interactions: a new kind of weak interaction. *J. Phys. Chem. A*, **2013**, *117*(1), 207-212.
[http://dx.doi.org/10.1021/jp3104535] [PMID: 23240981]

[27]   Sanders, J.; Kandrot, E. *CUDA by Example: An Introduction to General-purpose GPU Programming*; Tsinghua University Press: Boston, **2010**.

[28]   Cruz, J.C.; Hernández-Esparza, R.; Vázquez-Mayagoitia, Á.; Vargas, R.; Garza, J. Implementation of the molecular electrostatic potential over graphics processing units. *J. Chem. Inf. Model.*, **2019**, *59*(7), 3120-3127.
[http://dx.doi.org/10.1021/acs.jcim.8b00951] [PMID: 31145605]

[29]   Szabo, A.; Ostlund, N.S. *Modern Quantum Chemistry: Introduction to Advanced Electronic Structure Theory*; Dover: New York, **1996**.

[30]   Helgaker, T.; Jørgensen, P.; Olsen, J. *Molecular Electronic Structure Theory*; John Wiley & Sons, LTD: Chichester, **2000**.
[http://dx.doi.org/10.1002/9781119019572]

[31]   Pople, J.A.; Hehre, W.J. Computation of electron repulsion integrals involving contracted Gaussian basis functions. *J. Comput. Phys.*, **1978**, *27*, 161-168.
[http://dx.doi.org/10.1016/0021-9991(78)90001-3]

[32]   Gill, P.M. Molecular integrals Over Gaussian Basis Functions. *Adv. Quantum Chem.*, **1994**, *25*, 141-205.
[http://dx.doi.org/10.1016/S0065-3276(08)60019-2]

[33]   Yasuda, K. Two-electron integral evaluation on the graphics processor unit. *J. Comput. Chem.*, **2008**, *29*(3), 334-342.
[http://dx.doi.org/10.1002/jcc.20779] [PMID: 17614340]

[34]   Asadchev, A.; Allada, V.; Felder, J.; Bode, B.M.; Gordon, M.S.; Windus, T.L. Uncontracted Rys Quadrature Implementation of up to G Functions on Graphical Processing Units. *J. Chem. Theory Comput.*, **2010**, *6*(3), 696-704.
[http://dx.doi.org/10.1021/ct9005079] [PMID: 26613300]

[35]   Sandberg, J.A.R.; Rinkevicius, Z. An algorithm for the efficient evaluation of two-electron repulsion integrals over contracted Gaussian-type basis functions. *J. Chem. Phys.*, **2012**, *137*(23)234105
[http://dx.doi.org/10.1063/1.4769730] [PMID: 23267469]

[36]   Miao, Y.; Merz, K.M., Jr Acceleration of Electron Repulsion Integral Evaluation on Graphics

Processing Units via Use of Recurrence Relations. *J. Chem. Theory Comput.*, **2013**, *9*(2), 965-976.
[http://dx.doi.org/10.1021/ct300754n] [PMID: 26588740]

[37]   Kussmann, J.; Ochsenfeld, C. Hybrid CPU/GPU Integral Engine for Strong-Scaling Ab Initio
Methods. *J. Chem. Theory Comput.*, **2017**, *13*(7), 3153-3159.
[http://dx.doi.org/10.1021/acs.jctc.6b01166] [PMID: 28636392]

[38]   Kikuchi, O.; Nakajima, H.; Horikoshi, K.; Takahashi, O. Rapid evaluation of molecular electrostatic
potential maps by simple analytical functions - extension to compounds containing pi-electron
systems. Theochem-. *J. Mol. Struct.*, **1993**, *104*, 57-69.
[http://dx.doi.org/10.1016/0166-1280(93)87019-A]

[39]   Lee, M.; Leiter, K.; Eisner, C.; Knap, J. Atom-partitioned multipole expansions for electrostatic
potential boundary conditions. *J. Comput. Phys.*, **2017**, *328*, 344-353.
[http://dx.doi.org/10.1016/j.jcp.2016.10.012]

[40]   Weiner, P.K.; Langridge, R.; Blaney, J.M.; Schaefer, R.; Kollman, P.A. Electrostatic potential
molecular surfaces. *Proc. Natl. Acad. Sci. USA*, **1982**, *79*(12), 3754-3758.
[http://dx.doi.org/10.1073/pnas.79.12.3754] [PMID: 6285364]

[41]   Gasteiger, J.; Marsili, M. Iterative partial equalization of orbital electronegativity – a rapid access to
atomic charges. *Tetrahedron*, **1980**, *36*, 3219-3228.
[http://dx.doi.org/10.1016/0040-4020(80)80168-2]

[42]   García-Risueño, P.; Alberdi-Rodriguez, J.; Oliveira, M.J.T.; Andrade, X.; Pippig, M.; Muguerza, J.;
Arruabarrena, A.; Rubio, A. A survey of the parallel performance and accuracy of Poisson solvers for
electronic structure calculations. *J. Comput. Chem.*, **2014**, *35*(6), 427-444.
[http://dx.doi.org/10.1002/jcc.23487] [PMID: 24249048]

[43]   Kumar, A.; Gadre, S.R.; Mohan, N.; Suresh, C.H. Lone pairs: an electrostatic viewpoint. *J. Phys.
Chem. A*, **2014**, *118*(2), 526-532.
[http://dx.doi.org/10.1021/jp4117003] [PMID: 24372481]

[44]   Suresh, C.H.; Gadre, S.R. Electrostatic potential minimum of the aromatic ring as a measure of
substituent constant. *J. Phys. Chem. A*, **2007**, *111*(4), 710-714.
[http://dx.doi.org/10.1021/jp066917n] [PMID: 17249762]

[45]   Kumar, A.; Gadre, S.R. Exploring the Gradient Paths and Zero Flux Surfaces of Molecular
Electrostatic Potential. *J. Chem. Theory Comput.*, **2016**, *12*(4), 1705-1713.
[http://dx.doi.org/10.1021/acs.jctc.6b00073] [PMID: 26881455]

[46]   Boys, S.F.; Electronic wave functions, I. A general method of calculation for the stationary states of
any molecular system. *Proc. Roy. Soc.*, **1950**, *A200*, 542-554.
[http://dx.doi.org/10.1098/rspa.1950.0036]

[47]   Mcmurchie, L.E.; Davidson, E.R. One- and Two-electron Integrals over Cartesian Gaussian Functions.
J. Comput. Phys., **1978**, *26*, 218-231.
[http://dx.doi.org/10.1016/0021-9991(78)90092-X]

[48]   Rys, J.; Dupuis, M.; King, H.F. Computation of Electron Repulsion Integrals Using the Rys
Quadrature Method. *J. Comput. Chem.*, **1983**, *4*, 154-157.
[http://dx.doi.org/10.1002/jcc.540040206]

[49]   King, H.F. Strategies for Evaluation of Rys Roots and Weights. *J. Phys. Chem. A*, **2016**, *120*(46),
9348-9351.
[http://dx.doi.org/10.1021/acs.jpca.6b10004] [PMID: 27934243]

[50]   Shizgal, B.D. A novel Rys quadrature algorithm for use in the calculation of electron repulsion
integrals. *Comput. Theor. Chem.*, **2015**, *1074*, 178-184.
[http://dx.doi.org/10.1016/j.comptc.2015.10.023]

[51]   Gautschi, W. *Orthogonal Polynomials: computation and approximation*; Oxford University Press:
New York, **2004**.

[52]   Schwenke, D.W. On the computation of high orders Rys quadrature weights and nodes. *Comput. Phys. Commun.,* **2014**, *185*, 762-763.
[http://dx.doi.org/10.1016/j.cpc.2013.11.004]

[53]   Shizgal, B. *Spectral Methods in Chemistry and Physics. Applications to Kinetic Theory and Quantum Mechanics*; Wiley: Boston, **2015**.
[http://dx.doi.org/10.1007/978-94-017-9454-1]

[54]   Sagar, R.P.; Smith, V.H. On the calculation of Rys polynomials and quadratures. *Int. J. Quantum Chem.,* **1992**, *42*, 827-836.
[http://dx.doi.org/10.1002/qua.560420420]

[55]   Bowdler, H.; Martin, R.; Reinsch, C.; Wilkinson, J. Handbook series linear algebra-QR and QL Algorithms for Symmetric Matrices. *Numer. Math.,* **1968**, *11*, 293.
[http://dx.doi.org/10.1007/BF02166681]

[56]   Helgaker, T.; Jorgensen, P.; Olsen, J. *Molecular Electronic-Structure Theroy*; Wiley-VCH: Weinheim, **2000**.

[57]   White, C.A.; Johnson, B.G.; Gill, P.M.W.; Head-Gordon, M. The continuous fast multipole method. *Chem. Phys. Lett.,* **1994**, *230*, 8-16.
[http://dx.doi.org/10.1016/0009-2614(94)01128-1]

[58]   Lambrecht, D.S.; Ochsenfeld, C. Multipole-based integral estimates for the rigorous description of distance dependence in two-electron integrals. *J. Chem. Phys.,* **2005**, *123*(18)184101
[http://dx.doi.org/10.1063/1.2079967] [PMID: 16292893]

[59]   Lambrecht, D.S.; Doser, B.; Ochsenfeld, C. Rigorous integral screening for electron correlation methods. *J. Chem. Phys.,* **2005**, *123*(18)184102
[http://dx.doi.org/10.1063/1.2079987] [PMID: 16292894]

[60]   Maurer, S.A.; Lambrecht, D.S.; Flaig, D.; Ochsenfeld, C.; Ochsenfeld, C. Distance-dependent Schwarz-based integral estimates for two-electron integrals: reliable tightness vs. rigorous upper bounds. *J. Chem. Phys.,* **2012**, *136*(14)144107
[http://dx.doi.org/10.1063/1.3693908] [PMID: 22502501]

[61]   Tian, L.; Chen, F. *Multiwfn: A multifunctional wavefunction analyzer*; , **2012**, 33, pp. 580-592.

[62]   Panwalkar, V.; Neudecker, P.; Schmitz, M. The Nedd4-1 WW Domain Recognizes the PY Motif Peptide through Coupled Folding and Binding Equilibria. *Biochemistry,* **2016**, *55*(4), 659-674.
[http://dx.doi.org/10.1021/acs.biochem.5b01028] [PMID: 26685112]

Molecular Electron Density Theory: A New Theoretical Outlook on Organic Chemistry

Luis R. Domingo[1,*] and **Nivedita Acharjee**[2]

[1] *Department of Organic Chemistry, University of Valencia, Dr. Moliner 50, Burjassot, E-46100, Valencia, Spain*

[2] *Department of Chemistry, Durgapur Government College, West Bengal, India*

Abstract: Organic Chemistry has evolved continuously as the backbone for the sustainability of different disciplines such as medicinal chemistry, chemical biology, biochemistry, biotechnology, material science, polymers, and nanotechnology. The beauty of organic reactions lies in their unique structural framework, reactivity, and selectivity, interesting stuff for molecular modelling research. However, theoretical interpretations of organic reactions have not been able to keep pace with the ever-increasing efficiency of computational chemistry software along the last two decades. This is probably due to the popular use of the Frontier Molecular Orbital (FMO) theory to study the course of organic reactions during the last 40 years, in spite of its failure and criticism in several cases. In 2016, Domingo proposed a new theory, called **"Molecular Electron Density Theory (MEDT)"** to study molecular reactivity of organic reactions, which is backed up by the use of quantum chemical tools. This theory proposes the decisive role of electron density changes in the reactivity of organic molecules, being opposed to FMO concepts. MEDT has been successfully applied to rationalize the experimental outcome of several Diels-Alder reactions, sigmatropic rearrangements, electrocyclic reactions, and [3+2] cycloaddition reactions. MEDT covers the detailed analysis of Conceptual DFT (CDFT) indices, exploration of the Potential Energy Surface (PES), calculation of the global electron density transfer (GEDT), topological analysis of the Electron Localization Function (ELF) and Quantum Theory of Atoms in Molecules (QTAIM), and Non-Covalent Interactions (NCI) with the aid of molecular modelling software. MEDT correlates the changes in electron density along a reaction path with the activation energies and establishes the polar character associated with the reorganization of the molecular mechanism to reach a meaningful insight into the reactivity of organic molecules. MEDT can be applied to study the mechanism, reactivities and selectivities of organic reactions, particularly those showing chemo-, regio-, and stereoselectivities in the synthesis of biologically active products. This chapter aims to provide a detailed description of the basic theoretical concepts covered by MEDT to design a precise computational model of an organic reaction. Some applications of MEDT have also been illustrated in the concluding section for ready reference.

* **Corresponding author Luis R. Domingo:** Department of Organic Chemistry, University of Valencia, Dr.Moliner 50, Burjassot, E-46100 Valencia, Spain; E-mail: domingo@utopia.uv.es

Zaheer Ul-Haq and Angela K. Wilson (Eds.)

Keywords: Bond Critical Points, Bonding Evolution Theory, Conceptual DFT, Electron Density, Topology, Electron Localization Function, Global Electron Density Transfer, Intrinsic Reaction Coordinate, Laplacian, Molecular Electron Density Theory, Non-Covalent Interactions, Potential Energy Surface, Quantum Theory of Atoms in Molecules, Transition States Structures.

INTRODUCTION

Having evolved as a highly creative discipline of science, Organic Chemistry is widely applied in medicine [1, 2], nanotechnology [3], polymers [4], industrial synthesis [5], biochemistry [6], biotechnology [7], and cosmetics [8]. "structural beauty and wide applicability of organic molecules come from their" chemical framework, which defines their unique reactivity. Thus, the interpretation of organic reactivity is a tough challenge, crossing the scientific boundaries of physical, chemical, and mathematical theories. The ever-increasing speed and capabilities of computational resources have significantly reduced the time required to calculate mathematical functions, which has provided the mathematical solution for physical and chemical theories applied to study the energetics and reactivity of molecular systems.

All chemical changes are accompanied by the absorption or release of heat, and this intimate connection of matter and energy has been in the focus of studies to understand the course of a chemical reaction. The first link between the experimental energy barrier of a chemical reaction and the theoretical concept of "transition state" was put forward in the "Transition State Theory (TST)" [9]. The chemical changes are associated with changes in the bonding framework of a molecular system, which causes energy transformations along a reaction path. Thus, the interplay of theory and experiment is well understood when exploring the sequence of bonding changes along a chemical reaction. Understanding "chemical bonding" in its deepest aspect and the electronic structure of a molecule is, therefore, crucial to analyze the energetics of a chemical reaction.

The Lewis concept of chemical bonding [10] laid the foundation for the advent of two new theories, popularly used in Organic Chemistry, namely the Valence Bond Theory [11 - 13] (VBT) and Molecular Orbital Theory [14] (MOT). VBT and MOT are two approximate methods for the solution of Schrödinger's equation [15]. Within VBT, organic reactivity is explained by hybridisation, resonance, inductive effect, and other relevant concepts. MOT evolved as a unique concept proposing the movement of electrons along molecular orbitals (MOs) in a molecule. MOT considers the Born Oppenheimer's approximation [16], which considers a molecule as an entity and each electron moves under the influence of the potential field of all the nuclei, rather than being centered on one nucleus.

MOs are constructed by the linear combination of atomic orbitals [17] (LCAO) from each atom of the molecule. Huckel's Molecular Orbital Theory [18 - 20] is a popular approach used in Organic Chemistry, where MOs are built only from valence atomic orbitals. This concept was presented as a novel treatment of chemical reactivity and named Frontier Molecular Orbital (FMO) Theory by Fukui *et al.* in 1952 [21]. Many years later, the FMO theory gained serious interest, when it was applied by Woodward and Hofmann [22] in 1965 to interpret the electrocyclic ring opening of cyclobutenes. The FMO theory thus started gaining ground as a reliable theoretical approach for the interpretation of organic reactions, particularly cycloaddition reactions [23, 24]. The FMO concept was advocated in the 1960-70s by Woodward and Hoffmann [25], Fukui [26], and Pearson [27], which led to the use of FMOs for the interpretation of cycloadditions by Houk [28] in 1972-73. Failures of the FMO theory were also recognized [23], but such cases were labelled as uncommon, subject to special circumstances in the 1970s. The general failure of the FMO theory to interpret electrophilic aromatic substitutions, S_N2 reactions, and so-called "pericyclic reactions" was critically reviewed and questioned by M.J.S. Dewar [29] in 1989. The author pointed out two reasons for the popularity and general acceptance of the FMO theory in Organic Chemistry: (I) the FMO theory is claimed as the only alternative for the interpretation of "pericyclic reactions", which cannot be explained by any other concept [23]; (II) by using the FMO theory, calculations can be performed swiftly, and the reactivity of organic reactions can be interpreted without using any concept of Organic Chemistry, which naturally, led to the popularity of the FMO concept. Thus, although the failures of the FMO theory were recognized in many cases, it is applied as an easy approach reactivity to understand organic reactions, particularly "pericyclic reactions".

The growth of computational chemistry [30] in the 20th and 21st century has made it possible to explore the potential energy surface (PES) [31] defined under the TST [9]. Consequently, reactants, products, transition state structures (TSs) and intermediates can be characterized on the PES. Now, a proper quantum chemical approach is required to obtain the electronic energies of these stationary states. As a consequence, a new quantum mechanical modelling method based on the Hohenberg and Kohn theorem [32], called the Density Functional Theory (DFT) [33] was proposed using Kohn-Sham equations [34]. DFT expresses the ground state energy of a non-degenerated N electron system as a function of electron density (ρ). DFT has proved to be a major breakthrough in computational chemistry due to its accurate predictions of energy and reactivity, unlike *ab initio* calculations [35].

Electronic energies of the stationary points located on the PES of a reaction can be obtained precisely by DFT calculations, but a proper theoretical approach is needed to work out the molecular reactivity and, hence determine the sequence of bonding changes along the reaction path.

Numerous physical chemists have analysed electron density changes to interpret molecular reactivity since the 1980s, but organic chemistry remained at a standstill and derived FMO concepts of the 1960s & 1970s to interpret molecular reactivity. In 2014, into a review in which a new model for the C-C single bond formation in cycloaddition reactions was presented, Domingo [36] proposed a new idea to interpret the molecular reactivity in organic reactions based on the changes in electron density. This concept was proposed after the interpretation of a vast array of organic reactivity for 20 years, majority of them dedicated to Diels-Alder (DA) and [3+2] cycloaddition (32CA) reactions, which were considered as strong arguments in favour of the FMO theory in the 1960-70s [23]. The reactivity concept proposed by Domingo was called "**Molecular Electron Density Theory (MEDT)**" for the first time in 2015 to interpret the domino process for the nucleophilic attack of a carbenoid intermediate on carbonyl compounds to generate 2-iminofuran derivative [37]. MEDT [38] published in 2016 states that *the changes in electron density and not MO interactions are responsible for molecular reactivity*, and it opposes the FMO theory. Within MEDT, computational quantum calculations (CQC) are performed to study the changes in electron density along a chemical reaction, and the energies associated with these changes are calculated and interpreted. MEDT involves a complete coverage of quantum chemical principles to interpret the molecular reactivity from electron density changes. This chapter provides a detailed insight into these quantum-chemical principles employed within MEDT and their applications. Mainly, three computational software are used within MEDT: Gaussian suite of programs [39], the Topmod Package [40], Multiwfn [41], and the visualization software such as VMD [42], UCSF Chimera [43], *etc.*

CONCEPTUAL DENSITY FUNCTIONAL THEORY (CDFT) INDICES

For an N electron system, the DFT concept based on Hohenberg and Kohn theorem [32] expresses electron density, ρ_r, as the functional derivative of ground state energy, E, with respect to the external potential v_r,

$$\rho_r = \left(\frac{\delta E}{\delta v_r} \right)_N \tag{1}$$

Within the DFT framework, the reactivity of a molecular system depends on the changes in the number of electrons and external potential created by the nuclei.

This allows defining a series of reactivity descriptors, popularly known as the Conceptual DFT (CDFT) indices. The concept of "*Conceptual DFT*" dates back to the pioneering work of Parr [44], and was subsequently reviewed by Geerlings, Proft, and Langenaeker in 2003 [45], and recently, by Domingo *et al.* in 2016 [46]. While the global CDFT indices [46, 47] remain unaltered at the different molecular centers, the local CDFT indices present different values. Within the MEDT, the most relevant CDFT indices are analyzed to address the chemical behaviour of the reactants. The more commonly global and local CDFT indices used within MEDT are discussed below.

Global CDFT Indices

Electronic Chemical Potential (μ)

The electronic chemical potential μ of a molecular system was defined by Parr [44] in 1983 as the energy changes at the ground state with respect to the number of electrons, N, at a fixed external potential v_r. The electronic chemical potential μ quantifies the propensity of a system to exchange electron density (at its ground state) with the environment. The electronic chemical potential μ can be approximated by evaluating it directly from the ionization potential (I) and electron affinity (A), which can be expressed as

$$\mu \approx -\frac{(I+A)}{2} \tag{2}$$

Using Kohn-Sham equations [34] and Koopman's theorem [48] within the DFT framework, I and A can be approximated in terms of HOMO ($-E_{\text{HOMO}}$) and LUMO ($-E_{\text{LUMO}}$) energies respectively and hence the electronic chemical potential μ can be expressed

$$\mu \approx \frac{E_{\text{HOMO}} + E_{\text{LUMO}}}{2} \tag{3}$$

Chemical Hardness (η) and Softness (S)

The hard and soft acid-base (HSAB) principle was established by Pearson in 1963 [49], specifying the hard-hard and soft-soft interactions between Lewis acids and bases. In 1983, Parr [47] expressed chemical hardness η as the changes in electronic chemical potential μ of a system with respect to the number of electrons, N, at a constant external potential v_r. The chemical hardness η measures the resistance of a molecule to exchange electron density with the environment. The chemical hardness, denoted by η, is expressed as

$$\eta = \left(\frac{\partial \mu}{\partial N} \right)_{v_r} \tag{4}$$

η can be expressed in terms of ionization potential (I) and electron affinity (A) as

$$\eta \approx \frac{I - A}{2} \tag{5}$$

From Koopman's theorem [48], η can be expressed in terms of the frontier HOMO and LUMO energies as

$$\eta \approx \frac{E_{\text{LUMO}} - E_{\text{HOMO}}}{2} \tag{6}$$

Generally, the factor 1/2 is neglected, which reduces Eq (6) to

$$\eta \approx E_{\text{LUMO}} - E_{\text{HOMO}} \tag{7}$$

The chemical softness S is expressed as the inverse of chemical hardness η, and it is given as

$$S = \frac{1}{\eta} \tag{8}$$

Electrophilicity Index (ω)

Electrophilicity index ω was defined by Parr *et al.* [50] in 1999 as the energy stabilisation of a molecular system when an additional electronic charge is acquired by the system from the environment. The electrophilicity index ω is expressed as

$$\omega = \frac{\mu^2}{2\eta} \tag{9}$$

Thus, electrophilicity index ω includes two factors: (I) the feasibility of a system to acquire an electronic charge (described by μ^2), and (II) the system resistance to exchange electronic charge with the environment (described by η). High value of ω is obtained by a high μ value and a low η value of the molecular system, which characterizes a good electrophile.

The maximum number of electrons, ΔN_{max}, that can be acquired by a system, which is expressed as [50]

$$-\Delta N_{max} = \frac{\mu}{\eta} \tag{10}$$

In 2002, Domingo [51] defined the single electrophilicity scale at DFT/B3LYP/6-31G(d) level of theory to classify organic molecules participating in DA reactions as strong, moderate and marginal electrophiles. Strong electrophiles are characterized by $\omega > 1.50$ eV, while moderate electrophiles show ω values in the range 0.80 eV $\leq \omega \leq 1.50$ eV. Molecules with $\omega < 0.80$ eV are classified as marginal electrophiles. Consistent increase is observed in ΔN_{max} values with the increase in electrophilicity index, ω. In 2007 [52], Domingo computed electrophilicity indices of a series of organic molecules by two processes: (I) vertical ionization potentials and vertical electron affinities, and (II) HOMO/LUMO energies. Data obtained from these processes presented a good correlation, which identified the reasonable use of HOMO/LUMO energies to devise valuable electrophilicity scales. Domingo [53] reported in 2009 that electrophilicity index ω shows a good correlation with the computed activation energies of DA reactions.

In 2002 and 2003, Domingo and co-workers [51, 54] presented the quantitative characterization of the reagents involved in DA and 32CA reactions from the CDFT concept at DFT/B3LYP/6-31G(d) level of theory. Some of these reported results are shown in Table 1. At the top of Table 1 are found strong electrophilic species, which present high μ and η values, while at the bottom are found poor electrophilic species, which present low μ and high η values.

Table 1. DFT/B3LYP/6-31G(d) calculated electronic chemical potential (μ), chemical hardness (η), electrophilicity (ω) indices of some common organic molecules [51, 54].

Entry	Molecule	μ (au)	η (au)	ω (eV)
1	Ozone	-0.26	0.15	6.10
2	Tetracyanoethylene	-0.26	0.15	5.96
3	Nitroethylene	-0.20	0.20	2.61
4	Carbonyl oxide	-0.17	0.15	2.43
5	Acrolein	-0.16	0.19	1.84
6	Methyl acrylate	-0.16	0.23	1.51
7	Nitrous oxide	-0.18	0.32	1.37
8	Nitrone	-0.13	0.20	1.06
9	1,3-butadiene	-0.13	0.21	1.05
10	Cyclopentadiene	-0.11	0.20	0.83
11	Ethylene	-0.12	0.29	0.73
12	Nitrile oxide	-0.12	0.29	0.73
13	Azide	-0.12	0.30	0.66
14	Acetylene	-0.11	0.33	0.54
15	Methyl vinyl ether	-0.09	0.26	0.42

It is evident that the presence of the electron-withdrawing group increases μ value and hence increases the feasibility of the molecule to accept electrons. Following the electronegativity equalisation principle [55, 56], when two atoms A and B approach towards each other, the difference in electronegativity between them results in the flux of electron density. For two molecules undergoing polar reaction, the difference in electronic chemical potentials $\Delta\mu$ indicates the direction of electronic flux between them. Non-polar processes are associated with minimal differences in electronic chemical potentials $\Delta\mu$ at the ground state of the reagents.

Nucleophilcity Index (N)

Since 1998, the nucleophilicity index has been studied by several approaches [57 - 59]. In 2007, Gazquez [60] explained the propensity of a molecule to donate electrons by defining ω^- power and to accept electrons by ω^+ power, which are expressed in terms of the ionization potential I and electron affinity A of the molecular system as

$$\omega^+ = \frac{A^2}{2(I-A)} \tag{11}$$

$$\omega^- = \frac{I^2}{2(I-A)} \tag{12}$$

The relative nucleophilicity index N, proposed by Domingo *et al.* [61] in 2008, is expressed as

$$N = E_{\text{HOMO(Nucleophile)}} - E_{\text{HOMO(TCE)}} \tag{13}$$

Here, TCE denotes tetracyanoethylene, which is considered as the neutral species showing very low HOMO energy.

A different concept of nucleophilicity index was put forward in 2010 by Roy *et al.* [62], which proposed nucleophilicity index N' as the inverse of ω^- power. N' was subsequently defined as N'' to take care of the low values of N' (less than 1).

$$N'' = \frac{1}{\omega^-} \times 10 \tag{14}$$

Thus, the nucleophilicity index concept was defined by three different approaches: (I) relative nucleophilicity N of molecular systems with respect to tetracyanoethylene; (II) inverse of electrophilicity N', and (III) Roy's nucleophilicity index N''.

These three indices were compared by Domingo and Pérez [63] in 2011 for a series of organic molecules. A very good correlation was observed in all three cases for simple systems, such as *para* substituted phenols. However, for complex organic systems having two or more functional groups of different electronic nature, only the relative nucleophilicity index N showed a good correlation with the activation energies of polar organic reactions. Relative nucleophilicity index N of some common organic molecules are listed in Table **2**.

Table 2. DFT/B3LYP/6-31G(d) calculated Nucleophilicity (N) indices of some common organic molecules

Molecule	N (eV)
Aniline	3.72
Hydrazine	3.65
Phenol	3.16
Toluene	2.71
Benzene	2.42
Acetophenone	2.39
Benzaldehyde	2.17
Methanol	1.92
Nitrobenzene	1.53
Water	1.20

The molecular systems were classified as strong, moderate or marginal nucleophiles within the nucleophilicity scale [64]. Strong nucleophiles are characterized by N values greater than 3.00 eV, while moderate nucleophiles are characterized by N values in the range 2.00 eV $\leq N \leq$ 3.00 eV. Marginal electrophiles present N values lesser than 2.00 eV. The presence of electron-releasing groups increases the nucleophilicity index N, while molecular systems bearing electron-withdrawing groups show moderate to marginal nucleophilicity.

Local CDFT Indices

Fukui Functions

In 1984, Parr [65] defined Fukui function (denoted by $f(r)$) for a molecular system as the variation of electron density $\rho(r)$ at point r with respect to the changes in number of electrons N at a constant external potential $v(r)$.

$$f(\text{r}) = \left(\frac{\partial \rho(\text{r})}{\partial N} \right)_{v(\text{r})} \qquad (15)$$

Fukui function at point r of the molecular system was expressed by Parr and Yang [65] by using frozen core approximation as

$$f^{+}(\text{r}) \approx \rho_{\text{HOMO}} \text{ (For nucleophilic attack)} \qquad (16)$$

$$f^-(\mathrm{r}) \approx \rho_{\mathrm{LUMO}} \text{ (For electrophilic attack)} \tag{17}$$

Further, Yang and Mortier [66] approximated Fukui function by the finite difference method, which is expressed as

$$f^+(\mathrm{r}) \approx \rho_{N+1}(\mathrm{r}) - \rho_{N_0}(\mathrm{r}) \text{ (For nucleophilic attack)} \tag{18}$$

$$f^-(\mathrm{r}) \approx \rho_{N_0}(\mathrm{r}) - \rho_{N-1}(\mathrm{r}) \text{ (For electrophilic attack)} \tag{19}$$

where p_{N0}, ρ_{N-1} and ρ_{N+1}, are the atomic charges for species containing N (neutral), $N-1$ (cationic), and $N+1$ (anionic) number of electrons

The most electrophilic and nucleophilic centers in an organic molecule can be characterized by the calculation of local electrophilicity [67] ω_k index and local nucleophilicity [68] N_k index at site k. These indices are defined in terms of Fukui functions as

$$\omega_k = \omega f_k^+ \tag{20}$$

$$N_k = N f_k^- \tag{21}$$

Where ω and N are, respectively, the global electrophilicity and nucleophilicity indices, while f_K^+ and f_K^- are the Fukui functions for nucleophilic and electrophilic attacks.

Another descriptor, was proposed by Morell *et al.* [69] in 2005 to analyse the reactivity observed in chemical systems, which is given by

$$\Delta f(\mathrm{r}) = [f^+(\mathrm{r}) - f^-(\mathrm{r})] \tag{22}$$

Where $f^+(r)$ and $f^-(r)$ are the Fukui functions for nucleophilic and electrophilic attacks.

$\Delta f(r) > 0$ for an atomic site indicates the preferred site for a nucleophilic attack, whereas $\Delta f(r) < 0$ for an atomic site indicates the preferred site for an electrophilic attack. The index shows good agreement with experiments for activated and deactivated monosubstituted benzenes and can be employed to rationalize the Burgi-Dunitz trajectory on carbonyl compounds [69].

In 2012, Chattaraj and Domingo [70] proposed a single reactivity difference index R_k to characterize the local electrophilicity/nucleophilicity activation in complex organic molecules. The definition of R_k is given by the following three conditions to identity the centers with electrophilic ($R_k = +$n.nn) or nucleophilic ($R_k = -$n.nn) behavior as well as the ambiphilic ($R_k = \pm$n.nn) behavior in addition to eliminate the centers with marginal reactivity:

(a) If $(1 < \omega_k/N_k < 2)$ or $(1 < N_k/\omega_k < 2)$

      then $R_k \approx (\omega_k + N_k)/2$) $\Rightarrow$ ambiphilic ($R_k = \pm$n.nn)

(b) Else $R_k \_\approx (\omega_k - N_k)$

$$(23)$$

      where $R_k > 0$ $\Rightarrow$ electrophilic ($R_k = +$n.nn)

      and $R_k < 0$ $\Rightarrow$ nucleophilic ($R_k = -$n.nn)

(c) if $|R_k| < 0.1$, then $R_k = 0$

Together with the electrophilic and/or nucleophilic behavior of the center k given by the sign $(+, -, \pm)$, the magnitude of the R_k index n.nn accounts for the extent of the electronic/nucleophilic activation, a behavior that allows for the use of the R_k index as a measure of the molecular reactivity especially in polar processes. The R_k index has proved to be a useful index in the study of intramolecular reactions [71 - 73].

Parr Functions

When two reagents approach each other in a polar reaction, then, the two centre-to-centre interaction takes place between the most electrophilic and nucleophilic centers of the reagents [36]. On transfer of one electron from the nucleophile to the electrophile, the nucleophile is converted to a radical cation and the electrophile to a radical anion. Analysis of atomic spin densities (ASD) of the radical ions provides the distribution of electron density in the radical species. The ASD analysis determines the electron density changes at the reactive sites and is opposed to the FMO approach since the latter proposes nucleophile HOMO and electrophile LUMO interactions. In 2013, Domingo [74, 75] proposed a new local CDFT index for polar reactions, namely the electrophilic and nucleophilic Parr functions, obtained from the ASD analysis of the radical cation and radical anion.

$$P^+(\mathrm{r}) = \rho_s^{ra}(\mathrm{r}) \quad \text{(for nucleophilic attack)} \tag{24}$$

$$P^-(\mathrm{r}) = \rho_s^{rc}(\mathrm{r}) \quad \text{(for electrophilic attack)} \tag{25}$$

Where P^+ *(r)* and P^- *(r)* are the ASDs of radical cation and radical anion respectively.

The local electrophilicity ω_k index and local nucleophilicity N_k index can be calculated in terms of Parr functions as [74]:

$$\omega_k = \omega P_k^+ \tag{26}$$

$$N_k = N P_k^- \tag{27}$$

Domingo [74] presented a comparative analysis for the Parr functions, the Parr-Yang [65] (PY) Fukui functions, and the Yang-Mortier [66] (YM) Fukui functions at DFT/B3LYP/6-31G(d) level. The calculated values of propene are given in Fig. (**1**). The sums of P_k^+ and P_k^- of three carbon atoms are 0.95 and 0.99, respectively, which are comparable to the calculated sums of PY f_k^+ and f_k^- values of 0.92 and 0.94. However, the sums of YM f_k^+ and f_k^- values are 0.16 and 0.21, which will lead to serious errors in the local electrophilicity [67] ω_k and local nucleophilicity [68] N_k indices, used to predict the most electrophilic-nucleophilic centers, and consequently, the regioselectivity.

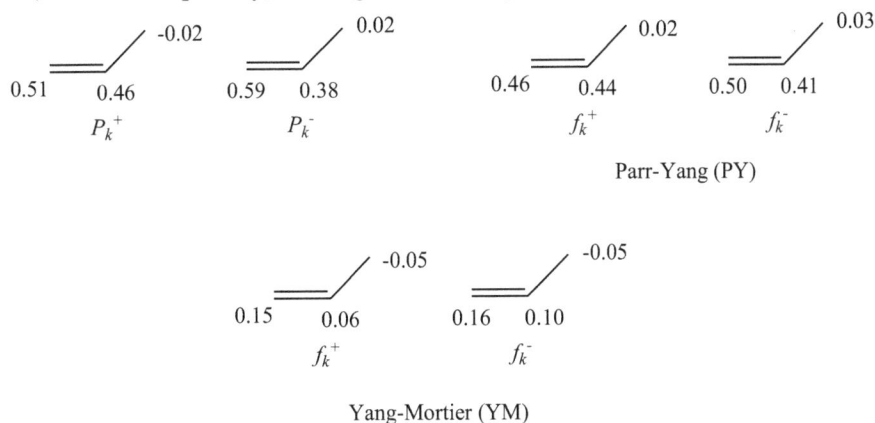

Parr-Yang (PY)

Yang-Mortier (YM)

Fig. (1). Fukui functions and Parr Functions of propene.

The Parr functions, PY Fukui functions, and YM Fukui functions for the electrophilic attack of acrolein are given in Fig. (**2**). The Parr functions and PY Fukui functions predict the β-conjugated and carbonyl carbons as the more electrophilic center of acrolein, while the YM Fukui function predicts the carbonyl oxygen as the most electrophilic center, which is contrary to the well-known regioselectivity observed in α,β-unsaturated carbonyl compounds. Thus, Parr functions supersede YM Fukui functions for the characterization of electrophilic and nucleophilic centers in organic molecules, and they are recommended for such calculations.

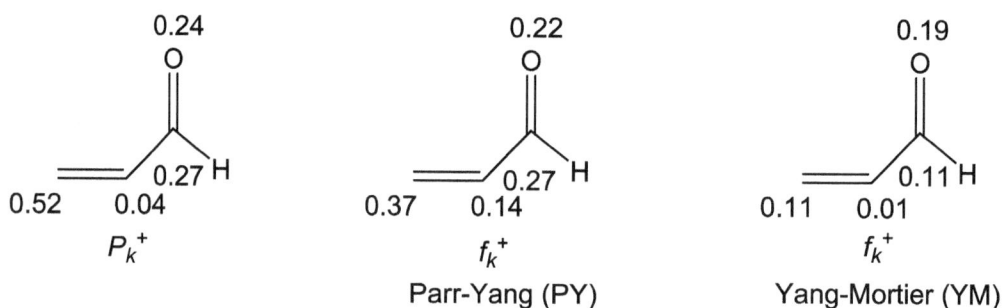

Fig. (2). Fukui functions and Parr Functions for electrophilic attack of acrolein.

Reactivity Indices for Free Radicals

In 2013, Domingo and Pérez [76] proposed a series of reactivity indices to predict the electrophilic and nucleophilic character of free radicals. The electrophilicity index ω^0 of free radicals is

$$\omega^0 = \frac{(\mu^0)^2}{2\eta^0} \tag{28}$$

Where μ° is the global chemical potential, and η° is the global hardness at the ground state of the free radical. μ° and η° are given by:

$$\mu^0 = \frac{E_{HOMO}^{\alpha^0} + E_{LUMO}^{\beta^0}}{2} \tag{29}$$

$$\eta^0 = E_{LUMO}^{\beta^0} - E_{HOMO}^{\alpha^0} \tag{30}$$

Where $E^{\alpha^{\circ}}{}_{HOMO}$ is the one electron energy of the HOMO in the α spin state of free radical and $E^{\beta^{\circ}}{}_{LUMO}$ is the one electron energy of the LUMO in the β spin state of free radical.

Relative nucleophilicity, N^0, of the free radical is expressed as

$$N^0 = E^{\alpha^0}_{HOMO} - E^{\alpha^0}_{HOMO}(DCM) \tag{31}$$

Where DCM is dicyanomethyl radical, which is considered as the reference for global nucleophilicity of free radicals.

The local radical Parr function P_k° at site k is expressed as the atomic spin density (ASD), ρ_k, of the neutral radical at site k

$$P_k^0 = \rho_k \tag{32}$$

The local ω°_k electrophilicity and N_k° nucleophilicity indices for free radicals are given by

$$\omega_k^0 = \omega P_k^0 \tag{33}$$

$$N_k^0 = N P_k^0 \tag{34}$$

DFT/B3LYP/6-31G(d) calculated reactivity indices for methyl free radicals are given in Table 3. The simplest methyl radical with $\omega^0 = 1.43$ eV is a moderate electrophile (0.80 eV $\le \omega \le 1.50$ eV) and a marginal nucleophile with $N^0 = 1.79$ eV ($N^0 < 2.00$ eV). Introduction of cyano group causes electrophilic activation with $\omega^0 = 3.79$ eV for cyanomethyl radical and $\omega^0 = 6.54$ eV for dicyanomethyl radical, while methoxy group decreases the electrophilic index with calculated $\omega^0 = 0.63$ eV for methoxymethyl radical. The presence of both cyano, as well as methoxy substituents, causes electrophilic as well as nucleophilic activation of the free radical (see Table 3). The electrophilicity ω^0 and nucleophilicity N^0 indices of free radicals show poor correlation ($R^2 = 0.67$) to each other. The trend of local indices does not show any dependence on the electron-releasing or electron-withdrawing substituents of the methyl radical. In the case of the cyanomethyl radical, the electrophilicity is concentrated on carbon C1, while the nucleophilicity power is also concentrated on the carbon C1 in the case of the methoxymethyl radical.

Table 3. CDFT indices of free radicals.

Entry	Free radical	μ^0 (eV)	η^0 (eV)	ω^0 (eV)	N^0 (eV)	ω^0_k(eV), C1	N^0_k (eV), C1
1		-3.70	4.80	1.43	1.79	1.65	2.07
2		-5.29	3.70	3.79	0.75	3.36	0.66
3		-6.35	3.08	6.54	0.00		
4		-2.24	3.96	0.63	3.67	0.60	3.48
5		-3.99	3.16	2.53	2.32		

Within the MEDT studies, the analysis of global and local CDFT indices at the ground state of reagents provides the initial comprehension of the polar character and direction of electronic flux along a reaction path.

LEWIS-LIKE STRUCTURE FROM TOPOLOGICAL ANALYSIS OF ELECTRON LOCALIZATION FUNCTION (ELF)

The Lewis's Valence theory [77] is the foundation for modern theories on chemical bonding. This theory assumes the possibility to identify a group of atoms spatially distributed in a molecule, and consequently, the regions with the maximum probability of finding a given group of electrons. In this context, the Electron Localization Function (ELF) concept introduced by Becke and Edgecombe [78] in 1990 presents a precise mathematical representation of the Lewis valence theory for simple and complex molecular systems. In 1994, Silvi and Savin [79] extended the ELF concept to propose three basin types of localization attractors: core, bonding, and non-bonding valence ELF attractors. Basins are obtained by the topological partitioning of ELF gradient fields. This classification leads to the identification of different regions in a chemical system. Core basins, labelled as C(A) surround the atomic nuclei. Valence basins can be monosynaptic [V(A)], disynaptic [V(A,B)], and trisynaptic [V(A,B,C)], depending on the number of atomic shells associated to them. The synaptic order of a valence basin denotes the number of core basins, which share a boundary with the valence basin. V(A) monosynaptic basins are associated with the presence of lone pairs, *pseudoradical* or carbenoid centers [80, 81], while V(A,B) disynaptic basins are associated with the bonding region between A and B. Valence basin populations, can be calculated by integrating the electron density over the volume of basins, thus allowing a quantitative analysis. A straightforward connection between electron density and molecular structure is established by the topological analysis of the ELF.

Domingo and Ríos-Gutiérrez [82, 83] applied ELF topological analysis to classify the three-atom-components (TACs) participating in 32CA reactions. Carbon atoms with a V(A) monosynaptic basin integrating at a population less than 1e are associated with *pseudoradical* centers, while those with a monosynaptic basin integrating more than 1 e for neutral molecules are associated with carbenoid centers [82]. TACs showing the presence of one *pseudoradical* center are classified as *pseudo(mono)radical* TAC, those showing the presence of two *pseudoradical* centers are classified as *pseudodiradical* TAC, and those with a carbenoid center are classified as carbenoid TAC. The absence of a *pseudoradical* or a carbenoid center classifies a TAC as zwitterionic. Lewis like structures can be proposed for chemical systems from the calculated valence basin populations, and not on the number of basins. The proposed Lewis like structures and TAC classifications are shown in Fig. (**3**). Simplest nitrile ylide [84] and nitrile imine [85] showing the presence of carbenoid centers at the carbon are classified as carbenoid type TACs. Simplest azomethineylide [80] and carbonyl ylide [80] and carbonyl ylide [81] showing the presence of two *pseudoradical* centers are classified as *pseudodiradical* TACs. Simplest diazoalkane [86] and azomethine imine [87] showing the presence of one *pseudoradical* center are classified as *pseudo(mono)radical* TACs. Finally, Simplest nitrone [88] and nitrile oxide [89] do not show either the presence of any *pseudoradical* or a carbenoid center, allowing their classification as zwitterionic TACs.

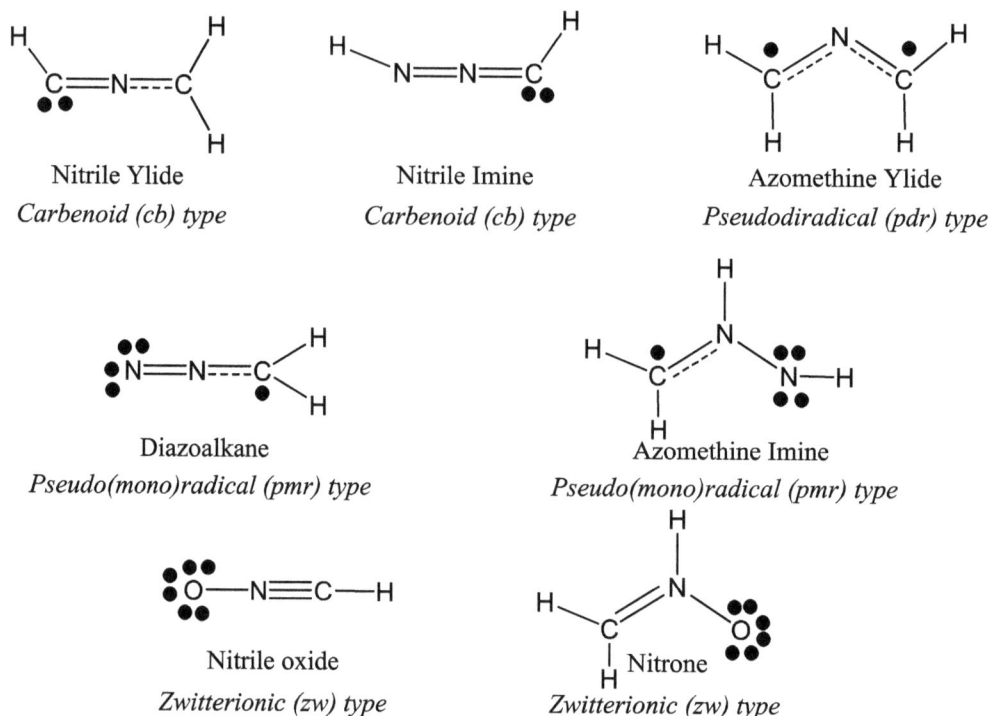

Fig. (3). Lewis-like bonding model and classification of three atom components (TACs) participating in 32CA reactions.

This classification of the TACs has been related to the activation energies of 32CA reactions. Domingo and Ríos-Gutiérrez [82, 83] calculated activation energy barrier for 32CA reaction of simplest azomethine ylide, azomethine imine, nitrile ylide, and nitrone to ethylene as 1.0, 7.7, 7.8, and 13.4 kcal/mol at MPWB1K/6-311G(d,p) level of theory. The activation energies increase in the order *pseudodiradical* < *pseudomonoradical* ≈ carbenoid < zwitterionic. The electronic structure of the TAC was related to its reactivity in 32CA reactions, thus defining the *pdr-type*, *pmr-type*, *cb-type*, and *zw-type* 32CA reactions [82, 83].

ELF localisation domains can be visualized for molecular systems using UCSF Chimera software [43]. MPWPW91/6-311G(d,p) optimized substituted ethylene derivatives and their ELF basin populations were reported in one of our earlier studies [90] in 2018. ELF localized domains of these optimized structures created by UCSF Chimera software are given in Fig. (**4**).

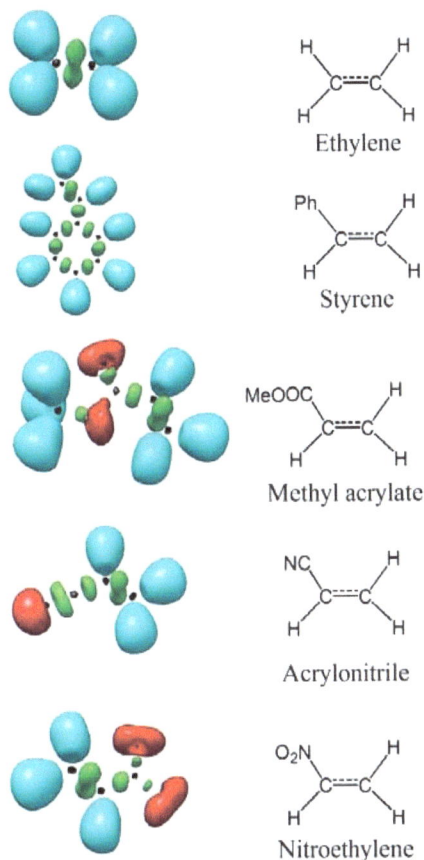

Fig. (4). ELF localization domains [Isovalue: 0.79] of monosubstituted ethylenes. Protonated basins are shown in blue, disynaptic basins are shown in green, monosynaptic basins are shown in red, and core basins are shown in black colours.

Monosynaptic basins (shown in red) in the case of acrylonitrile, nitroethylene, and methyl acrylate can be associated with lone pair regions associated with nitrogen and oxygen atoms. At MPWPW91/6-311G(d,p) level of theory, ethylene, styrene, methyl acrylate, acrylonitrile, and nitroethylene show the presence of disynaptic V(C4,C5) & V'(C4,C5) with the total integrating populations of 3.38e, 3.38e, 3.36e, 3.30e, and 3.48e, respectively, which can be associated with the underpopulated double bond in these compounds. This allows proposing the Lewis-like bonding model of these substituted ethylene derivatives (see Fig. **4**).

EXPLORING THE POTENTIAL ENERGY SURFACE (PES)

The mechanism of a chemical reaction is outlined in two stages: (a) Specification of the participating species; and (b) Identification of the elementary chemical step, and determining the stoichiometry and kinetics of the step. Fig. (**5**) shows an elementary chemical step in which the reactant (R) moves to product (P) passing through a TS.

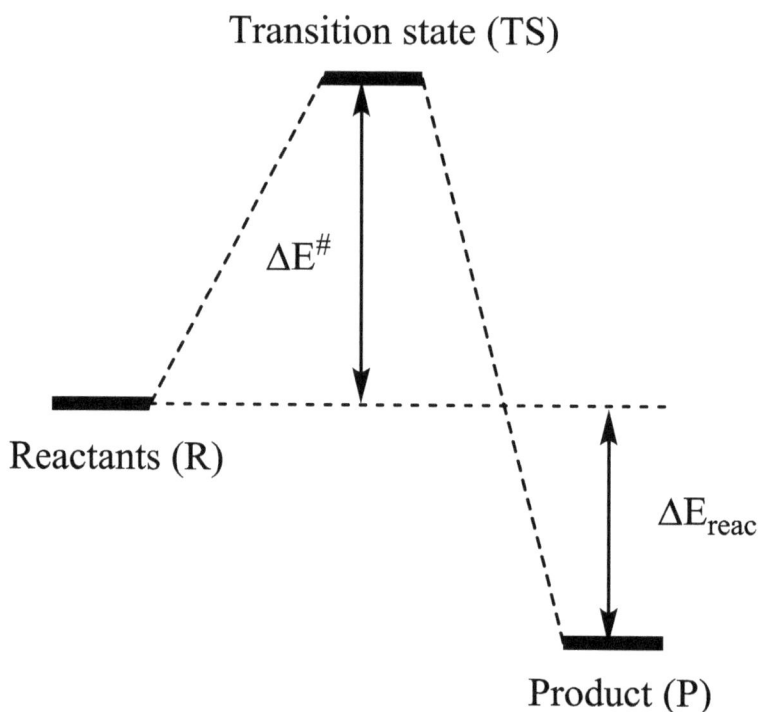

Fig. (5). Elementary step of a chemical reaction.

In computational chemistry, chemical reactivity is understood by modelling the configurations through which the reactants get transformed into products. Energy transformation along a chemical reaction is analyzed from the free energy profile obtained by finding the TSs that match with the experimental data. If the reaction is very exergonic, it will be irreversible, and consequently, the reaction will be kinetically controlled; in this case, the reaction product associated with the lowest TS energy will be the favoured. If the reaction is poor exergonic, the reaction will be reversible and thermodynamically controlled. In that case, the reaction product with the lowest energy will be favoured.

Using the Born-Oppenheimer approximation [16], the minimum electronic energy of a system associated with its stationary electronic state can be assumed to define the equilibrium arrangement of the bonded atoms in a system. With the advent of high-speed computers and simple but efficient chemical theories in the last decade, it has been possible to locate and characterize stationary electronic states of molecular systems. The study of reaction dynamics is based on the location of stationary states on the PES [31, 91]. The concept of PES, explained by Sutcliffe [91] in 2006, states each molecular geometry in a chemical reaction is associate with a potential energy. This allows a topological perspective of a chemical reaction, where molecular arrangements evolve along *"valleys"* and *"passes"* on the PES. Reactants, products, and intermediates are the minimum positions in the PES valley, while TS are saddled points of first order on the PES. A continuous curve connects two minima through a saddle point of index one, which defines the formulation of reaction coordinate [92]. The reaction minimum moves to the product minimum through TS and intermediates on the PES. Detailed exploration of the PES is performed under MEDT to extract the minima, maxima and saddle points for a chemical reaction. Location and characterization of TS is a key aspect of PES analysis, which is verified by the presence of a single imaginary frequency at the TS, and the intrinsic reaction coordinate [92] (IRC) study. A reaction path is obtained by following the minimum energy path from the TS to the reactants and the products. Fukui [92] stated that this path becomes IRC when mass-weighted coordinates are used. In 1990, Gonzalez and Schlegel [93, 94] defined a modified algorithm to follow the IRC in mass-weighted internal coordinates, which has been used in several MEDT studies [82 - 90] of organic reactions to verify the reaction path.

GLOBAL ELECTRONIC FLUX IN THE LOWERING OF ACTIVATION ENERGIES: THE GEDT CONCEPT

In 1996, Houk [95] studied the mechanism and activation parameters for the DA reaction between butadiene and ethylene. Computed activation enthalpies at RHF/6-31G(d), B3LYP/6-31G(d), and MP2/6-31G(d) levels are 47.4, 24.8, and 20.0 kcal/mol, respectively while the experimental value is 27.5 kcal/mol. It is interesting to note that the C-C bond distance in the optimized TSs obtained using these three computational methods was 2.2 Å, in spite of the wide variation in activation energies. Thus, we arrive at the question: *What causes a difference in the activation energies of the TSs with similar optimized geometries?*

Let us study two theoretical investigations carried out in this context. In 1996 [96], Sustmann and Sicking studied the DA reactions of substituted 1,3-butadiene systems to acrylonitrile, fumaronitrile and 1,1,-dicyanoethylene theoretically. Covalent interactions and ionic stabilizations of reactants were found to lower the activation energies. In 1999, Domingo [97] carried out DFT calculations for the DA reactions of nitroethylene and three ethylene derivatives of increased nucleophilicity. The decrease in activation energies with the increase in the nucleophilic character of the ethylene derivative was observed in that study. Thus, the influence of reactant polarity on the reaction course was qualitatively established. Finally, Domingo [98] provided a quantitative assessment of these findings in 2003 by correlating the activation parameters and global electrophilicity ω index of the DA reactions between cyclopentadiene and cyanoethylene series. In 2009, Domingo [99] reported a good correlation (R^2 = 0.99) between the experimental rate constants and charge transfer (CT) at the TSs associated with these reactions, thus proposing the polar DA reaction mechanism, followed for most of the DA reactions taking place experimentally. CTs were obtained with the Natural Bond Orbital method [100, 101]. Thus, the direct relationship between the activation energy and CT along synchronous and non-synchronous bond formation processes was now established and hence the role of electron density flux in the lowering of activation barrier was evident. Because the flux of electron density is a global process involving the two interacting molecules, and the electron density, and not charges, are transferred in a polar process, the flux of electron density at the TSs was termed by Domingo as the Global Electron Density Transfer [36] (GEDT). GEDT is calculated from the sum of the natural atomic charges (q), obtained by a Natural Population Analysis [100, 101] (NPA), of the atoms belonging to each framework (f) at the TSs.

$$\text{GEDT } (f) = \sum_{q \in f} q \qquad (35)$$

The sign of GEDT indicates the direction of the electron density flux in such a manner that positive values mean a flux from the considered framework to the other one. GEDT allows the classification of reactions as non-polar or polar processes. Non-polar reactions show GEDT values below 0.10 e, while polar processes are associated with GEDT values above 0.20 e.

More the GEDT is, more polar and faster the reaction is the role of GEDT in lowering the energy cost of polar cycloadditions was recently explained by Domingo *et al.* [102] in 2017. On going from reagents to TSs, the DA reactions involve the rupture of C-C double bonds present at the reagents. In polar processes, high GEDT results in the depopulation of C-C double bond regions, leading to the easy rupture of these C-C double bond. Consequently, polar reactions show lower activation energies and higher reaction rates compared to non-polar cycloadditions. The origin of GEDT can be attributed to the difference in electronic chemical potential μ [47] between the reactants, which causes the flux of electron density from the reactant with lower μ value to the reactant with higher μ value. Thus, the initial comprehension of electron density flux from the analysis of CDFT indices of the reactants is finally established by GEDT calculations at the TSs.

Very recently, Domingo and Ríos-Gutiérrez studied the nature of organic electron density transfer complexes (EDTC) within the MEDT [103]. The favourable nucleophilic/electrophilic interactions, which favour the GEDT towards the electrophile, are responsible for the formation of these species. Molecular complexes presenting a GEDT above 0.05e are classified as EDTCs. Analysis of the Parr functions at the separated reagents and the topological analysis of the electron density of the EDTCs allow the understanding of the subtle changes in the electronic structure of this significant class of molecular complexes, and consequently, their physical properties [103].

REVEALING THE MECHANISM AND ELECTRON DENSITY TRANSFER BY BONDING EVOLUTION THEORY

We have previously discussed the use of ELF [78, 79] to propose a Lewis-like structure of organic molecules. The Bonding Evolution Theory (BET) proposed by Krokidis [104] in 1997 represents a conjunction of ELF topological analysis [78, 79] and Thoms' Catastrophe theory [105]. The term *"catastrophe"*, which indicates a qualitative change observed in an object, when smooth change occurs in the parameters upon which it depends, was first introduced in 1972 by R. Thom [105]. The "catastrophe theory" became popular in 1976, when Zeeman [106] used this as one term for singularity theory and bifurcation theory in mathematics. A reaction path consists of ELF structural stability domains (SSDs), namely also phases, which are separated by turning points, sometimes called catastrophes. Identification of these catastrophes can reveal the changes in chemical rearrangements along a reaction path and hence the reaction mechanism. This is the framework of the BET study. This has been employed to study mechanism of numerous organic reactions [107] such as isomerisations [108], 32CA reactions [82, 83], S_N2 reactions [109], electron transfers [110], proton transfers [111], Nazarov cyclization [112], DA reactions [113], Cope rearrangement [114], etc. In the first step of BET, a topological analysis of the ELF at each IRC point is carried out [92 - 94], and then, the population of the ELF basins is analysed. As discussed above, IRC connects stationary points on the PES [91]. Thus, a family of configurations between reactant minimum and product minimum are connected by the IRC and therefore, following the electron density changes along the IRC path can reveal the mechanism of a chemical reaction. The topological analysis of ELF along the IRC yields the formation/disappearance of ELF basins. Changes in the number and nature of ELF basins denote the turning points or catastrophes which are associated with bond making/bond breaking, formation of lone pairs or *pseudoradical* centers [81], and other related electronic rearrangements. This allows the structuring of a plausible mechanism for the chemical reaction. As an example, we can consider the BET study for 32CA reaction of *C,N*-dimethyl nitrone and acrolein performed by Domingo *et al* [88] in 2018 at MPWB1K/6-311G(d,p) level of theory. The reaction can be differentiated into ten topological *phases* (see Fig. **6**).

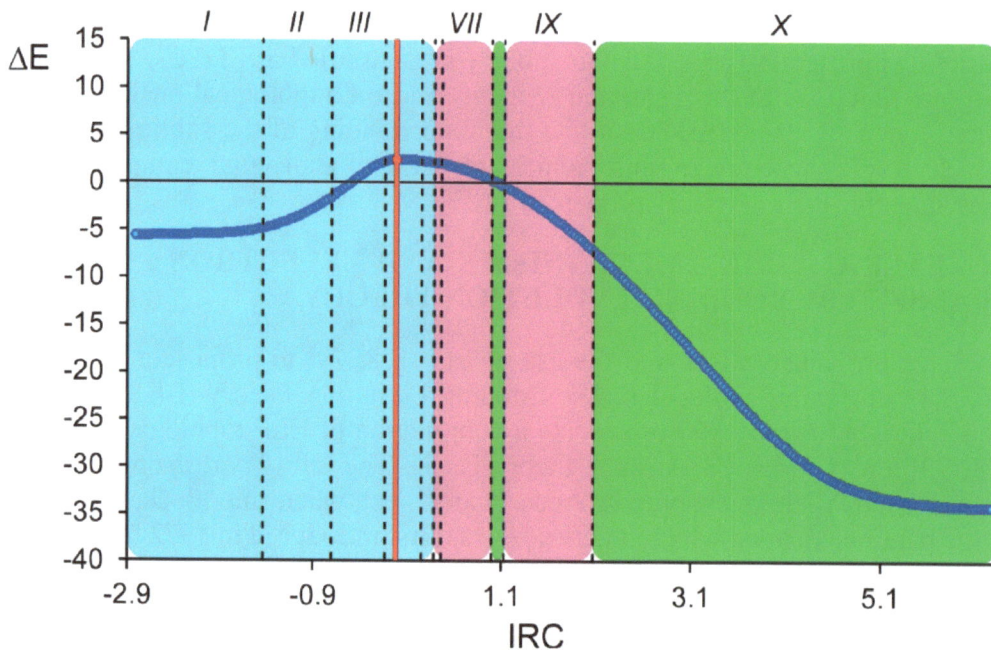

Fig. (6). ELF topological *phases* for polar 32CA reaction between C,N dimethyl nitrone and acrolein. The red point indicates the position of TS, black dashed lines separate the phases defined by points along the IRC, while coloured areas represent the different groups in which the reaction is topologically divided. Relative energies (ΔE, in kcal·mol^{-1}) are given with respect to the separated reagents.

A simple representation of the proposed molecular mechanism is shown in Fig. (**7**), which predicts the formation of new single bonds by the coupling of two *pseudoradical* centers. *Phases I – V* are characterized by the depopulation of the V(N2,C3) and V(C4,C5) disynaptic basins and formation of the V(N2) monosynaptic basin. These changes account for the rupture of the C4−C5 double bond of acrolein and the rupture of the N2−C3 double bond of nitrone leading to the formation of the N2 nitrogen lone pair at the nitrone framework. *Phases VI* and *VII* involve the formation of the V(C5) monosynaptic basin, which can be associated with the formation of the *pseudoradical* centre at C5 at the acrolein framework demanded for the formation of the second C3−C5 single bond. At *Phase VIII*, the formation of the first O1-C4 single bond is shown by the creation of the V(O1,C4) disynaptic basin. *Phases IX* involves the formation of the V(C3) monosynaptic basin, which can be associated with the *pseudoradical* centre at C3. Finally, the creation of the disynaptic V(C3,C5) basin in *Phase IX* shows the formation of the second C3-C5 single bond and to the molecular electronic relaxation associated with the formation of isoxazolidine.

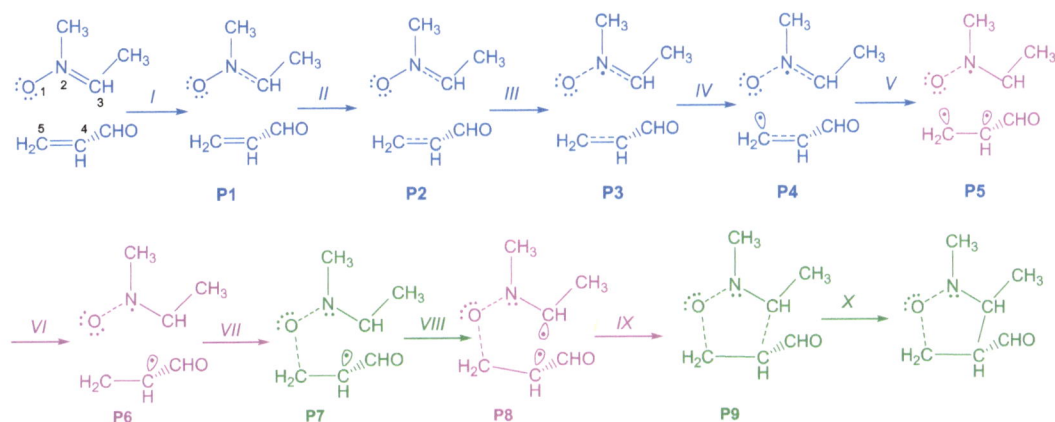

Fig. (7). A simple representation of the mechanism of 32CA reaction of *C,N*-dimethyl nitrone and acrolein predicted by Bonding Evolution Theory (BET) study. The different groups in which the BET is regrouped are given in different colours.

The BET procedure allows identifying the flow of electron density along the chemical reaction. This was explained by Silvi and co-workers [115] by correlating geometrical deformations and electron density transfer observed in the cyclization of 1,3-butadiene. Electron density transfer along the reaction path causes an increase in bond length and the adjacent bond is shortened to which the electron density transfer takes place. The formation of *pseudoradical* centers or lone pairs comes from the electron density of the adjacent bonds. This can be understood from ELF topological analysis performed in 2018 by Ayouchia *et al.* [116] for a 32CA reaction between nitrophenyl azide and methyl phenyl alkyne (see Fig. 8).

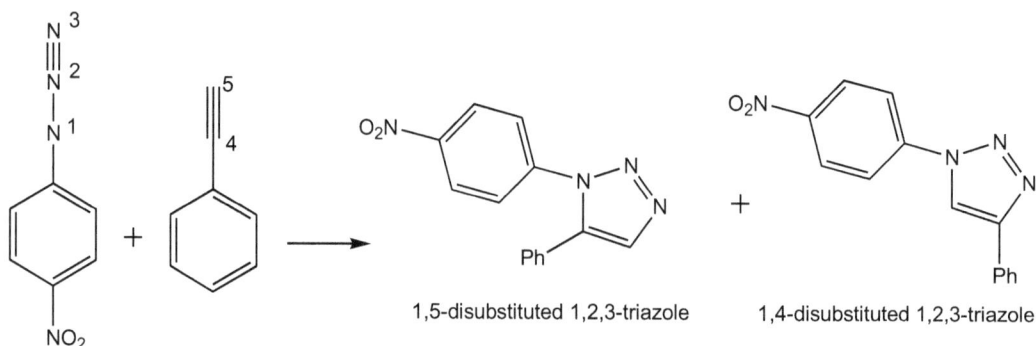

Fig. (8). 32CA reaction between nitrophenyl azide and methyl phenyl alkyne.

Table 4. ELF valence basin populations of selected IRC points along the favored 1,5-regioisomeric path for 32CA reaction between nitrophenyl azide and methyl phenyl alkyne

	P1	**P2**	**P3**	**P4**
V(N1,N2)	1.75	1.76	1.75	1.73
V(N2)	2.46	2.54	2.84	2.82
V(N2,N3)	2.48	2.42	2.08	2.07
V(C4,C5)	2.28	2.23	2.05	2.04
V'(C4,C5)	2.26	2.21	2.07	2.04
V(C4,C6))	2.52	2.55	2.29	2.24
V(N1)	3.33	3.31	2.30	1.00
V'(N1)			1.12	1.19
V(N3)	3.67	3.52	3.16	3.13
V'(N3)	0.28			
V(C5)	0.57			
V(N3,C5)		1.13	2.01	2.04
V(N1,C4)				1.31

The ELF valence basin populations of selected IRC points, **P1-P4**, along the favoured 1,5-regioisomeric path is shown in Table **4**. At **P1**, the monosynaptic V(N2) basin integrates at 2.46e, which increases to 2.54e at **P2**, while disynaptic V(N2,N3) basin population decreases from 2.48e at **P1** to 2.42e at **P2**. This indicates that N2 lone pair formation (associated with the V(N2) monosynaptic basin) derives electron density from the N2-N3 bond (associated with the V(N2,N3) disynaptic basin). Formation of the V'(N1) monosynaptic basin at **P3** derives electron density from the V(N1) monosynaptic basin, as the basin population of the V(N1) monosynaptic basin decreases from 3.31e at **P2** to 2.30e at **P3**. The formation of a new N1-C4 single bond is shown by the creation of the V(N1,C4) disynaptic basin, which derives its electron density from the V(N1) monosynaptic basin (decreasing from 2.30e at **P3** to 1.00e at **P4**). Thus, electron density flux is associated with the bonding changes along the reaction path.

REVEALING THE ELECTRON DENSITY DISTRIBUTION BY BADERS' THEORY OF ATOMS IN MOLECULES

When two nuclei interact, their electron density distribution is affected, which is used for the characterization of chemical bonding. Bader's Quantum Theory of Atoms-in-Molecules [117, 118] (QTAIM) characterizes chemical bonding from the charge density or electron density distribution. Under MEDT, QTAIM is used to reveal the nature of atomic interactions in the stationary states on the PES.

Topological Analysis of the Electron Density Distribution and Search for Critical Points

The electron density distribution in molecules shows some interesting features. The simple example of ethylene has been considered here to explain these features. Relief maps of the electron density distribution in ethylene and nitroethylene (optimized at MPWPW91/6-311G(d,p) level in our previous study [90]) are shown in Fig. (**9**). It is evident that the local maxima of electron density distribution exist at the nuclei sites and electron density decays away with increasing distance from the nuclei. The topology of electron density distribution can be analyzed quantitatively by considering the first derivative, $\nabla^2\rho$ (r_c). The electron density shows a maximum, a minimum, and a saddle point in space at points, which are called critical points (CPs). The first derivative of electron density vanishes at these points, thus, $\nabla^2\rho$ (r_c) , where r_c is the critical point.

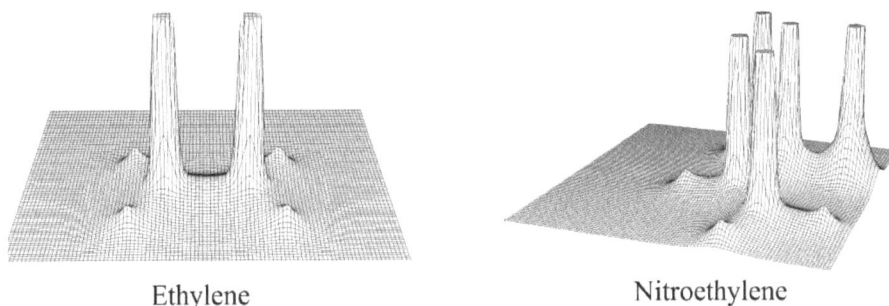

Ethylene   Nitroethylene

Fig. (9). Relief maps of electron density distribution in ethylene and nitroethylene.

Analogues to several mathematical interpretations, the characteristics of a CP r_c can be determined from the second derivative of ρ, denoted by $\nabla^2\rho$ (r_c). The second derivative of ρ is determined by the so-called Hessian of ρ, which can be diagonalized to obtain three eigenvalues, λ_1, λ_2 and λ_3, which are the principle axes of curvature. The sum of λ_1, λ_2 and λ_3 gives the Laplacian of electron density $\nabla^2\rho$ (r_c) , which is expressed as

$$\nabla^2 \rho(r_c) = \frac{\partial^2 \rho(r_c)}{dx^2} + \frac{\partial^2 \rho(r_c)}{dy^2} + \frac{\partial^2 \rho(r_c)}{dz^2} = \lambda_1 + \lambda_2 + \lambda_3 \tag{36}$$

For a molecular system, CPs are denoted as (rank, signature). Rank of a CP is equal to the number of non-zero eigenvalues and signature denotes the algebraic sums of the signs of the eigenvalues. With rank 3, there are four possible CPs of a chemical system: (3,-3), (3,-1), (3,+1), and (3,+3). At (3,-3) CP, all the curvatures are negative and ρ corresponds to a local maximum. (3,-3) CPs occur at the nuclei positions, which are shown in magenta colour in Fig. (**10**) for ethylene and nitroethylene. At (3,-1) CP, the two curvatures are negative, ρ is maximum in the plane defined by three axis and ρ is minimum at the CP along the third axis perpendicular to this plane. At (3,+1) CP, the two curvatures are positive, ρ is minimum in the plane defined by three axis and ρ is maximum at the CP along the third axis perpendicular to this plane. At (3,+3) CP, all the curvatures are positive and ρ corresponds to a local minimum.

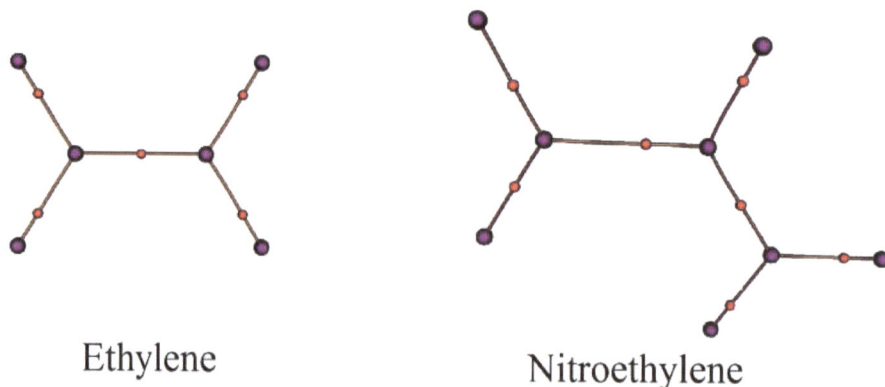

Ethylene   Nitroethylene

Fig. (10). Critical points in ethylene and nitroethylene; (3,-3) critical points are shown in magenta and (3,-1) critical points are shown in red colours.

(3,-1) CP exists between two nuclei. The gradient vector fiels of ρ in the molecular plane of ethylene and nitroethylene are shown in Fig. (**11**). The pair of gradient paths originate at (3,-1) CP and terminate at the two nuclei. This special trajectory connecting two neighbouring nuclei is called the *atomic interaction line*. If (3,-1) CP exists between two nuclei in an equilibrium geometry, then it indicates the accumulation of electron density between the nuclei and the atomic interaction line is called the *bond path*. (3,-1) CP is called the bond critical point or BCP.

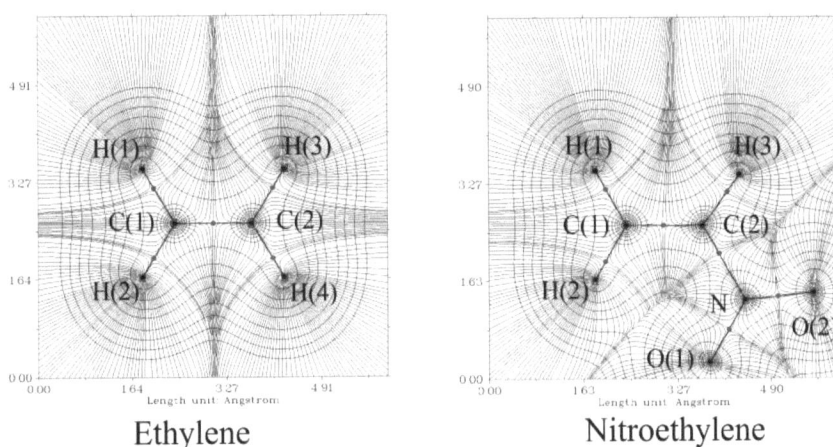

Ethylene   Nitroethylene

Fig. (11). Gradient vector field of electron density, ρ, in the molecular plane of ethylene and nitroethylene.

The search for CPs present in a molecule is checked for completion by Poincare Hopf rule given by

$$n- b + r- c =1 \qquad (37)$$

Where n, b, r, and c, respectively, denote the number of (3,-3), (3,-1), (3,+1), and (3,+3) critical points

The Sign of Laplacian of Electron Density

The Laplacian of electron density $\nabla^2\rho$ (r_c), at the BCPs is an important parameter to characterize atomic interactions. In 1984, Bader and Essen [118] characterized atomic interactions between a pair of atoms connected by a bond path under the following two categories:

Shared interactions: For a shared (covalent) interaction, the total electron density $\rho(r_c)$ at the BCP (located between the two atoms) shows the high value of the order of > 0.1au. It shows a negative Laplacian of electron density $\nabla^2\rho$ (r_c). The electron density is locally concentrated between the two atoms.

Closed shell interactions: For a closed shell interaction, the total electron density $\rho(r_c)$ at BCP is less than that for shared interaction. Here, the local charge is confined mainly to the valence shell of the atom with high electronegativity. Such interactions show positive Laplacian of electron density $\nabla^2\rho$ (r_c) over most of the internuclear region and it only shows negative value in the valence shell region of the more electronegative atom.

When a shared interaction takes place between two atoms, then the electron density $\rho(r_c)$ increases between the interacting atoms and the nodal surface of the Laplacian makes its way into the valence region of the more electropositive atom, gradually leading to the formation of a contiguous region of local charge concentration. As a result, the sign of Laplacian of electron density $\nabla^2\rho$ (r_c) changes from positive to negative.

NCI Gradient Isosurfaces

The Laplacian of electron density $\nabla^2\rho$ (r_c) can be decomposed into three eigenvalues (see Eq. **36**), λ_1, λ_2 and λ_3 of the electron density Hessian matrix. Non-covalent interactions (NCI) are identified by the sign of λ_2. For bonded NCIs, $\lambda_2 < 0$ in the interatomic region and if the atoms are in non-bonded contact, then $\lambda_2 > 0$.

In 2011, García *et al.* proposed the NCI plot [119] program to map and analyse NCI in molecular systems. In the NCI plot, the reduced density gradient [120] (RDG) is used to isolate non-covalent interactions in the real space. RDG is defined as

$$\text{RDG}(r) = \frac{1}{2(3\pi^2)^{1/3}} \frac{|\nabla\rho(r)|}{\rho(r)^{4/3}} \qquad (37)$$

NCI plot generates gradient isosurfaces which enclose the regions of the real space for the location of bonded NCI in the real space. Large negative values of sign $(\lambda_2)\rho$ indicate attractive interactions such as dipole-dipole or hydrogen bonding, while large positive values of sign $(\lambda_2)\rho$ are associated with NCI. Near zero values of sign $(\lambda_2)\rho$ indicate weak, van der Waals interactions. NCI gradient isosurfaces of TSs for 32CA reaction between nitrile oxide and α-santonin reported in our very recent MEDT study is given in Fig. (**12**) [121].

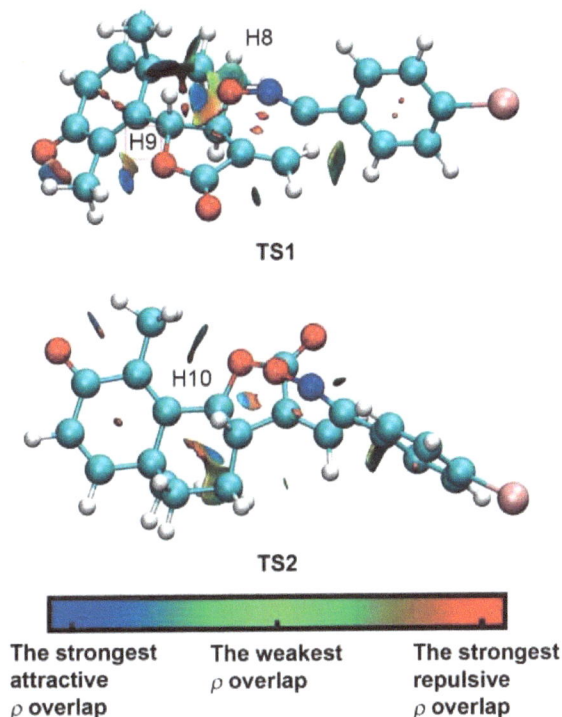

Fig. (12). NCI gradient isosurfaces of the TSs associated with the *ortho/syn/anti* reaction channel of 32CA reaction p-bromophenylnitrile oxide and α-santonin Surfaces are colored in the [−0.02, 0.02] a.u. range of sign(λ_2)ρ(isosurfaces) = 0.5 a.u.

QTAIM Topological Analysis Performed Under MEDT

Within MEDT, QTAIM topological analysis is carried out, together with the ELF analysis, to characterize the bond formation process along a chemical reaction, steps of which are given below:

- Searching of all CPs and bond paths present in the optimized stationary structures
- Calculations of the total electron density $\rho(r_c)$, Laplacian of electron density $\nabla^2\rho$ (r_c) and the total energy density H_c, at the (3,-1) CPs, known as BCPs, related to the formation of new bonds. In a 2017 MEDT study [102], Domingo *et al.* calculated these QTAIM parameters at the BCPs of TSs for the DA reaction between cyclopentadiene and ethylene. The calculated values of $\rho(r_c)$, $\nabla^2\rho$ (r_c) , and H_c at BCPs of the TS are 0.055 au, 0.046 au, and -0.012 au, respectively. These values indicate an accumulation of electron density between the

interacting nuclei at the atomic interaction line, and the positive Laplacian of electron density indicates weak non-covalent interactions. Thus, the TS is not presenting a bonding situation for this reaction and shared interaction is not present in the TSs

- Characterisation of the atomic interactions as covalent/non-covalent from the calculated $\rho(r_c)$ values at the BCPs and the corresponding signs of the Laplacian of electron density $\nabla^2\rho\ (r_c)$

- Identification of non-covalent interactions by visualization of NCI gradient isosurfaces at the TSs

APPLICATIONS OF MOLECULAR ELECTRON DENSITY THEORY

Diels-Alder Reactions

The mechanism of the DA reaction has been widely investigated since the 1960s. The mechanistic approaches are presented in Fig. (**13**). Woodward and Hoffmann [25] proposed a concerted mechanism for the DA reaction between 1,3-butadiene and ethylene, where electrons show a concerted movement around a circle, on a closed curve (see Fig. **13a**). An aromatic TS, initially proposed by Evans [122], was extended to this reaction by Dewar [123] in 1971 (see Fig. **13b**). Later in 1995, Houk [124] reviewed these proposed concepts. The mechanism received significant contribution after 2002 [36, 113, 125, 126], which allowed to rule out the "*closed curve*" concept, and proposed a symmetric rearrangement through the homolytic rupture of the C-C double bond at the first stage of the reaction (see Fig. **13c**). An unbound TS was proposed [126, 36] from the topological analysis of electron density (see Fig. **13d**).

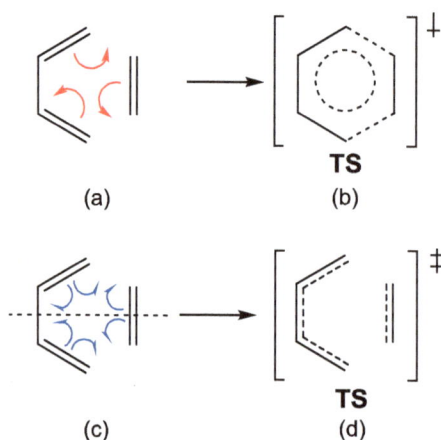

Fig. (13). Mechanistic approaches to the DA reaction.

Very recently in 2018, Domingo *et al.* [127] reported a MEDT study for the DA reaction between 1,3-butadiene and ethylene. The reaction was found to be non-polar, with high activation enthalpy of 23.2 kcal/mol, close to the experimental value of 27.5 kcal/mol. The sequential bonding changes are given in Fig. (**14**). The molecular mechanism can be differentiated into seven topological *phases*, which belong to four groups A, B, C, and D (see Fig. **14**). Group A involves the rupture of the three double bonds and accounts for the high activation energy of the TS belonging to the last *phase* of this group. Group B presents the formation of *pseudoradical* centers at the four interacting carbon atoms, while group C is associated with the synchronous formation of two new C-C single bonds. Finally, group D shows the formation of a C-C double bond with the release of molecular relaxation energy for the formation of cyclobutene. Thus, the MEDT study [127] provided a solid rationalization of the molecular mechanism with the characterization of the TS for the DA reaction, which is not addressed by the FMO concept.

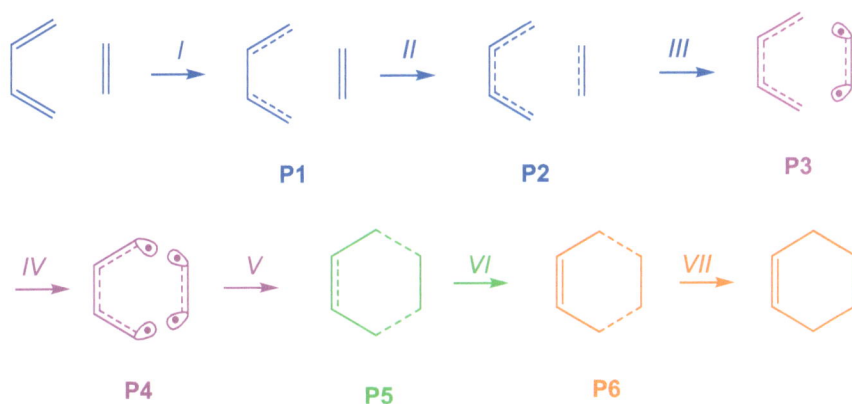

Fig. (14). Sequential bonding changes in the ELF topological groups for the DA reaction between 1,3-butadiene and ethylene. The different groups in which the BET is regrouped are given in different colours.

In 2017, a MEDT study [102] underlined the role of the GEDT in the decreasing of the activation energy barrier of polar DA reactions. The GEDT causes bonding changes in the reagents, resulting in depopulation of the C-C double bond in polar reactions. These electronic changes decrease the energy required for the rupture of C-C double bonds making the reaction easier. The non-polar DA reaction between cyclopentadiene and ethylene shows an activation enthalpy of 18.1 kcal/mol, while polar DA reaction between cyclopentadiene and tetracyanoethylene presents an activation enthalpy of 1.2 kcal/mol. Another MEDT study [128] explained the increase in activation energy of the non-polar DA reaction between 2-methylfuran and propene due to the loss of the aromatic character of 2-methylfuran along the reaction. MEDT has also been employed recently [129] to study the competitiveness of Lewis acid catalysed polar DA reaction and polar Alder-ene reaction. Low polar reactions favour the formation of the DA product, while the increase in polarity may kinetically favour the formation of Alder-ene adducts.

Electrocyclic Reactions and Sigmatropic Rearrangements

A MEDT study [127] was carried out to analyze the sequential bonding changes in the electrocyclic reaction of 1,3,5-hexatriene (see Fig. **15**). The molecular mechanism of this *pseudocyclic* reaction can be reorganized into four groups A, B, C, and D from the ELF topological characterization. Group A involves the rupture of three double bonds, while group B is associated with the formation of *pseudoradical* centers. Group C shows the formation of a new C-C single bond. Finally, C-C double bonds are formed in group D with the release of molecular relaxation energy.

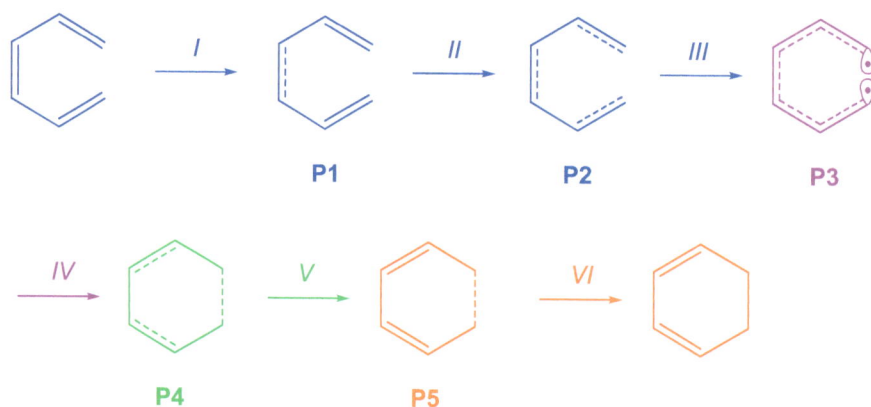

Fig. (15). Sequential bonding changes in the ELF topological groups of the electrocyclic reaction of 1,35-hexatriene. The different groups in which the BET is regrouped are given in different colours.

The molecular mechanisms of the [3, 3] sigmatropic rearrangement and the [1, 5] rearrangement of 1,5-hexadiene and the [1, 5] sigmatropic rearrangement of 1,3-pentadiene were also analyzed by MEDT studies [127] in 2018. The topological characterization of ELF along the reaction path indicated that these reaction involve homolytic C-C(H) bond ruptures and bond formations. Thus, the use of MEDT has given a quantum chemical proof to rule out the Woodward-Hoffmann [25] concept of concerted bond formation on a closed curve for "pericyclic reactions".

[3+2] Cycloaddition Reactions

32CA reactions are the most widely studied class of reactions within the MEDT. These reactions remain a versatile protocol for regio- and stereochemical synthesis of five-membered heterocycles of potent biological significance [130]. As commented on above, the TACs participating in 32CA reactions are classified into *pseudodiradical, pseudoradical,* carbenoid and zwitterionic by topological analysis of the electron density (see Fig. **3**) [82]. Depending on the electronic structure of the TAC, 32CA reactions have been classified into *pseudodiradical* (*pdr-type*), *pseudoradical* (*pmr-type*), carbenoid (*cb-type*), and zwitterionic (*zw-type*) with the activation energies increasing in the order, *pdr-type* < *pmr-type* ≈ *cb-type* < *zw-type* [82].

Simplest azomethine ylide [38, 80, 127] (**AY**) is classified as *a pseudodiradical* TAC and shows a symmetric bonding pattern. The *pseudodiradical* structure obtained from the topological analysis of ELF is different from the diradical species proposed by Firestone in the 1970s [131] for 32CA reactions. The singlet diradical species proposed by Firestone [131] demands unrestricted calculations, while the structure of *a pseudodiradical* TAC arises from the topological analysis of electron density obtained from restricted calculations. Simplest azomethine imine [87] (**AI**) is classified as *a pseudodiradical* TAC, while nitrone [88] (**NI**) is classified as *a zwitterionic* TAC. The mechanisms of the 32CA reactions of the simplest **AY** [80, 127], **AI** [87, 127], and **NI** [88, 127] to ethylene have been investigated within MEDT. The formation of the C-C single bonds along these reactions involves the C-to-C coupling of two *pseudoradical* centers formed at the two interacting nuclei during the reaction course (see Fig. **16**).

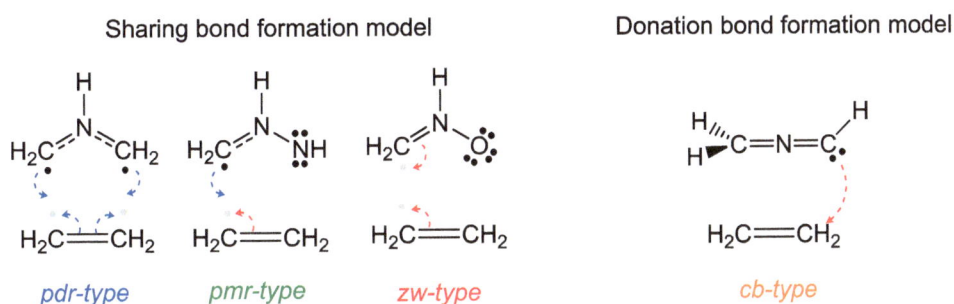

Fig. (16). Bonding changes associated with the formation of the new C–C single bonds along the four types of non-polar 32CA reactions revealed by MEDT studies.

The 32CA of **AY** with ethylene involves a homolytic rupture of the double bond of ethylene to generate two *pseudoradical* centers which no energy cost. On the other hand, for the 32CA of **NI**, the *pseudoradical* centers are generated by the rupture of C-N and C-C double bonds of **NI** and ethylene, which accounts for the high energy barrier of *zw-type* 32CA reaction of **NI**. Nitrile ylide [84] and nitrile imine [85] are classified as *cb*-type TACs and show a completely different reactivity in 32CA reactions. In these cases, the carbenoid centre donates the non-bonding electron density in new bond formation, which has a more decisive role to play than the coupling of *pseudoradical* centers at the interacting nuclei. Thus, MEDT provides a solid rationalization for the mechanism and reactivity of 32CA reactions.

MEDT has been applied to analyse several important 32CAs since 2016 [82]. Some of these reactions are given below in brief:

Reaction 1: 32CA reaction of nitrile ylide with electron-deficient ethylene (2016) [84].

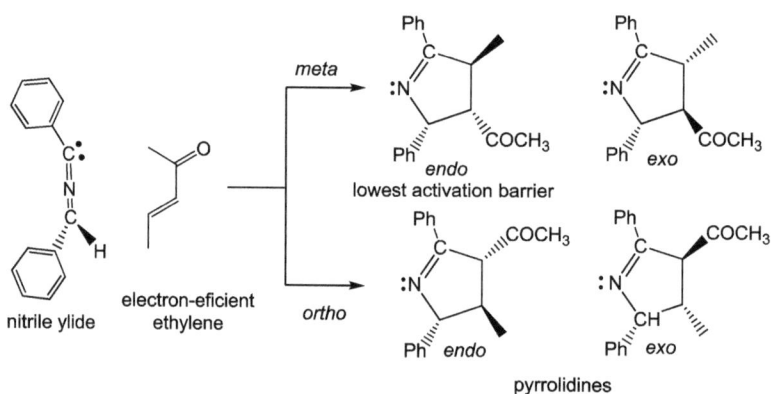

Fig. (17). 32CA reaction of nitrile ylide with electron-deficient ethylene.

Appealing conclusions obtained from this MEDT study (Fig. **17**):

- *cb-type* 32CA reaction
- Non-concerted *one-step* mechanism, low activation energy, strongly exothermic
- Preferred *meta*/*endo* reaction path, showing a low activation barrier in agreement with experiments
- The formation of the first C-C single bond starts with the donation of the lone pair of carbenoid carbon to β-conjugate carbon of the enone

Reaction 2: 32CA reaction of nitrone with ketenes (2017) [132].

Fig. (18). 32CA reaction of nitrone with ketenes.

Appealing conclusions obtained from this MEDT study (Fig. **18**):

- *zw-type* 32CA reactions
- The non-concerted *two-stage one-step* mechanism changes to a two-step mechanism for electrophilic ketenes
- A complete C=O chemoselective and regioselective reaction were predicted in agreement with the experimental outcome

Reaction 3: 32CA reaction of chiral azomethine ylides with β-nitrostyrene (2017) [133].

Fig. (19). 32CA reaction of chiral azomethine ylides with β-nitrostyrene.

Appealing conclusions obtained from this MEDT study (Fig. **19**):

- *pdr-type* 32CA reactions
- Non-concerted *two-stage one-step* mechanism with more advanced C1-C5 single bond formation
- *The meta/endo* reaction path was predicted as the most favourable one in agreement with experiments

Reaction 4: Domino reaction of 1-diazopropan-2-one and 1,1-dinitroethylene (2017) [86].

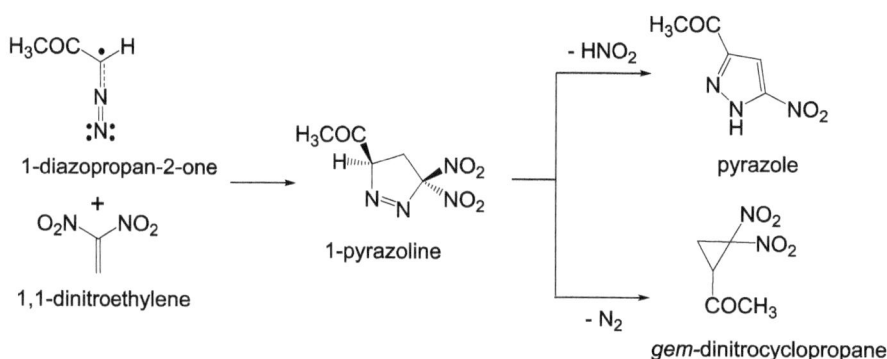

Fig. (20). Domino reaction of 1-diazopropan-2-one and 1,1-dinitroethylene.

Appealing conclusions obtained from this MEDT study (Fig. **20**):

- *pmr-type* 32CA reaction
- This domino process proceeds through two consecutive reactions. At first, a polar 32CA reaction between 1-diazopropan-2-one and 1,1-dinitroethylene to give 1-pyrazoline takes place. MEDT rationalizes the chemical conversion of 1-pyrazoline into the reaction products, a pyrazole and a *gem*-dinitrocyclopropane experimentally obtained

Reaction 5: 32CA reaction of nitrone and strained allene (1,2-cyclohexadiene) (2017) [134].

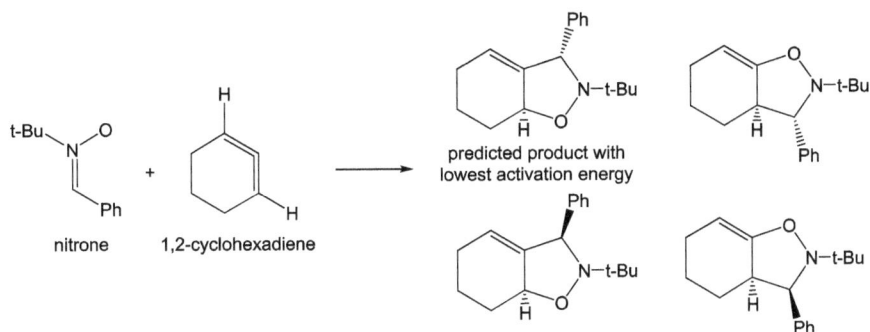

Fig. (21). 32CA reaction of nitrone and strained allene.

Appealing conclusions obtained from this MEDT study (Fig. **21**):

- Low *endo* stereo- and regioselective reaction
- Non-concerted *two-stage one-step* mechanism characterized by the initial attack of the central carbon of the strained cyclohexadiene on the carbon of the nitrone
- MEDT makes it possible to predict that 1,2-cyclohexadiene behaves as a radical species, which is responsible for its high reactivity compared to the simplest linear allene. The latter behaves as an ethylene derivative in a *zw-type* 32CA reaction with nitrone, presenting a high activation energy
- Geometrical predistortion of 1,2-cyclohexadiene, as Houk proposed [135], is not responsible for its high reactivity

Reaction 6: 32CA reaction of azomethine imine to *p*-nitrobenzylidene malononitrile (2018) [136].

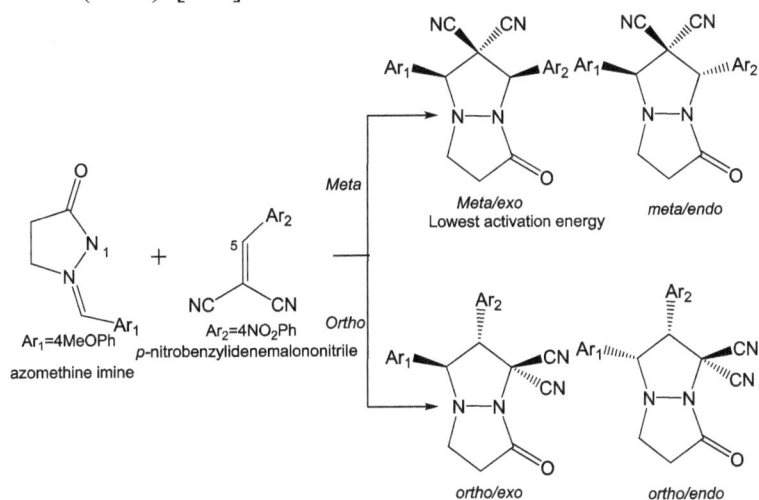

Fig. (22). 32CA reaction of azomethine imine to p-nitrobenzylidenemalononitrile.

Appealing conclusions obtained from this MEDT study (Fig. **22**):

- *zw-type* 32CA reaction
- *p*-nitrobenzylidene group increases the electrophilic character of the ethylene, resulting in a polar reaction
- Formation of the *meta/exo* product was predicted with the lowest energy barrier of 0.6 kcal/mol, in agreement with the experimental data
- Non-concerted *two-stage one-step* mechanism with the advanced formation of N1-C5 single bond
- [2n+2τ] mechanism characterized by donation of the N1 lone pair density to C5 carbon

Reaction 7: Intramolecular 32CA reaction of cyclic nitrone (2018) [137].

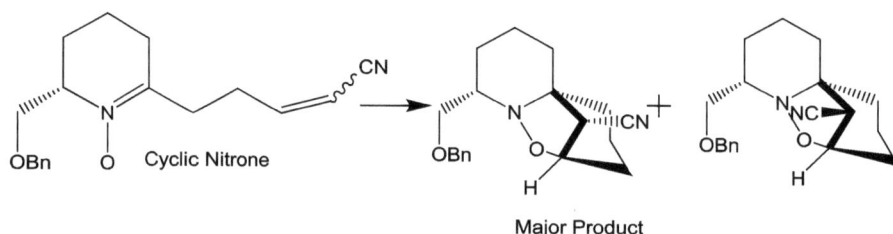

Fig. (23). Intramolecular 32CA reaction of cyclic nitrone.

Appealing conclusions obtained from this MEDT study (Fig. **23**):

- Non-concerted *one-step* mechanism
- 6,6,5-ring fused isoxazolidines are thermodynamically preferred over 6,6,5-ring bridged isoxazolidines
- Complete regio-, stereo-, and stereospecific process was predicted in agreement with experiments
- High activation energy due to the low electrophilic character of the nitrone, and the strain caused by the chain

Reaction 8: Copper metalation of 32CA reactions between azomethine ylides and electron-deficient ethylenes (2018) [138].

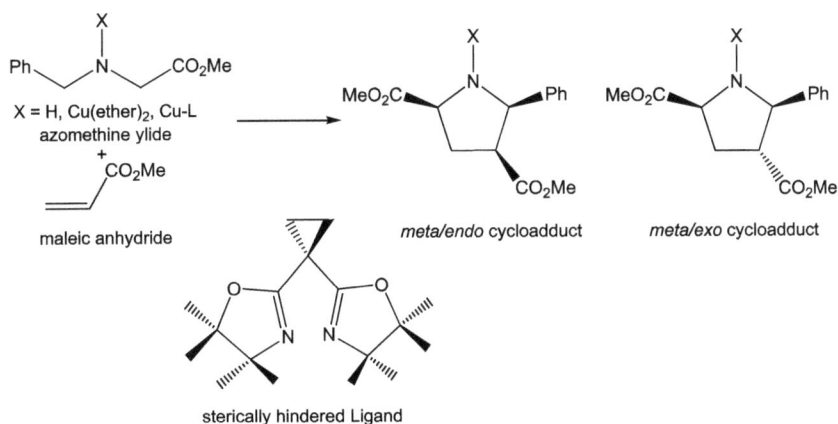

Fig. (24). Copper metalation of 32CA reactions between azomethine ylides and electron-deficient ethylenes.

Appealing conclusions obtained from this MEDT study (Fig. **24**):

- *pdr-type* 32CA reaction
- Non-concerted *two-stage one-step* mechanism
- Metalation produces no effect on reaction rate, but the regio- and stereoselectivities are significantly influenced by the size of the ligands used in the metalation

Reaction 9: 32CA reaction between trifluoroacetonitrile N-oxide and thioketones (2018) [139].

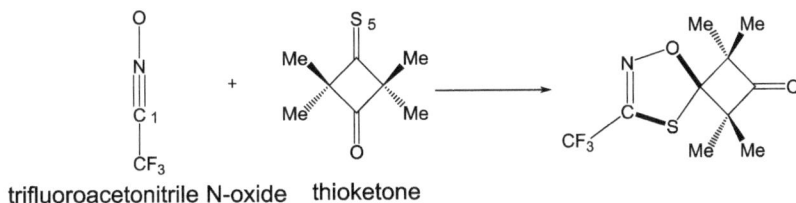

trifluoroacetonitrile N-oxide   thioketone

Fig. (25). 32CA reaction between trifluoroacetonitrile N-oxide and thioketones.

Appealing conclusions obtained from this MEDT study (Fig. **25**):

- *pmr-type* 32CA reaction
- Non-concerted *two-stage one-step* mechanism
- Complete chemo- and regioselective reaction predicted with the electrophilic attack of the C1 atom of nitrile oxide over S5 atom of the thioketone, in agreement with the experimental outcome

Reaction 10: 32CA reactions of *C,N*-dialkyl nitrones and ethylene derivatives (2018) [88].

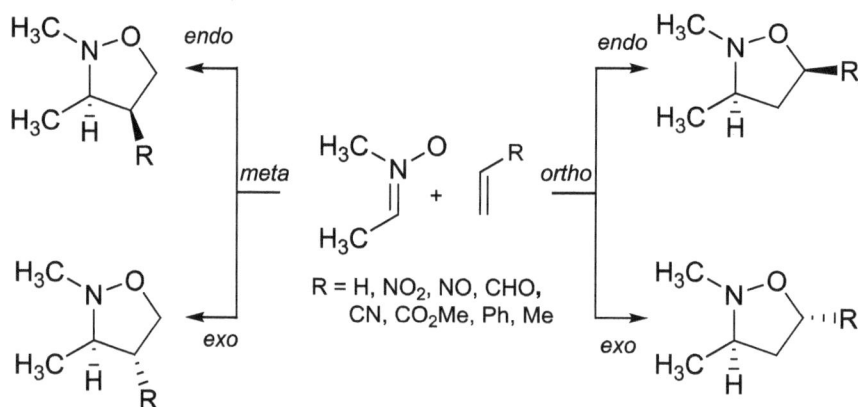

Fig. (26). 32CA reactions of *C,N*-dialkyl nitrones and ethylene derivatives.

Appealing conclusions obtained from this MEDT study (Fig. **26**):

- *zw-type* 32CA reactions
- Non-concerted *one-step* mechanism
- The non-polar 32CA reaction of nitrones with propene is *ortho/exo* selective, while the polar 32CA reaction with acrolein is *meta/endo* selective, in complete agreement with experiments
- MEDT makes it possible to rationalize the change in regioselectivity and molecular mechanism due to electrophilic activation of ethylene, which has not been addressed by standard computational models

Reaction 11: 32CA reaction of aryl azides and alkyne derivatives (2018) [116].

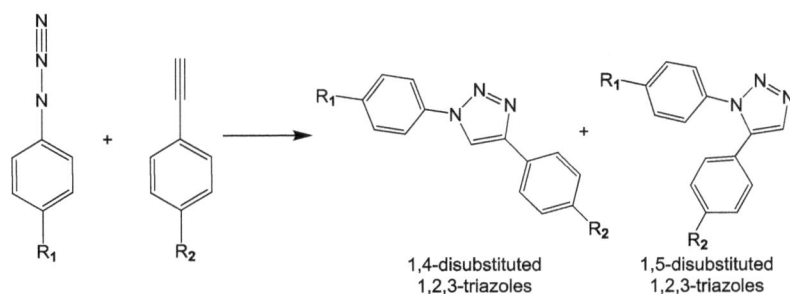

Fig. (27). 32CA reactions of aryl azides and alkyne derivatives.

Appealing conclusions obtained from this MEDT study (Fig. **27**):

- *zw-type* 32CA reactions
- Non-concerted *two-stage one-step* mechanism for the 1,5-regioisomeric reaction path, while the 1,4-regioisomeric reaction path takes place through *one-step* mechanism through a low asynchronous TS

Reaction 12: Synthesis of anticancer spiroisoxazolidine from 32CA reaction of α-santonin (2019) [121].

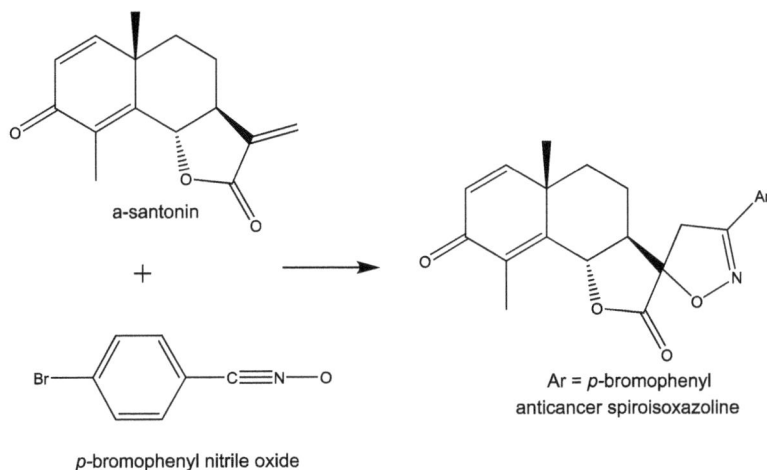

Fig. (28). Synthesis of anticancer spiroisoxazolidine from 32CA reaction of α-santonin.

Appealing conclusions obtained from this MEDT study (Fig. **28**):

- Non-polar *zw-type* 32CA reaction
- Non-concerted *one-step* mechanism
- *Ortho* regio- and chemoselective reaction, with a *syn* diastereofacial selectivity, was obtained in complete agreement with experiments

CONCLUSION

Although the historic conception of "vitalism" for organic compounds has been scientifically discredited, their wide applicability in medical and applied sciences has been acknowledged. This is attributed to their unique structural framework and reactivity. With the advent of "density functional theory" in the late 1980s and the ever-increasing speed of computers and software designing since the early 1990s, computational chemistry gained its position to interpret the mechanism and reactivity of chemical reactions. Physical chemists have performed numerous studies to understand the interactions and forces responsible for molecular reactivity. However, theoretical interpretations for molecular reactivity in organic reactions have remained standstill since the 1960s and derived concepts from the FMO theory in spite of the recognized failures and criticisms of this theory in several cases. The FMO concept gained superiority over other theoretical treatments for the last 40 years since it was claimed to be an easy and reliable theoretical model to study the course of "pericyclic reactions". After comprehensive theoretical investigations of more than 20 years dedicated to organic reactions, mainly DA and 32CA reactions, Domingo established the "Molecular Electron Density Theory (MEDT)" in 2016. This theory proposes that changes in electron density and not the molecular orbital interactions are responsible for molecular reactivity and rules out the previously claimed exclusive reliability of the FMO concept to study the course of "pericyclic reactions".

Within the MEDT study of a reaction, the propensity of electron density transfer and acceptance by the reactants through analysis of the CDFT indices, and the electronic structure of the reagents by a topological analysis of the ELF are first analysed. The PES is explored in detail to study the reaction course and associated activation parameters. The topological analysis of the ELF in conjunction with the catastrophe theory under the BET framework are perfromed to characterize the molecular mechanism. Finally, atomic interactions at the interatomic bonding regions are characterized by QTAIM analysis and NCI studies at the TSs. During the last three years, the complete coverage of these quantum-chemical principles within MEDT by the aid of molecular modelling software has provided a solid rationalization of the molecular mechanism and reactivity of several organic reactions in agreement with the experimental findings. The MEDT has thus proved to be a precise and reliable theoretical alternative to the FMO concept.

CONSENT FOR PUBLICATION

Not applicable.

CONFLICT OF INTEREST

The authors has no conflicts of interest, financial or otherwise, to declare.

ACKNOWLEDGEMENTS

Declare none.

REFERENCES

[1]   Silverman, R.B.; Holladay, M.W. *The Organic Chemistry of Drug Design and Drug Action,* 3rd ed; Academic Press: San Diego, CA, **2014**.

[2]   Rice, J.E. *Organic Chemistry Concepts and Applications for Medicinal Chemistry*; Academic Press: San Diego, CA, **2014**.

[3]   Grimsdale, A.C.; Müllen, K. The chemistry of organic nanomaterials. *Angew. Chem. Int. Ed. Engl.,* **2005**, *44*(35), 5592-5629.
        [http://dx.doi.org/10.1002/anie.200500805] [PMID: 16136610]

[4]   Saunders, K.J. *Organic Polymer Chemistry,* 2nd ed; Chapman and Hall Ltd: New York, **1988**.
        [http://dx.doi.org/10.1007/978-94-009-1195-6]

[5]   Weissermel, K.; Arpe, H-J. *Industrial Organic Chemistry,* 4th ed.; Weinhem: Wiley-VCH, **2003**.
        [http://dx.doi.org/10.1002/9783527619191]

[6]   Seager, S.L.; Slabaugh, M.R. *Organic and Biochemistry for Today,* 6th ed; Brooks/Cole Belmont: CA, **2008**.

[7]   Shinar, R.; Shinar, J. *Organic Electronics in Sensors and Biotechnology*; Mc Graw-Hill Education: New York, **2009**.

[8]   Rhein, L.D.; Schlossman, M.; O' Lenick, A. *Surfactants in Personal Care Products and Decorative Cosmetics,* 3rd ed.; CRC Press: Boca Raton, **2007**.

[9]   Trautz, M. Das gesetz der reaktionsgeschwindigkeit und der gleichgewichte in gasen. Bestätigung der additivität von Cv-3/2R. Neue bestimmung der integrationskonstanten und der moleküldurchmesser. *Z. Anorg. Allg. Chem.,* **1916**, *96*, 1-28.
        [http://dx.doi.org/10.1002/zaac.19160960102]

[10]   Lewis, G.N. The atom and the molecule. *J. Am. Chem. Soc.,* **1916**, *38*, 762-785.
        [http://dx.doi.org/10.1021/ja02261a002]

[11]   Slater, J.C. Directed valence in polyatomic molecules. *Phys. Rev. B.,* **1931**, *37*, 481-489.
        [http://dx.doi.org/10.1103/PhysRev.37.481]

[12]   Slater, J.C. Molecular energy levels and valence bonds. *Phys. Rev. B.,* **1931**, *38*, 1109-1144.
        [http://dx.doi.org/10.1103/PhysRev.38.1109]

[13]   Pauling, L. The metallic state. *Nature,* **1948**, *161*(4104), 1019-1020.
        [http://dx.doi.org/10.1038/1611019b0] [PMID: 18935081]

[14]   Mulliken, R.S. Spectroscopy, molecular orbitals, and chemical bonding. *Science,* **1967**, *157*(3784), 13-24.
[http://dx.doi.org/10.1126/science.157.3784.13] [PMID: 5338306]

[15]   Schrödinger, E. An undulatory theory of the mechanics of atoms and molecules. *Phys. Rev. B.,* **1926**, *28*, 1049-1070.
[http://dx.doi.org/10.1103/PhysRev.28.1049]

[16]   Born, M.; Oppenheimer, J.R. Zur Quantentheorie der Molekeln. *Ann. Phys.,* **1927**, *389*(20), 457-484. [On the Quantum Theory of Molecules]. [in German].
[http://dx.doi.org/10.1002/andp.19273892002]

[17]   Lennard-Jones, J.E. The electronic structure and the interaction of some simple radicals. *Trans. Faraday Soc.,* **1934**, *30*, 70-85.
[http://dx.doi.org/10.1039/tf9343000070]

[18]   Hückel, E. Quantum-theoretical contributions to the benzene problem. I. The electron configuration of benzene and related compounds. *Z. Phys.,* **1931**, *70*, 204-286.

[19]   Hückel, E. Quantum contributions to the benzene problem. II. *Z. Phys.,* **1931**, *72*, 310-337.

[20]   Hückel, E. Quantum contributions to the problem of aromatic and unsaturated compounds. III. *Z. Phys.,* **1932**, *76*, 628-648. [In German].

[21]   Fukui, K.; Yonezawa, T.; Shingu, H. A Molecular Orbital Theory of Reactivity in Aromatic Hydrocarbons. *J. Chem. Phys.,* **1952**, *20*, 722-725.
[http://dx.doi.org/10.1063/1.1700523]

[22]   Woodward, R.B.; Hoffmann, R. Stereochemistry of Electrocyclic Reactions. *J. Am. Chem. Soc.,* **1965**, *87*, 395-397.
[http://dx.doi.org/10.1021/ja01080a054]

[23]   Fleming, I. *Frontier Orbitals and Organic Chemical Reactions*; John Wiley & Sons: New York, **1976**.

[24]   Sauer, J.; Sustmann, R. Mechanistic Aspects of Diels-Alder Reactions: A Critical Survey. *Angew. Chem. Int. Ed. Engl.,* **1980**, *19*, 779-807.
[http://dx.doi.org/10.1002/anie.198007791]

[25]   Woodward, R.B.; Hoffmann, R. the conservation of Orbital Symmetry. *Angew. Chem. Int. Ed. Engl.,* **1969**, *8*, 781-932.
[http://dx.doi.org/10.1002/anie.196907811]

[26]   Fukui, K. Recognition of Stereochemical Paths by Orbital Interaction. *Acc. Chem. Res.,* **1971**, *4*, 57-64.
[http://dx.doi.org/10.1021/ar50038a003]

[27]   Pearson, R.G. Symmetry rules for predicting the course of chemical reactions. *Theoret. chim. Acta(Berl.),* **1970**, *16*, 107-110.
[http://dx.doi.org/10.1007/BF00572779]

[28]   Houk, K.N. Pericyclic Reactions and Orbital Symmetry. *Sur. Prog. Chem.,* **1973**, *6*, 113-208.
[http://dx.doi.org/10.1016/B978-0-12-610506-3.50010-8]

[29]   Dewar, M.J.S. A critique of frontier orbital theory. *J. Mol. Struct. THEOCHEM,* **1989**, *200*, 301-323.
[http://dx.doi.org/10.1016/0166-1280(89)85062-6]

[30]   Krylov, A.; Windus, T.L.; Barnes, T. Computational chemistry software and its advancement as illustrated through three grand challenge cases for molecular science. *J. Chem. Phys.,* **2018**, *149*, 180901-1-11.
[http://dx.doi.org/10.1063/1.5052551]

[31]   Moss, S.J.; Coady, C.J. Potential-energy surfaces and transition state theory. *J. Chem. Educ.,* **1983**, *60*, 455-461.

[http://dx.doi.org/10.1021/ed060p455]

[32]   Hohenberg, P.; Kohn, W. Inhomogeneous electron gas. *Phys. Rev.,* **1964**, *136*, B864-B871.
       [http://dx.doi.org/10.1103/PhysRev.136.B864]

[33]   Parr, R.G.; Yang, W. *Density functional theory of atoms and molecules*; Oxford University Press: New
       York, **1989**.

[34]   Kohn, W.; Sham, L.J. Self-consistent equations including exchange and correlation effects. *Phys. Rev.
       B.,* **1965**, *140*, A1133-A1138.
       [http://dx.doi.org/10.1103/PhysRev.140.A1133]

[35]   Hehre, W.J. Ab Initio Molecular Orbital Theory. *Acc. Chem. Res.,* **1976**, *9*, 399-406.
       [http://dx.doi.org/10.1021/ar50107a003]

[36]   Domingo, L.R. A new C–C bond formation model based on the quantum chemical topology of
       electron density. *RSC Advances,* **2014**, *4*, 32415-32418.
       [http://dx.doi.org/10.1039/C4RA04280H]

[37]   Ríos-Gutiérrez, M.; Domingo, L.R.; Pérez, P. Understanding the high reactivity of carbonyl
       compounds towards nucleophilic carbenoid intermediates generated from carbene isocyanides. *RSC
       Advances,* **2015**, *5*, 84797-84809.
       [http://dx.doi.org/10.1039/C5RA15662A]

[38]   Domingo, L.R. Molecular Electron Density Theory: A Modern View of Reactivity in Organic
       Chemistry. In: *Molecules*; , **2016**; 21, p. 1319(15 pages).

[39]   Frisch, M.J.; Trucks, G.W.; Schlegel, H.B.; Scuseria, G.E.; Robb, M.A.; Cheeseman, J.R.; Scalmani,
       G.; Barone, V. *Petersson, G.A.; Nakatsuji, H. et al. Gaussian 16*; Gaussian, Inc.: Wallingford, CT,
       USA, **2016**.

[40]   Noury, S.; Krokidis, X.; Fuster, F.; Silvi, B. Computational tools for the electron localization function
       topological analysis. *Comput. Chem.,* **1999**, *23*, 597-604.
       [http://dx.doi.org/10.1016/S0097-8485(99)00039-X]

[41]   Lu, T.; Chen, F. Multiwfn: a multifunctional wavefunction analyzer. *J. Comput. Chem.,* **2012**, *33*(5),
       580-592.
       [http://dx.doi.org/10.1002/jcc.22885] [PMID: 22162017]

[42]   Humphrey, W.; Dalke, A.; Schulten, K. VMD: visual molecular dynamics. *J. Mol. Graph.,* **1996**,
       14(1), 33-38.
       [http://dx.doi.org/10.1016/0263-7855(96)00018-5] [PMID: 8744570]

[43]   Pettersen, E.F.; Goddard, T.D.; Huang, C.C.; Couch, G.S.; Greenblatt, D.M.; Meng, E.C.; Ferrin, T.E.
       UCSF Chimera--a visualization system for exploratory research and analysis. *J. Comput. Chem.,* **2004**,
       25(13), 1605-1612.
       [http://dx.doi.org/10.1002/jcc.20084] [PMID: 15264254]

[44]   Parr, R.G.; Yang, W. Density-functional theory of the electronic structure of molecules. *Annu. Rev.
       Phys. Chem.,* **1995**, *46*, 701-728.
       [http://dx.doi.org/10.1146/annurev.pc.46.100195.003413] [PMID: 24341393]

[45]   Geerlings, P.; De Proft, F.; Langenaeker, W. Conceptual density functional theory. *Chem. Rev.,* **2003**,
       103(5), 1793-1874.
       [http://dx.doi.org/10.1021/cr990029p] [PMID: 12744694]

[46]   Domingo, L.R.; Ríos-Gutiérrez, M.; Pérez, P. Applications of the conceptual density functional theory
       indices to organic chemistry reactivity. *Molecules,* **2016**, *21*(6), 748.
       [http://dx.doi.org/10.3390/molecules21060748]

[47]   Parr, R.G.; Pearson, R.G. Absolute hardness: Companion parameter to absolute electronegativity. *J.
       Am. Chem. Soc.,* **1983**, *105*, 7512-7516.
       [http://dx.doi.org/10.1021/ja00364a005]

[48]   Koopmans, T. Über die Zuordnung vonWellenfunktionen und Eigenwerten zu den Einzelnen Elektronen Eines Atoms. *Physica,* **1933**, *1*, 104-113.
[http://dx.doi.org/10.1016/S0031-8914(34)90011-2]

[49]   Pearson, R.G. Hard and soft acids and bases. *J. Am. Chem. Soc.,* **1963**, *85*, 3533-3539.
[http://dx.doi.org/10.1021/ja00905a001]

[50]   Parr, R.G.; von Szentpaly, L.; Liu, S. Electrophilicity index. *J. Am. Chem. Soc.,* **1999**, *121*, 1922-1924.
[http://dx.doi.org/10.1021/ja983494x]

[51]   Domingo, L.R.; Aurell, M.J.; Pérez, P.; Contreras, R. Quantitative characterization of the global electrophilicity power of common diene/dienophile pairs in Diels-Alder reactions. *Tetrahedron,* **2002**, *58*, 4417-4423.
[http://dx.doi.org/10.1016/S0040-4020(02)00410-6]

[52]   Domingo, L.R.; Saéz, J.A.; Pérez, P. A comparative analysis of the electrophilicity of organic molecules between the computed IPs and EAs and the HOMO and LUMO energies. *Chem. Phys. Lett.,* **2007**, *438*, 341-345.
[http://dx.doi.org/10.1016/j.cplett.2007.03.023]

[53]   Domingo, L.R.; Sáez, J.A. Understanding the mechanism of polar Diels-Alder reactions. *Org. Biomol. Chem.,* **2009**, *7*(17), 3576-3583.
[http://dx.doi.org/10.1039/b909611f] [PMID: 19675915]

[54]   Pérez, P.; Domingo, L.R.; Aurell, M.J. Contreras, Quantitative characterization of the global electrophilicity pattern of some reagents involved in 1,3-dipolar cycloaddition reactions. *Tetrahedron,* **2003**, *59*, 3117-3125.
[http://dx.doi.org/10.1016/S0040-4020(03)00374-0]

[55]   Sanderson, R.T. Partial charges on atoms in organic compounds. *Science,* **1955**, *121*(3137), 207-208.
[http://dx.doi.org/10.1126/science.121.3137.207] [PMID: 17774721]

[56]   Sanderson, R.T. *Chemical Bonds and Bond Energy,* 2nd ed; Academic Press: New York, **1976**.

[57]   Roy, R.K.; Krishnamurti, S.; Geerlings, P.; Pal, S. Local softness and hardness based reactivity descriptors for predicting intra- and intermolecular reactivity sequences: Carbonyl compounds. *J. Phys. Chem. A,* **1998**, *102*, 3746-3755.
[http://dx.doi.org/10.1021/jp973450v]

[58]   Chattaraj, P.K.; Maiti, B.; Sarkar, U. Philicity: A unified treatment of chemical reactivity and selectivity. *J. Phys. Chem. A,* **2003**, *107*, 4973-4975.
[http://dx.doi.org/10.1021/jp034707u]

[59]   Contreras, R.; Andrés, J.; Safont, V.S.; Campodonico, P.; Santos, J.G. A theoretical study on the relationship between nucleophilicity and ionization potentials in solution phase. *J. Phys. Chem. A,* **2003**, *107*, 5588-5593.
[http://dx.doi.org/10.1021/jp0302865]

[60]   Gázquez, J.L.; Cedillo, A.; Vela, A. Electrodonating and electroaccepting powers. *J. Phys. Chem. A,* **2007**, *111*(10), 1966-1970.
[http://dx.doi.org/10.1021/jp065459f] [PMID: 17305319]

[61]   Domingo, L.R.; Chamorro, E.; Pérez, P. Understanding the reactivity of captodative ethylenes in polar cycloaddition reactions. A theoretical study. *J. Org. Chem.,* **2008**, *73*(12), 4615-4624.
[http://dx.doi.org/10.1021/jo800572a] [PMID: 18484771]

[62]   Pratihar, S.; Roy, S. Nucleophilicity and site selectivity of commonly used arenes and heteroarenes. *J. Org. Chem.,* **2010**, *75*(15), 4957-4963.
[http://dx.doi.org/10.1021/jo100425a] [PMID: 20670024]

[63]   Domingo, L.R.; Pérez, P. The nucleophilicity N index in organic chemistry. *Org. Biomol. Chem.,* **2011**, *9*(20), 7168-7175.

[http://dx.doi.org/10.1039/c1ob05856h] [PMID: 21842104]

[64] Jaramillo, P.; Domingo, L.R.; Chamorro, E.; Pérez, P. A further exploration of a nucleophilicity index based on the gas-phase ionization potentials. *J. Mol. Struct. THEOCHEM,* **2008**, *865*, 68-72.
[http://dx.doi.org/10.1016/j.theochem.2008.06.022]

[65] Parr, R.G.; Yang, W. Density functional approach to the frontier-electron theory of chemical reactivity. *J. Am. Chem. Soc.,* **1984**, *106*, 4049-4050.
[http://dx.doi.org/10.1021/ja00326a036]

[66] Yang, W.; Mortier, W.J. The use of global and local molecular parameters for the analysis of the gas-phase basicity of amines. *J. Am. Chem. Soc.,* **1986**, *108*(19), 5708-5711.
[http://dx.doi.org/10.1021/ja00279a008] [PMID: 22175316]

[67] Domingo, L.R.; Aurell, M.J.; Pérez, P.; Contreras, R. Quantitative characterization of the local electrophilicity of organic molecules. Understanding the regioselectivity on Diels-Alder reactions. *J. Phys. Chem. A,* **2002**, *106*, 6871-6875.
[http://dx.doi.org/10.1021/jp020715j]

[68] Pérez, P.; Domingo, L.R.; Duque-Noreña, M.; Chamorro, E. A condensed-to-atom nucleophilicity index. An application to the director effects on the electrophilic aromatic substitutions. *J. Mol. Struct. THEOCHEM,* **2009**, *895*, 86-91.
[http://dx.doi.org/10.1016/j.theochem.2008.10.014]

[69] Morell, C.; Grand, A.; Toro-Labbé, A. New dual descriptor for chemical reactivity. *J. Phys. Chem. A,* **2005**, *109*(1), 205-212.
[http://dx.doi.org/10.1021/jp046577a] [PMID: 16839107]

[70] Chattaraj, P.K.; Duley, S.; Domingo, L.R. Understanding local electrophilicity/nucleophilicity activation through a single reactivity difference index. *Org. Biomol. Chem.,* **2012**, *10*(14), 2855-2861.
[http://dx.doi.org/10.1039/c2ob06943a] [PMID: 22388637]

[71] Domingo, L.R.; Saéz, J.A.; Arnó, M. An ELF analysis of the C–C bond formation step in the N-heterocyclic carbene-catalyzed hydroacylation of unactivated C–C double bonds. *RSC Advances,* **2012**, *2*, 7127-7134.
[http://dx.doi.org/10.1039/c2ra21042h]

[72] Domingo, L.R.; Zaragozá, R.J.; Saéz, J.A.; Arnó, M. Understanding the mechanism of the intramolecular stetter reaction. A DFT study. *Molecules,* **2012**, *17*(2), 1335-1353.
[http://dx.doi.org/10.3390/molecules17021335] [PMID: 22301721]

[73] Soto-Delgado, J.; Aizman, A.; Contreras, R.; Domingo, L.R. On the catalytic effect of water in the intramolecular Diels–Alder reaction of quinone systems: a theoretical study. *Molecules,* **2012**, *17*(11), 13687-13703.
[http://dx.doi.org/10.3390/molecules171113687] [PMID: 23169266]

[74] Domingo, L.R.; Pérez, P.; Sáez, J.A. Understanding the local reactivity in polar organic reactions through electrophilic and nucleophilic Parr functions. *RSC Advances,* **2013**, *3*, 1486-1494.
[http://dx.doi.org/10.1039/C2RA22886F]

[75] Chamorro, E.; Pérez, P.; Domingo, L.R. On the nature of Parr functions to predict the most reactive sites along organic polar reactions. *Chem. Phys. Lett.,* **2013**, *582*, 141-143.
[http://dx.doi.org/10.1016/j.cplett.2013.07.020]

[76] Domingo, L.R.; Pérez, P. Global and local reactivity indices for electrophilic/nucleophilic free radicals. *Org. Biomol. Chem.,* **2013**, *11*(26), 4350-4358.
[http://dx.doi.org/10.1039/c3ob40337h] [PMID: 23685829]

[77] Lewis, G.N. *Valence and the Structure of Atoms and Molecules*; Dover Publications Inc: New York, **1966**.

[78] Becke, A.D.; Edgecombe, K.E. A simple measure of electron localization in atomic and molecular systems. *J. Chem. Phys.,* **1990**, *92*, 5397-5403.

[http://dx.doi.org/10.1063/1.458517]

[79]   Silvi, B.; Savin, A. Classification of chemical bonds based on topological analysis of electron localization functions. *Nature*, **1994**, *371*, 686-686.
[http://dx.doi.org/10.1038/371683a0]

[80]   Domingo, L.R.; Chamorro, E.; Pérez, P. Understanding the High Reactivity of the Azomethine Ylides in [3 + 2] Cycloaddition Reactions. *Lett. Org. Chem.*, **2010**, *7*, 432-439.
[http://dx.doi.org/10.2174/157017810791824900]

[81]   Domingo, L.R.; Sáez, J.A. Understanding the electronic reorganization along the nonpolar [3 + 2] cycloaddition reactions of carbonyl ylides. *J. Org. Chem.*, **2011**, *76*(2), 373-379.
[http://dx.doi.org/10.1021/jo101367v] [PMID: 21158474]

[82]   Ríos-Gutiérrez, M.; Domingo, L.R. Unravelling the Mysteries of the [3+2] Cycloaddition Reactions. *Eur. J. Org. Chem.*, **2019**, 267-282.
[http://dx.doi.org/10.1002/ejoc.201800916]

[83]   Rios-Gutierrez, M. *Unravelling [3+2] cycloaddition reactions through the molecular electron density theory*; ProQuest Dissertations Publishing: Universitat de Valencia (Spain), **2018**.

[84]   Domingo, L.R.; Ríos-Gutiérrez, M.; Noreña, M.D.; Chamorro, E.; Pérez, P. Understanding the carbenoid-type reactivity of nitrile ylides in [3+2] cycloaddition reactions towards electron-deficient ethylenes. A molecular electron density theory study. *Theor. Chem. Acc.*, **2016**, *135*, 160.
[http://dx.doi.org/10.1007/s00214-016-1909-6]

[85]   Ríos-Gutiérrez, M.; Domingo, L.R. The carbenoid-type reactivity of simplest nitrile imine from a molecular electron density theory perspective. *Tetrahedron*, **2019**, *75*, 1961-1967.
[http://dx.doi.org/10.1016/j.tet.2019.02.014]

[86]   Domingo, L.R.; Ríos-Gutiérrez, M.; Emamian, S. Understanding the domino reactions between 1-diazopropan-2-one and 1,1-dinitroethylene. A molecular electron density theory study of the [3+2] cycloaddition reactions of diazoalkanes with electron-deficient ethylenes. *RSC Advances*, **2017**, *7*, 15586-15595.
[http://dx.doi.org/10.1039/C7RA00544J]

[87]   Domingo, L.R.; Ríos-Gutiérrez, M. A Molecular Electron Density Theory Study of the Reactivity of Azomethine Imine in [3+2] Cycloaddition Reactions. *Molecules*, **2017**, *22*, 750 (20 Pages).

[88]   Domingo, L.R.; Ríos-Gutiérrez, M.; Pérez, P. A Molecular Electron Density Theory Study of the Reactivity and Selectivities in [3 + 2] Cycloaddition Reactions of C,N-Dialkyl Nitrones with Ethylene Derivatives. *J. Org. Chem.*, **2018**, *83*(4), 2182-2197.
[http://dx.doi.org/10.1021/acs.joc.7b03093] [PMID: 29350934]

[89]   Ndassa, I.M.; Adjieufack, A.I.; Ketcha, J.M.; Berski, S.; Gutiérrez, M.R. Understanding the Reactivity and Regioselectivity of [3+2] Cycloaddition Reactions between Substituted Nitrile Oxides and Methyl Acrylate. *A Molecular Electron Density Theory Study. Int. J. Quantum Chem.*, **2017**, *117*, e25451.
[http://dx.doi.org/10.1002/qua.25451]

[90]   Domingo, L.R.; Acharjee, N. [3+2] Cycloaddition Reaction of C-Phenyl-N-methyl Nitrone to Acyclic-Olefin-Bearing-Electron-Donating Substituent: A Molecular Electron Density Theory Study. *ChemistrySelect*, **2018**, *3*, 8373-8380.
[http://dx.doi.org/10.1002/slct.201801528]

[91]   Sutcliffe, B.T. The idea of a potential energy surface. *Mol. Phys.*, **2006**, *104*, 715-722.
[http://dx.doi.org/10.1080/00268970500418059]

[92]   Fukui, K. The Path of Chemical Reactions - The IRC Approach. *Acc. Chem. Res.*, **1981**, *14*, 363-368.
[http://dx.doi.org/10.1021/ar00072a001]

[93]   Gonzalez, C.; Schlegel, H.B. An improved algorithm for reaction path following. *J. Chem. Phys.*, **1989**, *90*, 2154-2161.
[http://dx.doi.org/10.1063/1.456010]

[94]   Gonzalez, C.; Schlegel, H.B. Reaction path following in mass-weighted internal coordinates. *J. Phys. Chem.*, **1990**, *94*, 5523-5527.
[http://dx.doi.org/10.1021/j100377a021]

[95]   Goldstein, E.; Beno, B.; Houk, K.N. Density Functional Theory Prediction of the Relative Energies and Isotope Effects for the Concerted and Stepwise Mechanisms of the Diels-Alder Reaction of Butadiene and Ethylene. *J. Am. Chem. Soc.*, **1996**, *118*, 6036-6043.
[http://dx.doi.org/10.1021/ja9601494]

[96]   Sustmann, R.; Sicking, W. Influence of Reactant Polarity on the Course of (4 + 2) Cycloadditions. *J. Am. Chem. Soc.*, **1996**, *118*, 12562-12571.
[http://dx.doi.org/10.1021/ja961391d]

[97]   Domingo, L.R.; Arnó, M.; Andrés, J. Influence of Reactant Polarity on the Course of the Inverse-Electron-Demand Diels-Alder Reaction. A DFT Study of Regio- and Stereoselectivity, Presence of Lewis Acid Catalyst, and Inclusion of Solvent Effects in the Reaction between Nitroethene and Substituted Ethenes. *J. Org. Chem.*, **1999**, *64*, 5867-5875.
[http://dx.doi.org/10.1021/jo990331y]

[98]   Domingo, L.R.; José Aurell, M.; Pérez, P.; Contreras, R. Origin of the synchronicity on the transition structures of polar Diels-Alder reactions. Are these reactions [4 + 2] processes? *J. Org. Chem.*, **2003**, *68*(10), 3884-3890.
[http://dx.doi.org/10.1021/jo020714n] [PMID: 12737567]

[99]   Domingo, L.R.; Sáez, J.A. Understanding the mechanism of polar Diels-Alder reactions. *Org. Biomol. Chem.*, **2009**, *7*(17), 3576-3583.
[http://dx.doi.org/10.1039/b909611f] [PMID: 19675915]

[100]   Reed, A.E.; Weinstock, R.B. Weinhold, Natural Population Analysis. F. *J. Chem. Phys.*, **1985**, *83*, 735-746.
[http://dx.doi.org/10.1063/1.449486]

[101]   Reed, A.E.; Curtiss, L.A.; Weinhold, F. Intermolecular interactions from a natural bond orbital, donor-acceptor viewpoint. *Chem. Rev.*, **1988**, *88*, 899-926.
[http://dx.doi.org/10.1021/cr00088a005]

[102]   Domingo, L.R.; Ríos-Gutiérrez, M.; Pérez, P. How does the global electron density transfer diminish activation energies in polar cycloaddition reactions? A Molecular Electron Density Theory study. *Tetrahedron*, **2017**, *73*, 1718-1724.
[http://dx.doi.org/10.1016/j.tet.2017.02.012]

[103]   Domingo, L.R.; Ríos-Gutiérrez, M. On the nature of organic electron density transfer complexes within molecular electron density theory. *Org. Biomol. Chem.*, **2019**, *17*(26), 6478-6488.
[http://dx.doi.org/10.1039/C9OB01031A] [PMID: 31218320]

[104]   Krokidis, X.; Noury, S.; Silvi, B. Characterization of Elementary Chemical Processes by Catastrophe Theory. *J. Phys. Chem. A*, **1997**, *101*, 7277-7282.
[http://dx.doi.org/10.1021/jp9711508]

[105]   Thom, R. *Stabilité Structurelle et Morphogénèse*; Intereditions: Paris, **1972**.

[106]   Zeeman, E.C. Catastrophe theory: a reply to Thom. In: Dynamical Syst. Warwick 1974, A. Manning (ed.), Lect.*Notes Math, 468, Springer-Verlag: New York, Heidelberg, Berlin, Tokyo*; , **1976**, p. 405 pp.. Zbl.307.58009.

[107]   Andrés, J.; Berski, S.; Domingo, L.R.; Polo, V.; Silvi, B. Describing the Molecular Mechanism of Organic Reactions by Using Topological Analysis of Electronic Localization Function. *Curr. Org. Chem.*, **2011**, *15*, 3566-3575.
[http://dx.doi.org/10.2174/138527211797636156]

[108]   Krokidis, X.; Silvi, B.; Alikhani, M.E. Topological characterization of the isomerization mechanisms in XNO (X=H, Cl). *Chem. Phys. Lett.*, **1998**, *292*, 35-45.

[http://dx.doi.org/10.1016/S0009-2614(98)00650-2]

[109] Polo, V.; Gonzalez-Navarrete, P.; Andres, J.; Silvi, B. An electron localization function and catastrophe theory analysis on the molecular mechanism of gas-phase identity SN2 reactions. *Theor. Chem. Acc.,* **2008**, *120*, 341-349.
[http://dx.doi.org/10.1007/s00214-008-0427-6]

[110] Krokidis, X.; Silvi, B.; Dezarnaud-Dandine, C.; Sevin, A. Topological study, using a coupled ELF and catastrophe theory technique, of electron transfer in the Li+Cl2 system. *New J. Chem.,* **1988**, *22*, 1341-1350.
[http://dx.doi.org/10.1039/a801838c]

[111] Berski, S.; Latajka, Z. Quantum chemical topology description of the hydrogen transfer between the ethynyl radical and ammonia (C2H. + NH3) – The electron localization function study. *Chem. Phys. Lett.,* **2006**, *426*, 273-279.
[http://dx.doi.org/10.1016/j.cplett.2006.06.049]

[112] Polo, V.; Andrés, J. Lewis Acid and Substituent Effects on the Molecular Mechanism for the Nazarov Reaction of Penta-1,4-dien-3-one and Derivatives. A Topological Analysis Based on the Combined Use of Electron Localization Function and Catastrophe Theory. *J. Chem. Theory Comput.,* **2007**, *3*(3), 816-823.
[http://dx.doi.org/10.1021/ct7000304] [PMID: 26627401]

[113] Berski, S.; Andrés, J.; Silvi, B.; Domingo, L.R. The Joint Use of Catastrophe Theory and Electron Localization Function to Characterize Molecular Mechanisms. A Density Functional Study of the Diels-Alder Reaction between Ethylene and 1,3-Butadiene. *J. Phys. Chem. A,* **2003**, *107*, 6014-6024.
[http://dx.doi.org/10.1021/jp030272z]

[114] Polo, V.; Andrés, J. A joint study based on the electron localization function and catastrophe theory of the chameleonic and centauric models for the Cope rearrangement of 1,5-hexadiene and its cyano derivatives. *J. Comput. Chem.,* **2005**, *26*(14), 1427-1437.
[http://dx.doi.org/10.1002/jcc.20272] [PMID: 16082660]

[115] Andrés, J.; Berski, S.; Silvi, B. Curly arrows meet electron density transfers in chemical reaction mechanisms: from electron localization function (ELF) analysis to valence-shell electron-pair repulsion (VSEPR) inspired interpretation. *Chem. Commun. (Camb.),* **2016**, *52*(53), 8183-8195.
[http://dx.doi.org/10.1039/C5CC09816E] [PMID: 27218123]

[116] Ayouchia, H.B. El.; Lahoucine, B.; Anane, H.; Ríos-Gutiérrez, M.; Domingo, L.R.; Stiriba, S.E. Experimental and Theoretical MEDT Study of the Thermal [3+2] Cycloaddition Reactions of Aryl Azides with Alkyne Derivatives. *ChemistrySelect,* **2018**, *3*, 1215-1233.
[http://dx.doi.org/10.1002/slct.201702588]

[117] Bader, R.F.W. *Atoms in Molecules: A Quantum Theory*; Oxford: Clarendon Press: USA, **1990**.

[118] Bader, R.F.W.; Essén, H. The characterization of atomic interactions. *J. Chem. Phys.,* **1984**, *80*, 1943-1960.
[http://dx.doi.org/10.1063/1.446956]

[119] Contreras-García, J.; Johnson, E.R.; Keinan, S.; Chaudret, R.; Piquemal, J.P.; Beratan, D.N.; Yang, W. NCIPLOT: a program for plotting non-covalent interaction regions. *J. Chem. Theory Comput.,* **2011**, *7*(3), 625-632.
[http://dx.doi.org/10.1021/ct100641a] [PMID: 21516178]

[120] Johnson, E.R.; Keinan, S.; Mori-Sánchez, P.; Contreras-García, J.; Cohen, A.J.; Yang, W. Revealing noncovalent interactions. *J. Am. Chem. Soc.,* **2010**, *132*(18), 6498-6506.
[http://dx.doi.org/10.1021/ja100936w] [PMID: 20394428]

[121] Domingo, L.R.; Ríos-Gutiérrez, M.; Acharjee, N. A Molecular Electron Density Theory Study of the Chemoselectivity, Regioselectivity and Diastereofacial Selectivity in the Synthesis of an Anti-Cancer Spiro-Isoxazoline derived from α-Santonin. *Molecules,* **2019**, *24*, 832 (14 pages).
[http://dx.doi.org/10.3390/molecules24050832]

[122] Evans, M.G. The activation energies of reactions involving conjugated systems. *Trans. Faraday Soc.,* **1939**, *35*, 824-834.
[http://dx.doi.org/10.1039/tf9393500824]

[123] Dewar, M.J.S. Aromaticity and pericyclic rections. *Angew. Chem. Int. Ed. Engl.,* **1971**, *10*, 761-776.
[http://dx.doi.org/10.1002/anie.197107611]

[124] Houk, K.N.; Gonzalez, J.; Li, Y. Pericyclic Reaction Transition States: Passions and Punctilios, 1935-1995. *Acc. Chem. Res.,* **1995**, *28*, 81-90.
[http://dx.doi.org/10.1021/ar00050a004]

[125] Domingo, L.R. State of the Art of the Bonding Changes along the Diels-Alder Reaction between Butadiene and Ethylene: Refuting the Pericyclic Mechanism. *Org. Chem. Curr. Res.,* **2013**, *2*, 120.

[126] Domingo, L.R.; Ríos-Gutiérrez, M.; Chamorro, E.; Pérez, P. Aromaticity in Pericyclic Transition State Structures? A Critical Rationalisation Based on the Topological Analysis of Electron Density. *ChemistrySelect,* **2016**, *1*, 6026-6039.
[http://dx.doi.org/10.1002/slct.201601384]

[127] Domingo, L.R.; Ríos-Gutiérrez, M.; Silvi, B.; Pérez, P. The mysticism of pericyclic reactions: A contemporary rationalisation of organic reactivity based on electron density analysis. *Eur. J. Org. Chem.,* **2018**, 1107-1120.
[http://dx.doi.org/10.1002/ejoc.201701350]

[128] Hallooman, D.; Cudian, D.; Ríos-Gutiérrez, M.; Rhyman, L.; Alswaidan, I.A.; Elzagheid, M.I.; Domingo, L.R.; Ramasami, P. Understanding the Intramolecular Diels-Alder Reactions of N-Susbtituted N-allyl-furfurylamines. An MEDT Study. *ChemistrySelect,* **2017**, *2*, 9736-9745.
[http://dx.doi.org/10.1002/slct.201702136]

[129] Domingo, L.R.; Gutiérrez, M.R.; Pérez, P. A Molecular Electron Density Theory Study of the Competitiveness of Polar Diels-Alder and Polar Alder Ene Reactions. *Molecules,* **2018**, *23*, 1913.
[http://dx.doi.org/10.3390/molecules23081913]

[130] Padwa, A.; Pearson, W.H., Eds. *Synthetic Application of 1,3-Dipolar Cycloaddition Chemistry Toward Heterocycles and Natural Products*; Wiley: New York, **2002**.
[http://dx.doi.org/10.1002/0471221902]

[131] Firestone, R.A. Mechanism of 1,3-dipolar cycloadditions. *J. Org. Chem.,* **1968**, *33*, 2285-2290.
[http://dx.doi.org/10.1021/jo01270a023]

[132] Ríos-Gutiérrez, M.; Darù, A.; Tejero, T.; Domingo, L.R.; Merino, P. A molecular electron density theory study of the [3 + 2] cycloaddition reaction of nitrones with ketenes. *Org. Biomol. Chem.,* **2017**, *15*(7), 1618-1627.
[http://dx.doi.org/10.1039/C6OB02768G] [PMID: 28120980]

[133] Nasri, L.; Gutiérrez, M.R.; Nacereddine, A.K.; Djerourou, A.; Domingo, L.R. A Molecular Electron Density Theory study of [3+2] cycloaddition reactions of chiral azomethine ylides with ß-nitrostyrene. *Theor. Chem. Acc.,* **2017**, *136*, 104 (12 Pages).
[http://dx.doi.org/10.1007/s00214-017-2133-8]

[134] Domingo, L.R.; Gutiérrez, M.R.; Pérez, P. A Molecular Electron Density Theory Study of the [3+2] Cycloaddition Reaction of Nitrones with Strained Allene. *RSC Advances,* **2017**, *7*, 26879-26887.
[http://dx.doi.org/10.1039/C7RA01916E]

[135] Barber, J.S.; Styduhar, E.D.; Pham, H.V.; McMahon, T.C.; Houk, K.N.; Garg, N.K. Nitrone Cycloadditions of 1,2-Cyclohexadiene. *J. Am. Chem. Soc.,* **2016**, *138*(8), 2512-2515.
[http://dx.doi.org/10.1021/jacs.5b13304] [PMID: 26854652]

[136] Gutiérrez, M.R.; Nasri, L.; Nacereddine, A.K.; Djerourou, A.; Domingo, L.R. A molecular electron density theory study of the [3 + 2] cycloaddition reaction between an azomethine imine and electron deficient ethylenes. *J. Phys. Org. Chem.,* **2018**, *31* e3830
[http://dx.doi.org/10.1002/poc.3830]

[137] Domingo, L.R.; Gutiérrez, M.R.; Adjieufack, A.I.; Ndassa, I.M.; Nouhou, C.N.; Mbadcam, J.K. Molecular Electron Density Theory Study of Fused Regioselectivity in the Intramolecular [3+2] Cycloaddition Reaction of Nitrones. *ChemistrySelect,* **2018**, *3*, 5412-5420.
       [http://dx.doi.org/10.1002/slct.201800224]

[138] Domingo, L.R.; Ríos-Gutiérrez, M.; Pérez, P. A Molecular Electron Density Theory Study of the Role of the Copper Metalation of Azomethine Ylides in [3 + 2] Cycloaddition Reactions. *J. Org. Chem.,* **2018**, *83*(18), 10959-10973.
       [http://dx.doi.org/10.1021/acs.joc.8b01605] [PMID: 30052439]

[139] Emamian, S.; Lu, T.; Domingo, L.R.; Saremi, L.H.; Gutiérrez, M.R. A Molecular Electron Density Theory study of the chemo- and regioselective [3+2] cycloaddition reactions between trifluoroacetonitrile N-oxide and thioketone. *Chem. Phys.,* **2018**, *501*, 128-137.
       [http://dx.doi.org/10.1016/j.chemphys.2017.12.019]

CHAPTER 6

Frontier Molecular Orbital Approach to the Cycloaddition Reactions

Anjandeep Kaur[*]

Department of Chemistry, Government Mohindra College Patiala, Punjab, 147001, India

Abstract: The frontier molecular orbital (FMO) theory has provided a powerful model for the qualitative understanding of reactivity and regioselectivity of the Cycloaddition reactions, based on the electronic properties of isolated reactants. The 1,3-dipolar cycloaddition reactions are explicable by the frontier molecular orbital approach, which is based on the assumption that bonds are formed by a flow of electrons from the highest occupied molecular orbital (HOMO) of one reactant to the lowest unoccupied molecular orbital (LUMO) of another. But the tricky part here was to decide which molecule supplied the HOMO and which supplied the LUMO. Furthermore, the computational chemistry techniques like $HOMO_{dipole}$-$LUMO_{dipolarophile}$/$HOMO_{dipolarophile}$-$LUMO_{dipole}$ energy gaps, electronic chemical potentials (μ), electrophilicity indices (ω) and the charge capacities (ΔN_{max}) are useful in indicating whether the reactions are under normal or inverse electron demand conditions. Also, the relative kinetics of cycloaddition reactions can be rationalized by utilizing $HOMO_{dipole}$-$LUMO_{dipolarophile}$ energy gaps and ΔN_{max}, from which it was found that increasing electron-withdrawing power of the dipolarophile substituents, the energy gap decreases and, thus, reactions with the same dipole became faster in Normal electron demand cycloadditions, while the reverse occurs in case of inverse electron demand conditions.

Keywords: Cycloaddition, Chemical Potential, Charge Capacity, Diastereo-selectivity, Dipolarophile, Electrophilicity, Enantioselectivity, Inverse Electron Demand Reactions, Normal Electron Demand Reactions, Regioselectivity, Stereochemistry.

INTRODUCTION

The language of chemistry is replete with orbitals, whether they be hybrid orbitals (in valence-bond models) or linear combination of atomic orbitals (in molecular orbital models). The power of the orbital concept cannot be avoided: almost everything in chemistry can be described, at a qualitative level by the appropriate use of an orbital model. The 'Orbital' concept, which was established by many

[*] **Corresponding author Dr. Anjandeep Kaur:** Department of Chemistry, Government Mohindra College Patiala, Punjab, 147001, India, E-mail: kauranjandeep@gmail.com

Zaheer Ul-Haq and Angela K. Wilson (Eds.)

scientists, was initially used to construct the wave function of molecules through which molecular properties were usually interpreted [1]. In 1950's, Fukui developed Frontier molecular orbital theory on the basis of which he proposed that the pair of π-electrons occupying the highest orbital, which is referred to as frontier electrons, plays a decisive role in chemical activation of molecules. Instead of thinking about the total electron density, he focused on two particular orbitals, which act as an essential part in a wide range of chemical reactions and referred these orbitals as Highest occupied molecular orbitals (HOMO) and Lowest unoccupied molecular orbitals (LUMO) [2 - 4]. Based upon this FMO approach, he provided a simple and clear picture of theoretical interpretation of many reactions and 'The Overlap and Orientation' principle of Mulliken for the formation of molecular complexes [5, 6]. The utility of Fukui's study was further broadened by the role of symmetry orbitals pointed by Diels and Alder in the cycloaddition reactions. Based upon the symmetry of molecular orbitals, Woodward and Hoffmann proposed a principle called 'The conservation of orbital symmetry' to explain stereochemistry in pericyclic reactions [7 - 9]. The century-old class [10 - 13] of 1,3-dipolar cycloaddition (1,3-DC) is a well-studied field of organic chemistry. These reactions have a wide application in the synthesis of many pharmaceutically important compounds and intermediates [14, 15].

In the beginning, the concept of 1,3-dipolar cycloadditions was given by Smith in 1938 [16], but the generalization given by Huisgen, in 1960, made it widely applicable [17]. The research in the field of 1,3-dipolar cycloaddition reactions is utilized in almost every area of chemistry, from material chemistry [18] to drug discovery [19], indicating its diversity. 1,3-Dipolar cycloaddition is considered as an important reaction in synthetic organic chemistry where the two organic molecules *i.e.* a 1,3-dipole **I** and a dipolarophile **II,** join to give a five-membered heterocycle **III** (Scheme 1). A number of simple and complex heterocycles of immense importance for the industrial and academic world can be prepared from simple and easily accessible molecules by 1,3-dipolar cycloaddition reaction. These reactions are commonly used as a key step in several organic syntheses [20].

As the control of stereochemistry is the major challenge in cycloaddition reactions, therefore, in recent years, such reactions have entered a new stage of development. Nowadays, controlling the regioselectivity, enantioselectivity, and diastereoselectivity of the 1,3-dipolar cycloaddition reactions is the main challenge. Their stereochemistry can be brought under control by either controlling the reaction by a metal complex acting as a catalyst or by selecting the appropriate substrate [21 - 23].

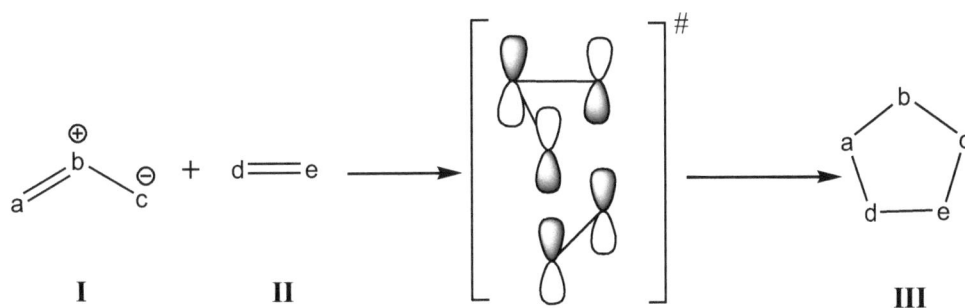

Scheme 1. Basic schematic representation of 1,3-dipolar cycloaddition reactions.

1,3-DIPOLE/YLIDE

Curtius was considered as a pioneer in the history of 1,3-dipoles with the discovery of diazoacetic ester, in 1883 [24]. The 1,3-dipole or an ylide has a positive and a negative charge spread over three atoms and possesses 4π-electrons [25]. Generally, the elements of groups 14, 15, and 16 are included in 1,3-dipoles among which nitrogen, carbon, and oxygen are the predominant ones and due to this limitation on the central atom of the dipole, a limited number of structures can be made by combination and permutations of these three atoms.

The elements of the higher row like sulfur and phosphorus, can also be included in 1,3-dipoles, but only a few cycloaddition reactions involving such types of dipoles have been published. The ylides or 1,3-dipoles can be represented by two octet-structures and two sextet structures. In the case of octet structure, the positive charge is present on the central atom while the two terminal atoms have a negative charge distributed over them. In sextet-structure, two among the four π-electrons are localized at the central atom. This structure explains the ambivalence of the 1,3-dipole but contributes little to resonance hybrid's electron distribution. The ambivalent nature of 1,3-dipole plays a major role in comprehending the regiochemistry, reactivity, and mechanism of 1,3-dipolar cycloadditions.

Classification of 1,3-Dipoles

1,3-Dipoles were categorized by Huisgen into two categories *i.e.* allyl anion type and propargyl/allenyl anion type 1,3-dipoles. They are also referred to as sp^2 and sp hybridized 1,3-dipoles, respectively.

Allyl Anion Type 1,3-Dipoles

The allyl anion type 1,3-dipoles are bent in nature and have four π-electrons in parallel to three *p*-orbitals, which are perpendicular to the plane of the dipole

(**Scheme 2**). Two octet structures, where all three centers have an electron octet and two sextet structures where the atom *a* or *c* has an electron sextet, can be drawn. Nitrogen, oxygen, or sulfur can be the central atom *b* Table **1**.

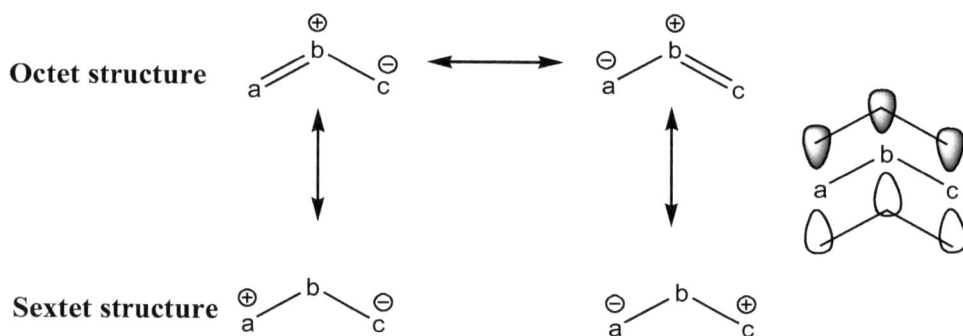

Octet structure

Sextet structure

Scheme 2. The allyl anion type 1,3-dipoles.

Table 1. Allyl anion type 1,3-dipoles.

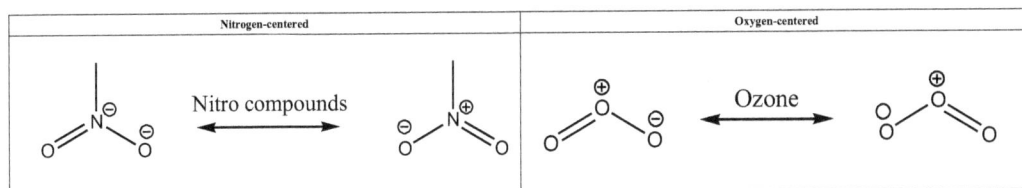

Propargyl/Allenyl Anion Type 1,3-Dipoles

These dipoles have an extra π-bond in comparison to allyl anion type dipoles perpendicular to their molecular plane (Scheme **3**). Due to this extra π-bond, propargyl/allenyl anion type 1,3-dipoles are of linear nature. The central atom *b* is generally nitrogen (Table **2**).

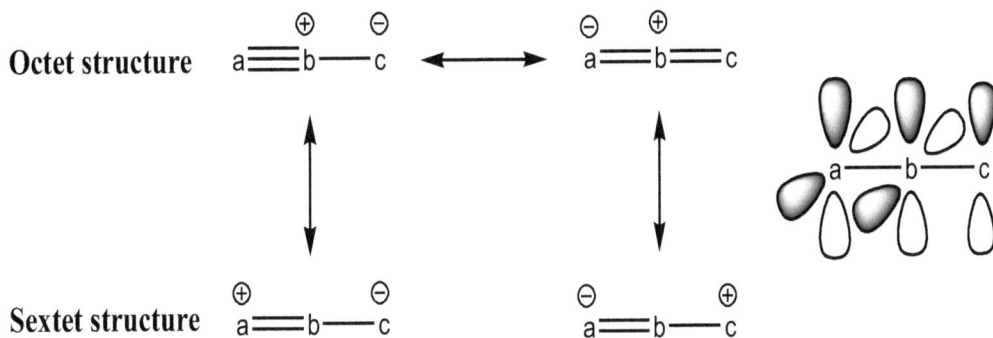

Scheme 3. The propargyl/allenyl anion type dipoles.

Table 2. Propargyl/Allenyl anion type 1,3-dipoles.

The Dipolarophile

The 2π-component is termed as dipolarophile in the case of 1,3-dipolar cycloaddition reactions. The dipolarophiles can have substituted double or triple bond, such as C≡C [26], C=C [27], C≡N [28], C=N [29], C=O [20] and C=S [30]. Such a large number of dipolarophiles make the 1,3-dipolar cycloadditions to be a very resourceful and beneficial reaction in the synthesis of numerous heterocyclic compounds. The dipolarophiles can have mono-, di-, tri- or tetra substitution on the π-bond, moreover, this π-bond may be conjugated [31], isolated or part of a cumulene [32] system. However, the tri- and tetrasubstituted dipolarophiles are quite less reactive in nature due to steric factors.

Mechanism of 1,3-Dipolar Cycloaddition

During 1960's, the mechanism of 1,3-dipolar cycloaddition reactions was a much debated subject. Huisgen proposed a synchronous-concerted mechanism [33, 34], but Firestone was in favor of a stepwise-diradical mechanism [35, 36] (Scheme 4).

(A) Huisgen's concerted mechanism

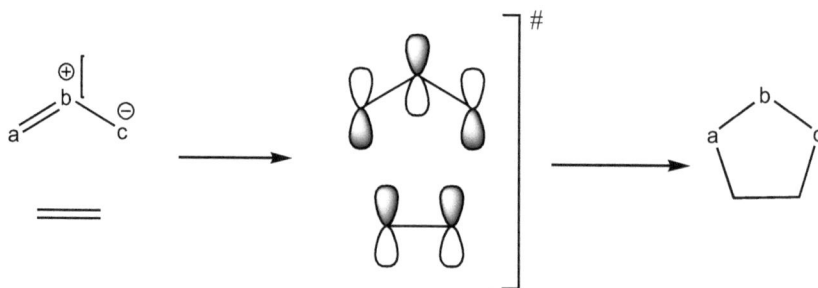

(B) Firestone's stepwise mechanism

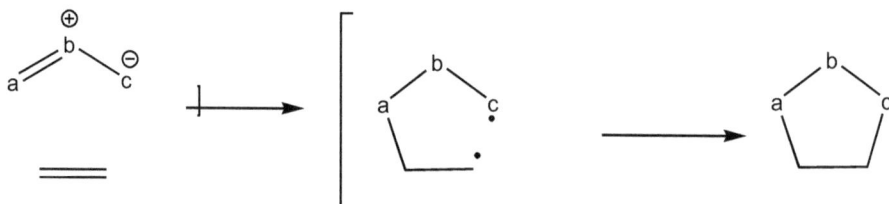

Scheme 4. Schematic representation of the mechanism of cycloaddition reactions **(A)** Concerted mechanism **(B)** Stepwise mechanism.

The concerted mechanism of Huisgen was vehemently supported by the 'cis' nature of the additions, *i.e.* the substituents on both the reactants got their geometrical relationships preserved in the product. But according to Firestone in its stepwise mechanism, the stereospecific nature of the reaction is due to a greater energy barrier for rotation around the single bond in the intermediate diradical than the activation energy for the ring closure, which also explains the cis-stereospecificity of the cycloaddition. In 1985, Houk worked with Firestone to study the cycloaddition of *p*-nitrobenzonitrile oxide to cis- and trans-dideuterioethylene (Scheme **5**) [37]. The cycloaddition of the benzonitrile oxide with cis-dideuterioethylene produced the cis adduct only. If this reaction has taken place through a diradical mechanism, then both the cis- and trans-products would have formed because rotation about single bonds to deuterated primary radical centers in diradicals is very fast in comparison to the cyclization process. Consistent results were obtained for trans-dideuterioethylene.

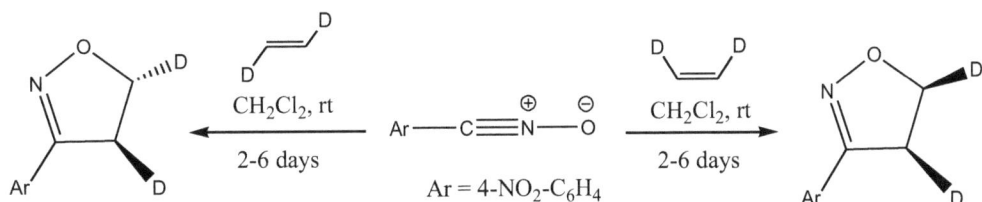

Scheme 5. 1,3-dipolar cycloaddition of *p*-nitrobenzonitrile oxide to cis- and trans-dideuterioethylene.

The arguments on the mechanism of 1,3-dipolar cycloaddition reactions are still unresolved. But it is generally concluded that these reactions are concerted but not synchronous. This means that in the transition state, the rate of formation of the two new bonds is not the same (Fig. **1**), or that one bond is formed faster than the other. As shown in Fig. (**1**) the bond a-e is more developed in comparison to the bond c-d, therefore, the bond a-e is formed earlier than the bond c-d. Thus, it can be concluded that there is a simultaneous bond formation but with different speeds. Hence it is considered that the mechanism of cycloaddition reactions is 'Concerted but asynchronous' and the transition state is considered as evidence for the concerted reaction mechanism.

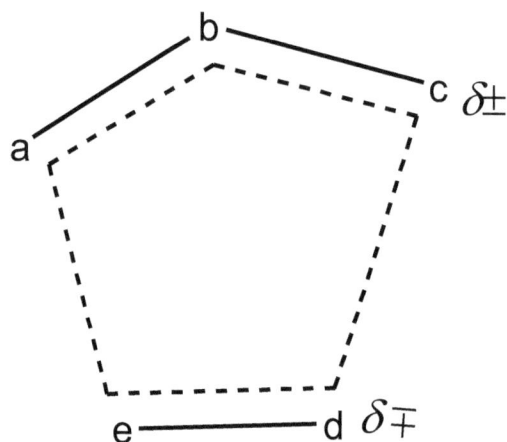

Fig. (1). Asynchronous mechanism.

Pedron *et al.* studied the mechanism of cycloadditions of nitrones with 1,2-diaza-1 ,3-dienes, by using density functional theory (DFT) methods and found that the reaction took place through an asynchronous concerted transition state rather than a stepwise mechanism, which occurs *via* a higher energy pathway in comparison to concerted process [38]. Jasinski and coworkers performed DFT calculations on 1,3-dipolar cycloaddition reactions of conjugated nitroalkanes to benzonitrile *N*-oxides and shed light on its one-step mechanism through asynchronous transition state [39].

The influence of hydrogen bonding on the transition state was studied by Tamilmani *et al. via* DFT investigations on cycloaddition reaction on hydroxy-ortho-quinodimethane with fumaric acid. The transition state of cycloaddition was stabilized by hydrogen bonding between the hydroxyl group of diene and carboxylic group of dienophile, which led to the syn product [40].

There is certainly an impact on the stability of the transition state due to steric factors, which further influence the stereoselectivity of the reaction. In previous research studies involving the cycloaddition of *N*-aryl maleimides with *C,N*-diphenylnitrones, two diastereomeric isoxazolidines were produced with high selectivity, the *exo-endo* ratio depends upon the secondary interactions due to π-π stacking between the aromatic rings as well as on the substituent's position on the *C*-phenyl ring of the *C,N*-diphenylnitrones [41]. When a *p*-OCH$_3$ group was present, then the secondary interactions dominated (Fig. **2**) to lead the *endo* product but the *exo-endo* isomeric ratio was reversed when the -OCH$_3$ group occupied the *ortho* or *meta* position on the *C*-phenyl ring. This happened because of the steric interactions arising due to the closer approach of methoxy group to the oxygen atom of carbonyl moiety, which led to the dominance of *exo* over *endo* isomer (Fig. (**3**).

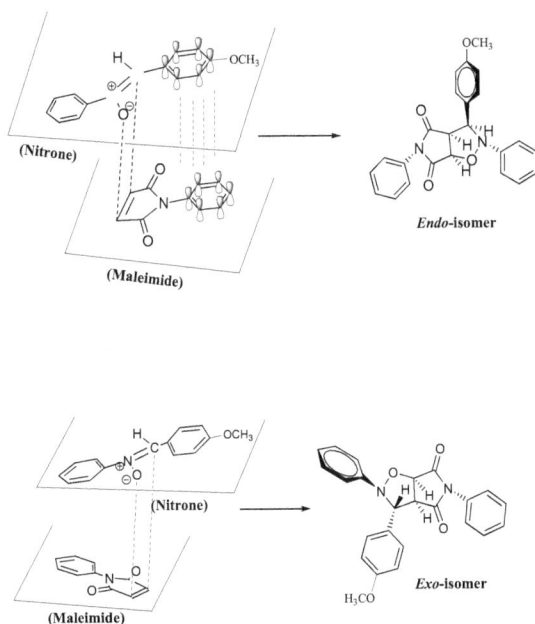

Fig. (2). Secondary interactions leading to *Endo*-isomer as a major product.

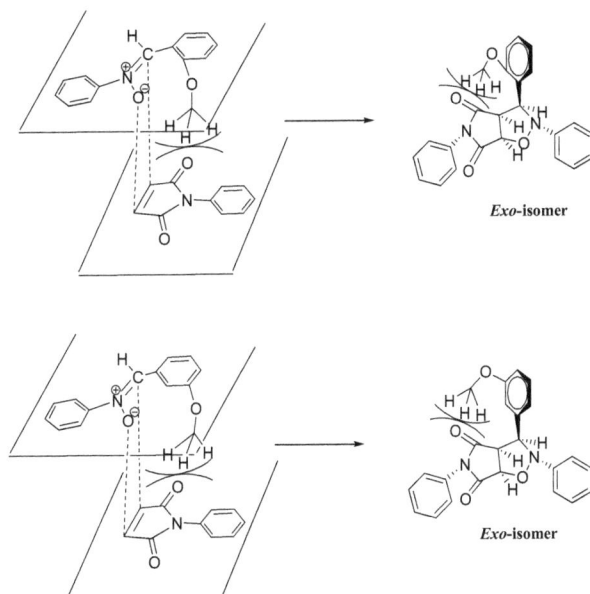

Fig. (3). Steric interactions leading to the dominance of *Exo*-isomer as a major product.

Regioselectivity in 1,3-Dipolar Cycloaddition

As the transition state of 1,3-dipolar cycloadditions is under the control of frontier orbital coefficients, hence, the regioselectivity of such reactions can be rationalized by frontier orbital theory [42]. In cycloaddition reactions, regioselectivity is discerned when unsymmetrically substituted components participate in the reaction, as in the case of a Diels-Alder reaction in which both the reactants, *i.e.* alkene and diene, are substituted unsymmetrically making the two regioisomers possible (Scheme **6**).

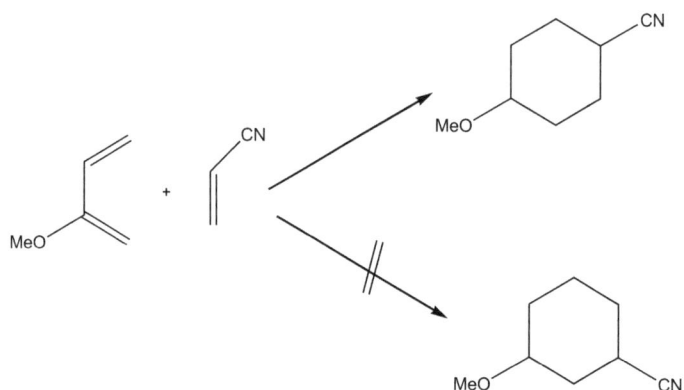

Scheme 6. Diels-Alder reaction between an unsymmetrically substituted diene and alkene.

The regioselectivity of the reactions can be predicted with the help of FMO theory and by analyzing the HOMO-LUMO energy gaps between the dipole and dipolarophile [43, 44]. In the case of cycloaddition reactions, two types of interactions, *i.e.* $HOMO_{dipole}$-$LUMO_{dipolarophile}$ and $HOMO_{dipolarophile}$-$LUMO_{dipole}$, can exist and the result is predicted by the overlap of orbitals with the largest coefficients. The governing electronic interaction is due to the combination between the largest HOMO and the largest LUMO. Therefore, the regioselectivity is directed by the atoms that have the largest orbital HOMO and LUMO coefficients [45, 46].

Considering the case of diazomethane cycloaddition to methyl acrylate and methyl cinnamate, it can ve seen that the diazomethane carbon has the largest HOMO while the largest LUMO is possessed by the terminal olefinic carbon of methyl acrylate. Hence, in this cycloaddition, we get the substitution regioselectively at the C-3 position (Scheme **7**). In the case of cycloaddition of diazomethane with methyl cinnamate, among the phenyl and carboxyl group, the better electron-withdrawing group is the carboxyl one, which makes the β-carbon to be the most electrophilic. Thus, the product has a phenyl group on C-4 and a carboxyl group on C-3 positions regioselectively.

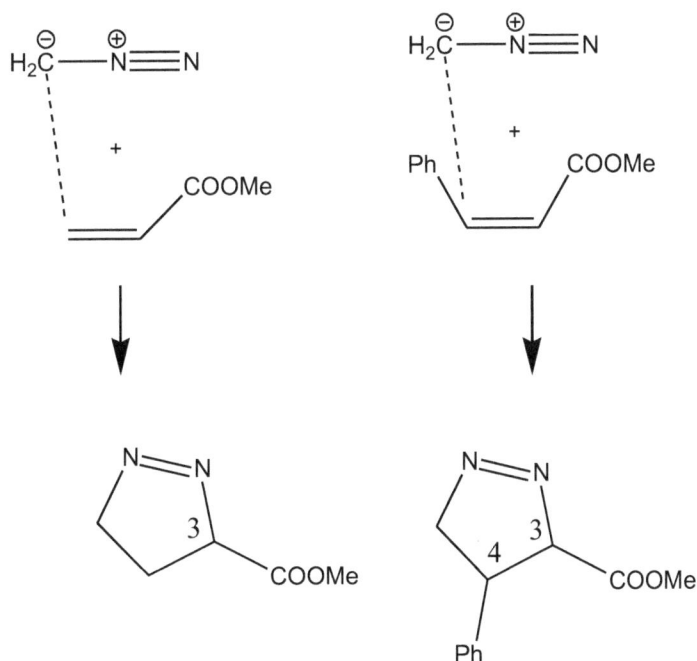

Scheme 7. Cycloaddition reaction of Diazomethane with methyl acrylate and methyl cinnamate.

Padwa and coworkers elucidated the regioselective cycloaddition of carbonyl ylide (generated *in situ* by using $Rh_2(OAc)_4$ catalyst in benzene solvent) with methyl propiolate and methyl propargyl ether [47, 48]. The $HOMO_{dipole}$-$LUMO_{dipolarophile}$ interaction dominated in the cycloaddition of carbonyl ylide with methyl propiolate. Here, the carbon of carbonyl ylide (proximal to carbonyl group) and the terminal alkyne carbon of methyl propiolate have the largest orbital coefficients, hence resulting in the single regioisomer in the product. But in the case of cycloaddition of carbonyl ylide with methyl propargyl ether, $HOMO_{dipolarophile}$-$LUMO_{dipole}$ interaction dominated. Here, the carbon distal to the carbonyl group of the carbonyl ylide and the terminal alkyne carbon of methyl propargyl ether, which possessed the largest orbital coefficients, resulted in single isomer in the product (Scheme **8**).

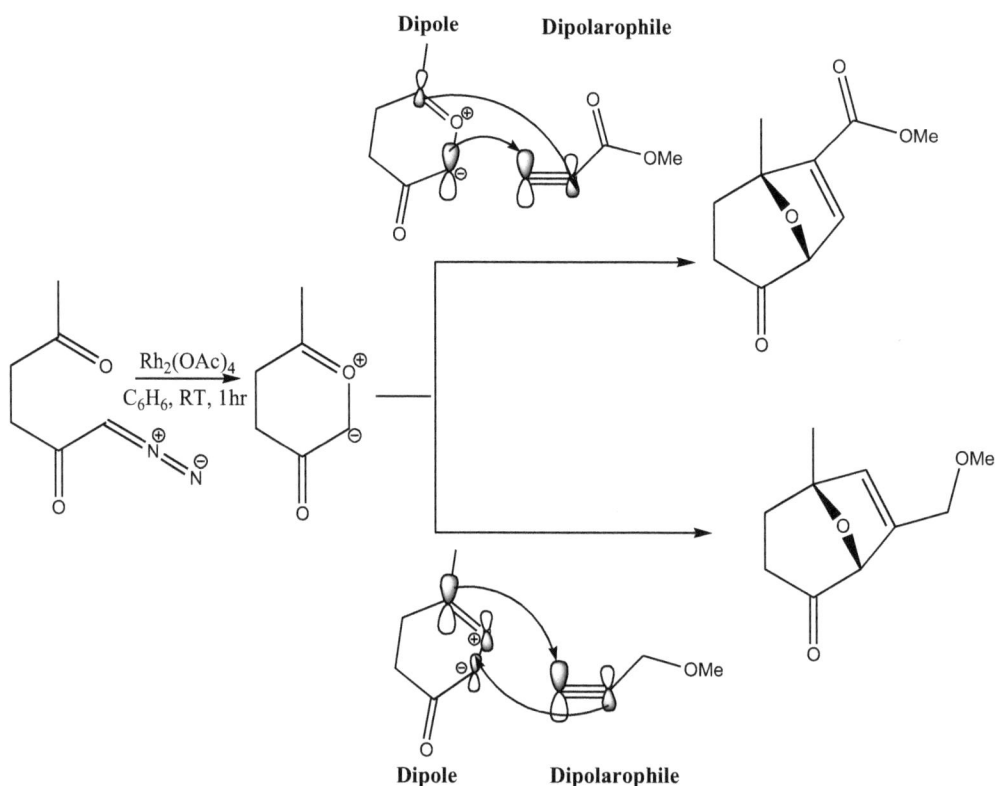

Scheme 8. Regioselective cycloaddition of Carbonyl ylide to methyl propiolate and methyl propargyl ether.

Based on predominant FMO interactions, Sustmann classified 1,3-dipolar cycloaddition reactions into three types: type-I, type-II, and type-III [49, 50]. In type-I reactions, the interaction is of HOMO$_{dipole}$ with LUMO$_{dipolarophile}$. These reactions are also termed as *'normal electron demand'* or *'HOMO controlled' reactions*. Fig. (**4**) shows symmetry allowed molecular orbitals (MO) interactions. In this figure, the low energy interactions are shown by the solid arrow and high energy interactions by the dotted arrow. These types of cycloadditions are accelerated by the presence of electron-donating groups (EDG) on 1,3-dipole as the HOMO of dipole becomes electron-rich and its energy rises towards the LUMO of dipolarophile. While the electron-withdrawing groups (EWG) on dipolarophile lowers the energy of the LUMO towards the HOMO of the dipole. Hence, the HOMO-LUMO gap in both cases gets diminished.

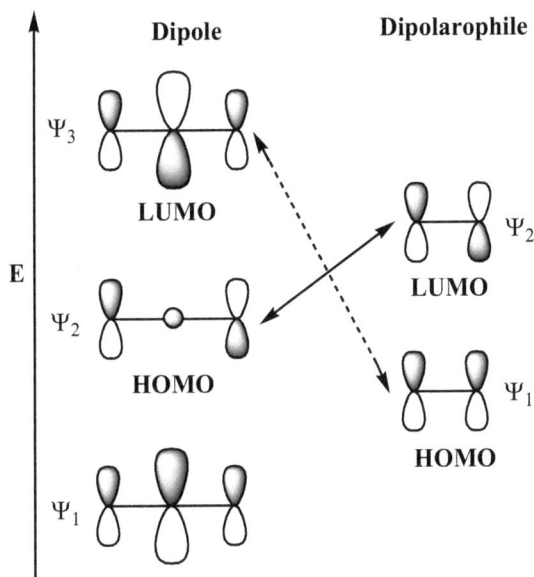

Fig. (4). Symmetry allowed interactions of Type-I 1,3-dipolar cycloadditions.

In type-II reactions, since FMO energies of the dipole and alkene are similar, both HOMO-LUMO interactions are to be considered. Adding either an EDG or EWG to the dipole or dipolarophile can accelerate these reactions (Fig. **5**). The cycloaddition reactions of azides belong to this category.

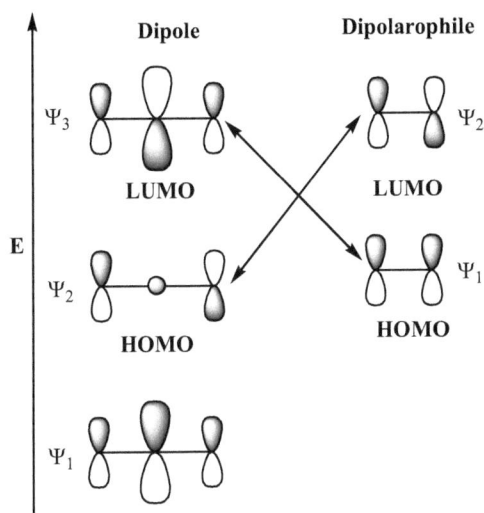

Fig. (5). Symmetry allowed interactions of Type-II 1,3-dipolar cycloadditions.

The 1,3-dipoles in type-III reactions react opposite to that of type-I 1,3-dipoles. The type-III reactions interact between $LUMO_{dipole}$ and $HOMO_{dipolarophile}$ (Fig. 6) hence the EDGs on the dipolarophile and EWGs on the dipole accelerate these reactions. These reactions are also called *'inverse electron demand'* or *'LUMO controlled'* reactions. The dipoles in Type-III reactions are considered to be electrophilic in nature because they react with electron-rich dipolarophiles more efficiently.

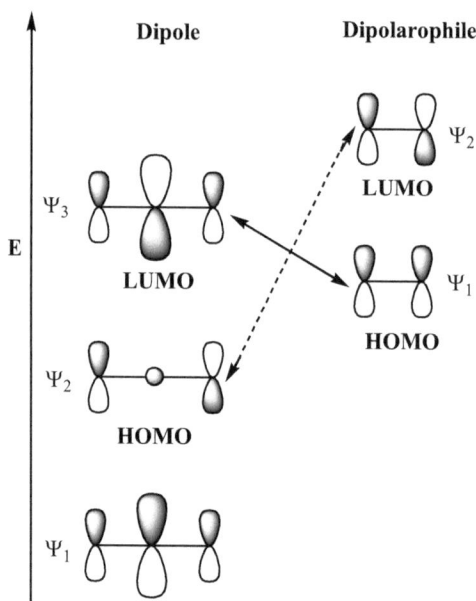

Fig. (6). Symmetry allowed interactions of Type-III 1,3-dipolar cycloadditions.

The azomethine ylide and azomethine imine cycloaddition reactions are examples of type-I reactions, while the cycloadditions of nitrone are normally considered as type-II reactions. Nitrile oxide cycloadditions are generally classified as borderline to type-III reactions while the cycloaddition of ozone and nitrous oxide is a typical example of type-III reaction. The FMO energies can change relatively by introducing either electron-withdrawing or electron-donating groups on the dipole or dipolarophile, which can further change the type of cycloaddition reaction, for example, the reaction of *N*-methyl-*C*-phenylnitrone with methyl acrylate is controlled by $HOMO_{dipole}$-$LUMO_{dipolarophile}$ interaction but the reaction of the same nitrone with methyl vinyl ether is under the control of $LUMO_{dipole}$-$HOMO_{dipolarophile}$ interaction.

Sustmann classification was applied to cycloaddition reactions of Azomethine ylide **3,** generated *in situ via* decarboxylative condensation of isatin **1** and sarcosine **2** with the exocyclic double bond of *N*-aryl-3-benzylidenepyrrolid-ne-2,5-diones **4a-i** dipolrophiles to substantiate whether the cycloaddition reaction is of NED or not (Scheme **9**).

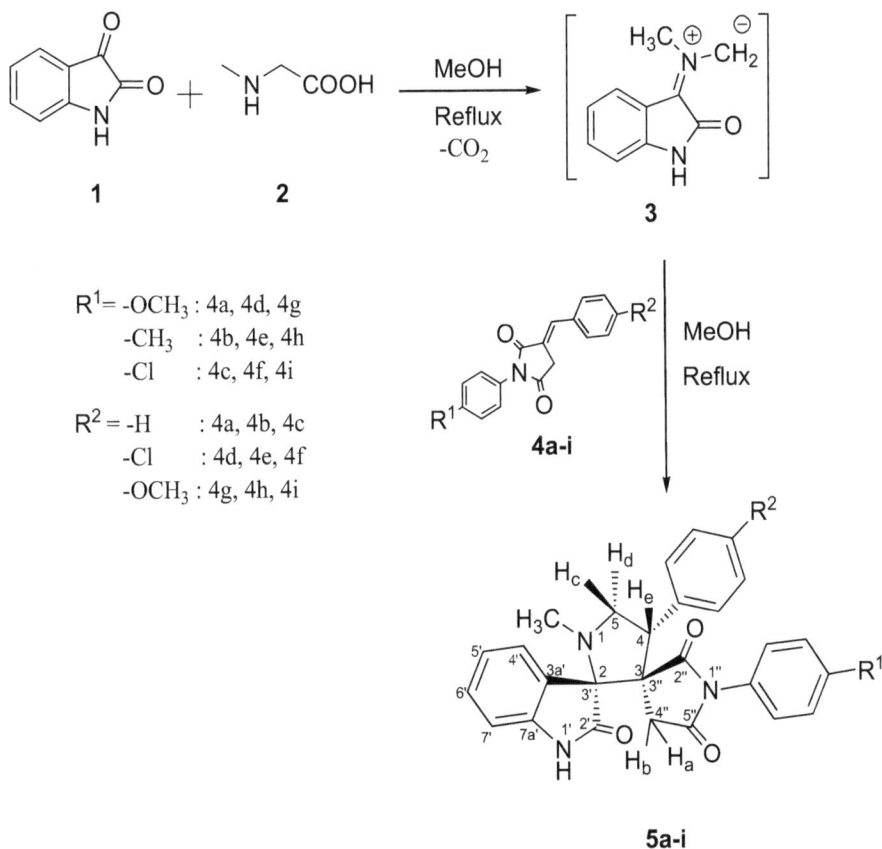

R^1= -OCH$_3$: 4a, 4d, 4g
      -CH$_3$: 4b, 4e, 4h
      -Cl   : 4c, 4f, 4i

R^2 = -H   : 4a, 4b, 4c
      -Cl   : 4d, 4e, 4f
      -OCH$_3$: 4g, 4h, 4i

Scheme 9. Synthesis of dispiro-oxindole-pyrrolidines from Azomethine ylide and differently substituted *N*-aryl-3-benzylidenepyrrolidine-2,5-diones.

As it is evident that the bond formation in 1,3-dipolar cycloaddition reactions is controlled by the flow of electrons from the highest occupied molecular orbital (HOMO) of one reactant to the lowest unoccupied molecular orbital (LUMO) of another, it is quite tricky to decide which molecule will supply the HOMO and which the LUMO. The energy of HOMO is directly related to ionization potential, which decides the susceptibility of the molecule towards the attack by electrophiles. On the other hand, the energy of LUMO is related to electron affinity and it characterizes the molecule's susceptibility towards the attack by nucleophiles.

The energies of HOMO and LUMO of 1,3-dipole **3** and differently substituted dipolarophiles **4a-i** along with the FMO energy gaps (eV) were calculated by using the Schrödinger (9.3) software (Table **3**). As the values of HOMO$_{dipole}$-LUMO$_{dipolarophile}$ energy gaps were lesser than the corresponding HOMO$_{dipolarophile}$-LUMO$_{dipole}$ energy gaps, hence it was concluded that the reaction took place in a normal electron-demand fashion or it is under the control of HOMO$_{dipole}$-LUMO$_{dipolarophile}$ interaction.

Table 3. The Frontier orbital energies (eV) for various dipolarophiles (4a-i) and 1,3-dipole (3) and at Schrödinger (9.3) theoretical level

Reactants	HOMO (eV)	LUMO (eV)	HOMO$_{Dipole}$-LUMO$_{Dipolarophile}$ (eV)	HOMO$_{Dipolarophile}$-LUMO$_{Dipole}$ (eV)
4a	-8.549	-1.030	6.604	8.014
4b	-8.746	-1.082	6.552	8.211
4c	-9.120	-1.143	6.491	8.585
4d	-8.620	-1.343	6.291	8.085
4e	-8.453	-1.494	6.140	7.918
4f	-9.257	-1.398	6.236	8.722
4g	-8.485	-0.936	6.698	7.950
4h	-8.773	-0.946	6.688	8.238
4i	-8.906	-1.189	6.445	8.371
3	-7.635	-0.535	-	-

Parr *et al.* [51] defined the global electrophilicity power as follows:

$$\omega = \mu^2/2\eta \tag{1}$$

The global electrophilicity power measures the energy stabilization of the system when it acquires an additional electronic charge ΔN_{max} from the environment. In Eq. (1) μ and η were the electronic chemical potential and the chemical hardness of the ground state of atoms and molecules, respectively. Here, μ describes the charge transfer pattern of the system in its ground state geometry and η describes the resistance to the change. The formula to calculate μ, in terms of electron energies of FMO HOMO and LUMO, ε_{HOMO} and ε_{LUMO}, is given by [52]:

$$\mu = \varepsilon_{HOMO} + \varepsilon_{LUMO}/2 \tag{2}$$

It has also been possible to give a quantitative representation to the chemical hardness concept introduced by Pearson [53] as:

$$\eta = \varepsilon_{LUMO} - \varepsilon_{HOMO} \qquad (3)$$

A good nucleophile is characterized by a large value of μ and a small value of η. The maximum amount of electronic charge, which an electrophile can accept, is given by [51]:

$$\Delta N_{max} = -\mu / \eta \qquad (4)$$

The Eq. (1) defines the quantity ω which describes the propensity of the system to acquire additional electronic charge from the environment while the quantity ΔN_{max} describes the charge capacity of the molecule defined in Eq. (4).

Table **4** delineated the values of the electronic chemical potentials (μ), chemical hardness η, global electrophilicities (ω), and the charge capacities (ΔN_{max}) of the reactants. It is evident from the results that the charge transfer took place from the dipole to the dipolarophiles because electronic chemical potentials μ (opposite of electronegativity of a species) of the dipole **3** is larger than that of all the dipolarophiles **4a-i.**

Table 4. Chemical hardness (η in eV), electronic chemical potential (μ in eV), global electrophilicity (ω in eV) and charge capacities (ΔN_{max}) for dipolarophile 4a-i and dipole 3 systems.

Reactants	η	μ	ω	ΔN_{max}
4a	7.518	-4.789	1.525	0.637
4b	7.664	-4.914	1.575	0.641
4c	7.977	-5.132	1.650	0.643
4d	7.277	-4.981	1.705	0.684
4e	6.959	-4.973	1.777	0.714
4f	7.859	-5.328	1.806	0.677
4g	7.549	-4.711	1.470	0.624
4h	7.827	-4.859	1.508	0.620
4i	7.717	-5.047	1.650	0.654
3	7.100	-4.085	1.175	0.575

The measure of stabilization of a species that arises from accepting charge, *i.e.* electrophilicity index ω, of the dipole **3**, is less than that of all the dipolarophiles **4a-i** and ΔN_{max} value of dipole was less than all the dipolarophiles. Consequently dipolarophiles behaved as electrophile, whereas dipole acted as a nucleophile. Thus the reaction between **4a-i** and **3** belonged to normal electron demand (NED) type. And the analysis of HOMO/LUMO energy gaps, electronic chemical potentials and global electrophilicities all indicated a normal electron demand character for the reaction.

Reactivity

The reactivity of 1,3-dipoles towards various dipolarophiles varies immensely. These cycloaddition reactions are influenced by the interactions between HOMO and LUMO of both the reactants and a small energy gap between the HOMO and LUMO indicates a stronger interaction.

Electron donating substituents on either the dipole or the dipolarophile increase the energy of both the HOMO and LUMO, while the reverse is experienced with the electron-withdrawing substituents. Conjugating groups tend to decrease the HOMO-LUMO gap of a molecule by raising the energy of HOMO and decreasing the energy of LUMO. The type of substituents can, thus, increase or decrease the rate of a reaction by either decreasing or increasing the FMO energy gap [54, 55].

The main dominant interaction in normal electron demand reaction is $HOMO_{dipole}$-$LUMO_{dipolarophile}$, for example, a reaction between dipole and an electron-deficient dipolarophile. The electron-withdrawing group (EWG) on dipolarophile will lower the energy of FMOs of dipolarophile relative to the dipolarophile without EWG. This lowering in the energy of $LUMO_{dipolarophile}$ will lead to a decrease in the energy difference between E_{HOMO} of dipole and E_{LUMO} of dipolarophile having EWG, compared to the interaction in the absence of EWG on the dipolarophile. The decreased energy gap between the interacting FMOs leads to faster reaction rates (Fig. **7**).

In the case of inverse electron demand reaction, for example, the reaction between dipole and an electron-rich dipolarophile, the FMO interaction that governs the course of reaction is $HOMO_{dipolarophile}$-$LUMO_{dipole}$ interaction [56, 57]. Here, the frontier molecular orbitals of dipolarophile have higher energies than frontier molecular orbitals of the dipole. The presence of electron-donating group (EDG) on dipolarophile will raise the energy of HOMO of dipolarophile and will lead to a decrease in the energy difference between E_{HOMO} of dipolarophile and E_{LUMO} of the dipole. The decreased energy gap between FMOs is responsible for the dominating interaction, which will cause an enhanced rate of 1,3-dipolar cycloaddition (Fig. **8**).

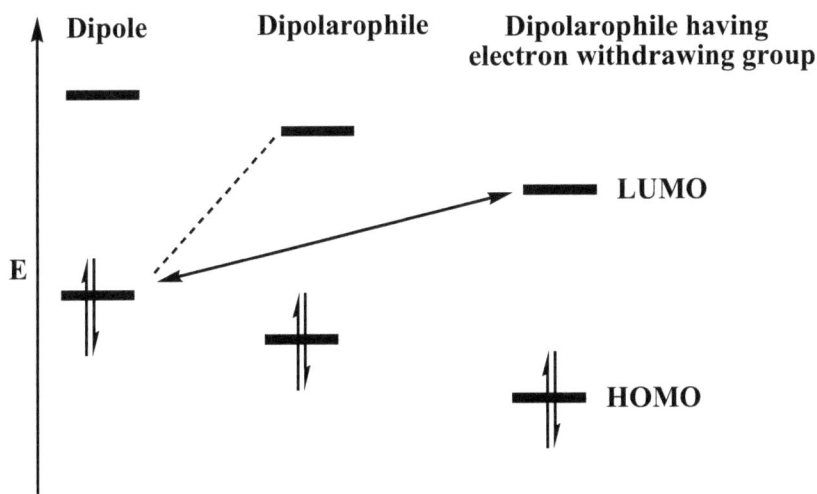

Fig. (7). HOMO-LUMO interaction in Normal Electron Demand reaction.

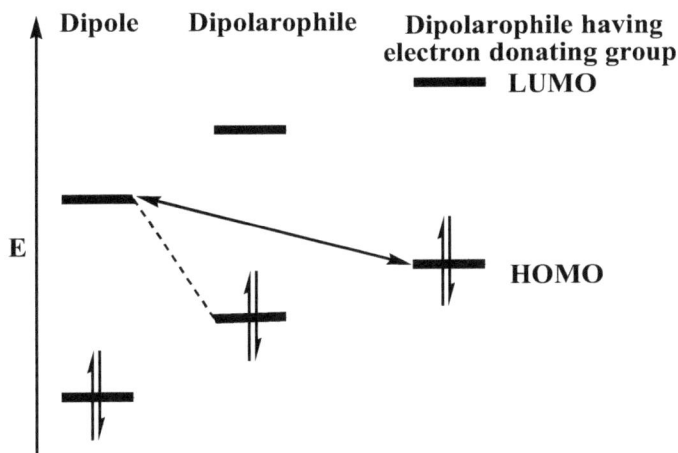

Fig. (8). HOMO-LUMO interaction in Inverse Electron Demand reaction.

The relative reactivities of different dipolarophiles in Scheme (**9**) *i.e.*, *N*-aryl-3-benzylidinepyrrolidine-2,5-dions **4a-i** in 1,3-dipolar cycloaddition reactions with azomethine ylide **3**, were correlated with the $HOMO_{dipole}$-$LUMO_{dipolarophile}$ energy gaps, which were helpful in determining the kinetic aspects of the reaction and, thus, helped in the selection of the best candidate for the reaction. Time taken by various substituted reactants to undergo cycloaddition was explained by the correlation with their $HOMO_{dipole}$-$LUMO_{dipolarophile}$ energy gaps (Table **5**). The results indicated that the electron-donating substituents on C_4-phenyl and benzylidene rings decreased the reactivity of the dipolarophiles by raising the LUMO energy, thus increasing the $HOMO_{dipole}$-$LUMO_{dipolarophile}$ energy gap, while electron-withdrawing substituents increased the reactivity by decreasing the energy of LUMO, thus, decreasing the energy gap.

Table 5. The Frontier orbital energy gaps (eV) between 1,3-dipole (3) and dipolarophiles (4a-i) and time for completion of reaction with various dipolarophiles.

Entry	Compounds	$HOMO_{Dipole}$-$LUMO_{Dipolarophile}$(eV)	Time (h)
1	**5a**	6.882	4.00
2	**5b**	7.016	4.50
3	**5c**	6.750	3.00
4	**5d**	6.981	4.25
5	**5e**	6.799	3.75
6	**5f**	6.618	2.50
7	**5g**	6.606	2.50
8	**5h**	6.942	4.25
9	**5i**	6.772	3.25

CONCLUSION

Here, it has been concluded that the reaction followed a normal electron demand path as indicated by the data of electronic chemical potentials (μ), HOMO/LUMO energy gaps, charge capacities (ΔN_{max}) and electrophilicity indices (ω). Also, the values of $HOMO_{dipole}$-$LUMO_{dipolarophile}$ energy gaps are helpful in rationalizing the kinetics of this cycloaddition, as it was found that with the increasing electron-withdrawing power of the dipolarophile substituents, the reaction became faster because the energy gap between $HOMO_{dipole}$-$LUMO_{dipolarophile}$ is decreased.

CONSENT FOR PUBLICATION

Not applicable.

CONFLICT OF INTEREST

The authors has no conflicts of interest, financial or otherwise, to declare.

ACKNOWLEDGEMENTS

Declared none.

REFERENCES

[1]   Parr, R.G. *The Quantum Theory of Molecular Electronic Structure*; Johns Hopkins University: Benjamin, New York, **1963**.

[2]   Fukui, K.; Yonezawa, T.; Shingu, H. A molecular orbital theory of reactivity in aromatic hydrocarbons. *J. Chem. Phys.,* **1952**, *20*, 722-725.
      [http://dx.doi.org/10.1063/1.1700523]

[3]   Fukui, K.; Yonezawa, T.; Nagata, C. Theory of substitution in conjugated molecules. *Bull. Chem. Soc. Jpn.,* **1954**, *27*, 423-427.
      [http://dx.doi.org/10.1246/bcsj.27.423]

[4]   Fujimoto, H.; Fukui, K. Molecular Orbital Theory of Chemical Reactions. *Adv. Quantum Chem.,* **1972**, *6*, 177-201.
      [http://dx.doi.org/10.1016/S0065-3276(08)60545-6]

[5]   Mulliken, R.S. Molecular complexes and their spectra. VI. Some problems and new developments. *Recl. Trav. Chim. Pays Bas,* **1956**, *75*, 845-852.
      [http://dx.doi.org/10.1002/recl.19560750720]

[6]   Fukui, K.; Yonezawa, T.; Nagata, C.; Shingu, H. Molecular Orbital Theory of Orientation in Aromatic, Heteroaromatic, and Other Conjugated Molecules. *J. Chem. Phys.,* **1954**, *22*, 1433-1442.
      [http://dx.doi.org/10.1063/1.1740412]

[7]   Woodward, R.B.; Hoffmann, R. Selection rules for concerted cycloaddition reactions. *J. Am. Chem. Soc.,* **1965**, *87*, 2046-2048.
      [http://dx.doi.org/10.1021/ja01089a050]

[8]   Woodward, R.B.; Hoffmann, R. Stereochemistry of electrocyclic reactions. *J. Am. Chem. Soc.,* **1965**, *87*, 395-397.
      [http://dx.doi.org/10.1021/ja01080a054]

[9]   Woodward, R.B.; Hoffmann, R. The conservation of orbital symmetry. *Angew. Chem. Int.,* **1969**, *8*, 781-853.
      [http://dx.doi.org/10.1002/anie.196907811]

[10] Bonin, M.; Chauveau, A.; Micouin, L. Asymmetric 1,3-Dipolar Cycloadditions of Cyclic Stabilized Ylides Derived from Chiral 1,2-Amino Alcohols. *Synlett,* **2006**, 2349-2363.
      [http://dx.doi.org/10.1055/s-2006-949626]

[11] Husinec, S.; Savic, V. Chiral catalysts in the stereoselective synthesis of pyrrolidine derivatives *via* metallo-azomethine ylides. *Tetrahedron Asymmetry,* **2005**, *16*, 2047-2061.
      [http://dx.doi.org/10.1016/j.tetasy.2005.05.020]

[12] Tufariello, J.J. Alkaloids from Nitrones. *Acc. Chem. Res.,* **1979**, *12*, 396-403.
      [http://dx.doi.org/10.1021/ar50143a003]

[13] Ohta, A.; Dahl, K.; Raab, R.; Geittner, J.; Huisgen, R. Diazodiphenylmethane and Monosubstituted Butadienes: Kinetics and a New Chapter of Vinylcyclopropane Chemistry. *Helv. Chim. Acta,* **2008**, *91*, 783-804.
      [http://dx.doi.org/10.1002/hlca.200890081]

[14] Benedetti-Doctorovich, V.; Burgess, E.M.; Lambropoulos, J.; Lednicer, D.; Van Derveer, D.; Zalkow, L.H. Synthesis of 2-methyl-(Z)-4-(phenylimino)naphth[2,3-d]oxazol-9-one, a monoimine quinone with selective cytotoxicity toward cancer cells. *J. Med. Chem.,* **1994**, *37*(5), 710-712.
      [http://dx.doi.org/10.1021/jm00031a023] [PMID: 8126711]

[15] Pereira, E.R.; Sancelme, M.; Voldoire, A.; Prudhomme, M. Synthesis and antimicrobial activities of 3-N-substituted-4,5-bis(3-indolyl)oxazol-2-ones. *Bioorg. Med. Chem. Lett.,* **1997**, *7*, 2503-2506.
      [http://dx.doi.org/10.1016/S0960-894X(97)10007-5]

[16] Smith, L.I. Aliphatic Diazo Compounds, Nitrones, and Structurally Analogous Compounds. Systems

Capable of Undergoing 1,3-Additions. *Chem. Rev.,* **1938**, *23*, 193-285.
[http://dx.doi.org/10.1021/cr60075a001]

[17]   Pellissier, H. Asymmetric 1,3-dipolar cycloadditions. *Tetrahedron,* **2007**, *63*, 3235-3285.
       [http://dx.doi.org/10.1016/j.tet.2007.01.009]

[18]   Zhu, J.; Lines, B.M.; Ganton, M.D.; Kerr, M.A.; Workentin, M.S. Efficient synthesis of isoxazolidine-
       tethered monolayer-protected gold nanoparticles (MPGNs) *via* 1,3-dipolar cycloadditions under high-
       pressure conditions. *J. Org. Chem.,* **2008**, *73*(3), 1099-1105.
       [http://dx.doi.org/10.1021/jo702398r] [PMID: 18181644]

[19]   Raj, A.A.; Raghunathan, R. A novel entry into a new class of spiroheterocyclic framework:
       regioselective synthesis of dispiro[oxindole-cyclohexanone]pyrrolidines and dispiro[oxindole-
       hexahydroindazole]pyrrolidines. *Tetrahedron,* **2001**, *57*, 10293-10298.
       [http://dx.doi.org/10.1016/S0040-4020(01)01042-0]

[20]   Krasiński, A.; Radić, Z.; Manetsch, R.; Raushel, J.; Taylor, P.; Sharpless, K.B.; Kolb, H.C. *In situ*
       selection of lead compounds by click chemistry: target-guided optimization of acetylcholinesterase
       inhibitors. *J. Am. Chem. Soc.,* **2005**, *127*(18), 6686-6692.
       [http://dx.doi.org/10.1021/ja043031t] [PMID: 15869290]

[21]   Huisgen, R. 1,3◻Dipolar Cycloadditions. Past and Future. *Angew. Chem. Int. Ed. Engl.,* **1963**, *2*, 565-
       598.
       [http://dx.doi.org/10.1002/anie.196305651]

[22]   Mita, T.; Ohtsuki, N.; Ikeno, T.; Yamada, T. Enantioselective 1,3-dipolar cycloaddition of nitrones
       catalyzed by optically active cationic cobalt(III) complexes. *Org. Lett.,* **2002**, *4*(15), 2457-2460.
       [http://dx.doi.org/10.1021/ol026079p] [PMID: 12123350]

[23]   Ohtsuki, N.; Kezuka, S.; Kogami, Y.; Mita, T.; Ashizawa, T.; Ikeno, T. Enantioselective 1,3-dipolar
       cycloaddition reactions between nitrones and α-substituted α,β-unsaturated aldehydes catalyzed by
       chiral cationic cobalt(III) complexes. *Synthesis,* **2003**, 1462-1466.

[24]   Ohtsuki, N.; Kezuka, S.; Mita, T.; Kogami, Y.; Ashizawa, T.; Ikeno, T.; Yamada, T. Enantioselective
       1,3-Dipolar Cycloaddition Reactions between Nitrones and α-Substituted α,β-Unsaturated Aldehydes
       Catalyzed by Chiral Cationic Cobalt(III). *Complexes. Bull. Chem. Soc. Jpn.,* **2003**, *76*, 2197-2207.
       [http://dx.doi.org/10.1246/bcsj.76.2197]

[25]   McNaught, A.D.; Wilkinson, A. *Compendium of Chemical Terminology,* 2nd ed; Blackwell Scientific
       Publications: Oxford, **1997**.

[26]   Huisgen, R. Cycloadditions-Definition, Classification, and Characterization. *Angew. Chem. Int. Ed.
       Engl.,* **1968**, *7*, 321-328.
       [http://dx.doi.org/10.1002/anie.196803211]

[27]   Sirion, U.; Bae, Y.J.; Lee, B.S.; Chi, D.Y. Ionic Polymer Supported Copper(I): A Reusable Catalyst
       for Huisgen's 1,3-Dipolar Cycloaddition. *Synlett,* **2008**, 2326-2330.

[28]   Hang, X.C.; Chen, Q.Y.; Xiao, J.C. 1,3-dipolar cycloaddition of difluoro(methylene)cyclopropanes
       with nitrones: Efficient synthesis of 3,3-difluorinated tetrahydropyridinols. *Synlett,* **2008**, 1989-1992.

[29]   Huisgen, R.; Stangl, H.; Sturm, J.; Wagenhofer, H. 1,3-Dipolare Additionen mit Nitril-yliden. *Angew.
       Chem.,* **1962**, *74*, 31-31.
       [http://dx.doi.org/10.1002/ange.19620740111]

[30]   Novikov, M.S.; Khlebnikov, A.F.; Egarmin, M.A.; Shevchenko, M.V.; Khlebnikov, V.A.; Kostikov,
       R.R.; Vidovic, D. Regioselectivity of the 1,3-dipolar cycloaddition of fluorinated fluoren-9-iminium
       ylides to heteroelement-containing dipolarophiles: Experimental and quantum-chemical study. *Russ. J.
       Org. Chem.,* **2006**, *42*, 1800-1812.
       [http://dx.doi.org/10.1134/S1070428006120086]

[31]   Huisgen, R.; Langhals, E. 1,3◻Dipolar cycloadditions of diphenyldiazomethane to thioketones: Rate
       measurements disclose thiones to be superdipolarophiles. *Heteroatom Chem.,* **2006**, *17*, 433-442.

[http://dx.doi.org/10.1002/hc.20262]

[32]   Yoo, C.L.; Olmstead, M.M.; Tantillo, D.J.; Kurth, M.J. Synthesis of 2H-pyrroles *via* the 1,3-dipolar cycloaddition reaction of nitrile ylides with acrylamides. *Tetrahedron Lett.,* **2006**, *47*, 477-481.
       [http://dx.doi.org/10.1016/j.tetlet.2005.11.066]

[33]   Huisgen, R. On the Mechanism of 1,3-dipolar cycloadditions. *Reply. J. Org. Chem.,* **1968**, *33*, 2291-2297.
       [http://dx.doi.org/10.1021/jo01270a024]

[34]   Huisgen, R. Concerted nature of 1,3-dipolar cycloadditions and the question of diradical intermediates. *J. Org. Chem.,* **1976**, *41*, 403-419.
       [http://dx.doi.org/10.1021/jo00865a001]

[35]   Firestone, R.A. On the Mechanism of 1,3-Dipolar Cycloadditions. *J. Org. Chem.,* **1968**, *33*, 2285-2290.
       [http://dx.doi.org/10.1021/jo01270a023]

[36]   Firestone, R.A. Orientation in 1,3-Dipolar Cycloadditions according to the Diradical Mechanism. Partial formal charges in the Linnett structures of the diradical intermediate. *J. Org. Chem.,* **1972**, *37*, 2181-2191.
       [http://dx.doi.org/10.1021/jo00978a027]

[37]   Houk, K.N.; Firestone, R.A.; Munchausen, L.L.; Mueller, P.H.; Arison, B.H.; Garcia, L.A. Stereospecificity of 1,3-dipolar cycloadditions of p-nitrobenzonitrile oxide to cis- and trans-dideuterioethylene. *J. Am. Chem. Soc.,* **1985**, *107*, 7227-7228.
       [http://dx.doi.org/10.1021/ja00310a105]

[38]   Pedron, M.; Delso, I.; Tejero, T.; Merino, P Stereospecificity of 1,3-dipolar cycloadditions of p-nitrobenzonitrile oxide to cis- and transdideuterioethylene *J. Am. Chem. Soc.,* **1985**, *107*, 7227-7228.

[39]   Jasiński, R.; Jasińska, E.; Dresler, E. A DFT computational study of the molecular mechanism of [3 + 2] cycloaddition reactions between nitroethene and benzonitrile N-oxides Radomir Jasiński & Ewa Jasińska & Ewa Dresler. *J. Mol. Model.,* **2017**, *23*, 1-9.
       [http://dx.doi.org/10.1007/s00894-016-3185-8]

[40]   Tamilmani, V.; Daul, C.A.; Robles, J.L.; Bochet, C.G.; Venuvanalingam, P. Hydrogen bond stabilization in Diels-Alder transition states: The cycloaddition of hydroxy-ortho-quinodimethane with fumaric acid and dimethylfumarate. *Chem. Phys. Lett.,* **2005**, *406*, 355-359.
       [http://dx.doi.org/10.1016/j.cplett.2005.03.017]

[41]   Kaur, A.; Singh, B. 1,3-Dipolar Cycloaddition Reactions Leading to the Synthesis of New 2,3, 5-Triaryl-4H,2,3,3a,5,6,6a-hexahydropyrrolo[3,4-d]isoxazole-4,6-diones. *J. Heterocycl. Chem.,* **2014**, *51*, 1421-1429.
       [http://dx.doi.org/10.1002/jhet.1838]

[42]   Fleming, I. *Frontier Orbitals and Organic Chemical Reactions*; Wiley-VCH: London, **1976**.

[43]   Emamian, S.R.; Asgari, S.A.; Zahedi, E. Mechanism and regioselectivity of 1,3-dipolar cycloaddition reactions of sulphur-centred dipoles with furan-2,3-dione: A theoretical study using DFT. *J. Chem. Sci.,* **2014**, *126*, 293-302.
       [http://dx.doi.org/10.1007/s12039-013-0540-5]

[44]   Houk, K.N.; Rondan, N.G.; Santiago, C.; Gallo, C.J.; Gandour, R.W.; Griffin, G.W. Theoretical studies of the structures and reactions of substituted carbonyl ylides. *J. Am. Chem. Soc.,* **1980**, *102*, 1504-1512.
       [http://dx.doi.org/10.1021/ja00525a006]

[45]   Pierluigi, C.; Houk, K.N. Geometries of nitrilium betaines. The clarification of apparently anomalous reactions of 1,3-dipoles. *J. Am. Chem. Soc.,* **1976**, *98*, 6397-6399.
       [http://dx.doi.org/10.1021/ja00436a062]

[46]   Pierluigi, C.; Gandour, R.W.; Hall, J.A.; Deville, C.G.; Houk, K.N. A derivation of the shapes and

energies of the molecular orbitals of 1,3-dipoles. Geometry optimizations of these species by MINDO/2 and MINDO/3. *J. Am. Chem. Soc.,* **1976**, *99*, 385-392.

[47]   Padwa, A.; Fryxell, G.E.; Lin, Z. Tandem cyclization-cycloaddition reaction of rhodium carbenoids. Scope and mechanistic details of the process. *J. Am. Chem. Soc.,* **1990**, *112*, 3100-3109.
[http://dx.doi.org/10.1021/ja00164a034]

[48]   Padwa, A.; Weingarten, M.D. Cascade processes of metallo carbenoids. *Chem. Rev.,* **1996**, *96*(1), 223-270.
[http://dx.doi.org/10.1021/cr950022h] [PMID: 11848752]

[49]   Sustmann, R. A simple model for substituent effects in cycloaddition reactions. I. 1,3-dipolar cycloadditions. *Tetrahedron Lett.,* **1971**, •••, 2717-2720.
[http://dx.doi.org/10.1016/S0040-4039(01)96961-8]

[50]   Sustmann, R. Orbital energy control of cycloaddition reactivity. *Pure Appl. Chem.,* **1974**, *40*, 569-593.
[http://dx.doi.org/10.1351/pac197440040569]

[51]   Parr, R.G.; Szentpaly, L.V.; Liu, S. Electrophilicity Index. *J. Am. Chem. Soc.,* **1999**, *121*, 1922-1924.
[http://dx.doi.org/10.1021/ja983494x]

[52]   Parr, R.G.; Yang, W. *Density Functional Theory of Atoms and Molecules*; Oxford University: New York, **1989**.

[53]   Parr, R.G.; Pearson, R.G. Absolute hardness: companion parameter to absolute electronegativity. *J. Am. Chem. Soc.,* **1983**, *105*, 7512-7516.
[http://dx.doi.org/10.1021/ja00364a005]

[54]   Houk, K.N.; Luskus, L.J. Influence of steric interactions on endo stereoselectivity. *J. Am. Chem. Soc.,* **1971**, *93*, 4606-4607.
[http://dx.doi.org/10.1021/ja00747a052]

[55]   Fuente, M.C.; Dominguez, D. Normal electron demand Diels–Alder cycloaddition of indoles to 2,3-dimethyl-1,3-butadiene. *Tetrahedron,* **2011**, *67*, 3997-4001.
[http://dx.doi.org/10.1016/j.tet.2011.04.030]

[56]   Wiessler, M.; Waldeck, W.; Kliem, C.; Pipkorn, R.; Braun, K. The Diels-Alder-reaction with inverse-electron-demand, a very efficient versatile click-reaction concept for proper ligation of variable molecular partners. *Int. J. Med. Sci.,* **2009**, *7*(1), 19-28.
[PMID: 20046231]

[57]   Bodwell, G.J.; Pi, Z.; Pottie, I.R. Electron deficient dienes. 2. one step synthesis of a coumarin-fused electron deficient diene and its inverse electron demand diels-alder reactions with enamines. *Synlett,* **1999**, *4*, 477-479.
[http://dx.doi.org/10.1055/s-1999-2645]

SUBJECT INDEX

A

Activation 68, 111, 113, 117, 118, 129, 206
  enthalpy 206
Adenomatous polyposis coli (APC) 113, 120
  tumour suppressor 113
ADME properties and binding 36
ADMET 17, 18, 31, 32
  analysis 31, 32
  properties 17, 18, 32
Allergic asthma 95
Alzheimer's disease 98, 120
Apoptosis 98, 113, 119
Arthritis 93
ASINEX database 36
Asinex platinum 82
Atherosclerosis 94
Atomic spin densities (ASDs) 185, 186, 188
Azomethine imine cycloaddition reactions 241

B

Bader's quantum theory 200
Bank of molecules of the Amazon (BMA) 38, 39
Bayesian theory 125
BIA evaluation software 43
Binding site 9, 11, 32, 83, 89, 100, 128
  analysis 32
  internal protein 9
Binding 6, 40
  strength 40
  thermodynamics 6
Bond 65, 69, 92, 199, 202, 228, 233, 234
  acidic peptide 92
  attack-mediated peptide 65
  reactive peptide 69
  triple 233
Bonding evolution theory (BET) 175, 196, 198, 206, 207
Bonding pattern consistency 36
Born-Oppenheimer approximation 175, 193
Brainstem encephalitis 96

Burgi-Dunitz trajectory 185

C

Cambridge structural database (CSD) 18
cAMP response element-binding protein-1 118
Cancer 37, 93, 101, 111, 118, 120, 123, 127, 132, 150
  breast 127
  gastrointestinal 127
  liver 37
Canonical pathway 113, 114, 120
  conservative 114
  independent 114
Canonical signalling pathway 113
Carnosic acid 131
Caspase-1 65, 98
  inhibitors 98
Chagas disease 65, 68, 70
Charge 26, 27, 195, 200, 228, 244, 247
  capacities 228, 244, 247
  density 26, 27, 200
  distribution 159, 160
  transfer (CT) 27, 195, 244
Chembridge 43, 47, 83
  and Asnex databases 83
  libraries 43, 47
Chemical
  behaviour 178
  changes 175
  composition 41
Chemical hardness 91, 178, 179, 181, 243, 244
  expressed 178
  concept 244
  structures of praziquantel and oxamniquine drugs 91
Chemoselective reaction 217
Chikungunya 63, 65, 88
  fever 88
  virus 63, 65, 88
Complete chemo- and regioselective reaction 215

S

Salivary gland tumour 121
SARS-CoV-2 44, 45, 46
    proteases 44
Schizophrenia 121, 123
Schrödinger's equation 175
Schrodinger software 43
Screening 2, 3, 6, 10, 12, 16, 17, 19, 21, 31,
    36, 47, 48, 64, 66, 160
  high-throughput 2, 10, 16, 17, 21, 64, 66
Secreted frizzled-related proteins (SFRPs) 120
SHAPE-GAUSS score function 74
Signalling 112, 113, 114, 116, 122, 123, 124,
    126
    activity 113
    cascades 112
    intracellular 116
    pathway impact analysis (SPIA) 123, 126
    upregulating mTOR 114
Signalling pathways 112, 113, 114, 119, 127
    canonical Wnt 114, 127
    dependent 113, 117
    dependent WNT 114
    intercellular 112
    non-canonical Wnt 119
Signal 20, 124, 125, 180
    transduction pathway databases 125
    electrophilicity scale 180
    gene analysis methods 124
    gene-based analysis method 124
    target drugs 20
Software 9, 13, 15, 16, 22, 39, 43, 45, 47, 83,
    174, 217, 243
    commercial 15
    discovery studio visualizer 45
    iGEMDOCK 47
    molecular modelling 174, 217
Sore throat 28
Structural stability domains (SSDs) 196
Structure 3, 4, 7, 8, 9, 10, 12, 13, 16, 21, 26,
    64, 66
    activity relationships (SAR) 12
    based computational methods 4
    based drug design (SBDD) 3, 4, 7, 8, 9, 10,
    12, 13, 16, 21, 26, 64, 66
Synthesis 13, 37, 89, 93, 97, 101, 114, 174,
    208, 216, 229, 233, 242
    stereochemical 208
    viral RNA 89

T

Tankyrase 111, 130
Thioredoxin glutathione reductase (TGR) 91
Thoms' catastrophe theory 196
Tools 2, 3, 6, 13, 19, 22, 28, 31, 43, 48, 64,
    65, 124, 125, 126
    efficient drug discovery 2
    elements mathematical 13
    integrated protein annotation 31
    intuitive bioinformatics 126
    proteomic data management 124
    quantum mechanics 28
    refinement 22
    toxicity predictions 48
Topological analysis 174, 189, 190, 196, 200,
    208, 217
    applied ELF 190
    of electron localization function 189
Toxicity 2, 5, 15, 19, 32, 101
    biological 2
Traditional chinese medicine database
    (TCMD) 30
Trypanosoma cruzi 63, 65, 70
Trypsinogen 68
Tuberous sclerosis protein 114

U

Universal force field (UFF) 24

V

Vaccine 1, 2, 29, 33, 37, 42, 48, 88
    effective 1, 37
    efficient 29, 48
    emergency 29
    novel 48
Valence 175, 190
    basins 190
    bond theory 175
Valence basin populations 190
    calculated 190
Valproic acid 46
Vertical 156, 180
    electron affinities 180
    ionization potentials 180
    recurrence relations 156

www.ingramcontent.com/pod-product-compliance
Lightning Source LLC
Chambersburg PA
CBHW050820220326
41598CB00006B/266